Oracle 12c 数据库基础教程

赵卫东　刘永红　于　曦　主编
李　立　鄢　涛　　　副主编

科学出版社
北京

内 容 简 介

Oracle 数据库是性能优异的数据库系统之一，拥有很大的市场占有率，在大中型企业的信息管理系统中有广泛的应用。Oracle 数据库管理和应用系统开发是国内外高等院校信息技术及相关专业的必修或者选修课程。Oracle 12c 版本是 Oracle 产品的最新产物，是当前企业级数据库开发的首选数据库版本。本教材按工程认证的标准，以 CDIO 的理念为基础，以工程案例开发为手段，全面提升数据库管理和开发能力。Oracle 知识点繁多，本书作者根据多年的教学及项目开发经验，在教材中将 Oracle 12c 的主要知识点进行了科学的归类和划分，制定了科学的学习路径，同时每章还配有课后练习和训练任务，力求使读者学习的时候可以循序渐进，深入浅出，提高学习效率。

本教材可以作为大学本科有关课程的专业教材，也可以作为广大 Oracle 数据库管理员和开发人员的参考资料。

本书还配有教学课件、全部数据库设计以及程序的脚本和代码，读者可以在出版社网站下载。

图书在版编目(CIP)数据

Oracle 12c 数据库基础教程 / 赵卫东，刘永红，于曦主编. —北京：科学出版社，2017.7
 ISBN 978-7-03-053896-3

Ⅰ.①O… Ⅱ.①赵… ②刘… ③于… Ⅲ.①关系数据库系统-教材 Ⅳ.①TP311.138

中国版本图书馆 CIP 数据核字 (2017) 第 161200 号

责任编辑：冯 铂 / 责任校对：韩雨舟
封面设计：墨创文化 / 责任印制：罗 科

科 学 出 版 社 出版
北京东黄城根北街16号
邮政编码：100717
http://www.sciencep.com

成都锦瑞印刷有限责任公司 印刷
科学出版社发行 各地新华书店经销
*

2017 年 7 月第 一 版　开本：787×1092　1/16
2017 年 7 月第一次印刷　印张：28.25
字数：580 千字
定价：59.00 元
(如有印装质量问题，我社负责调换)

前　　言

2016年6月2日，我国顺利成为《华盛顿协议》正式成员，这是促进我国工程师按照国际标准培养、提高工程技术人才培养质量的重要举措，是推进工程师资格国际互认的基础和关键，对我国工程技术领域应对国际竞争、走向世界具有重要意义。

本书是"以项目驱动为核心的工程实践能力培养系列教材"中的一门，全书按工程认证的标准，以CDIO的理念为基础，将一个实际项目的开发融入全书各个章节、各个知识点。通过本书的学习，让读者能快速理解并掌握各项知识点，全面提升面对复杂工程问题时的分析、解决和实际编码能力。

本门课程对应工程认证12项毕业要求指标点的关系如下：

能力点	覆盖度
1.工程知识：能够将数学、自然科学、工程基础和专业知识用于解决复杂工程问题	★★★
2.问题分析：能够应用数学、自然科学和工程科学的基本原理，识别、表达、并通过文献研究分析复杂工程问题，以获得有效结论	★★★★
3.设计/开发解决方案：能够设计针对复杂工程问题的解决方案，设计满足特定需求的系统、单元(部件)或工艺流程，并能够在设计环节中体现创新意识，考虑社会、健康、安全、法律、文化以及环境等因素	★★★★★
4.研究：能够基于科学原理并采用科学方法对复杂工程问题进行研究，包括设计实验、分析与解释数据、并通过信息综合得到合理有效的结论	★★★
5.使用现代工具：能够针对复杂工程问题，开发、选择与使用恰当的技术、资源、现代工程工具和信息技术工具，包括对复杂工程问题的预测与模拟，并能够理解其局限性	★★★★
6.工程与社会：能够基于工程相关背景知识进行合理分析，评价专业工程实践和复杂工程问题解决方案对社会、健康、安全、法律以及文化的影响，并理解应承担的责任	★★★
7.环境和可持续发展：能够理解和评价针对复杂工程问题的专业工程实践对环境、社会可持续发展的影响	★★★
8.职业规范：具有人文社会科学素养、社会责任感，能够在工程实践中理解并遵守工程职业道德和规范，履行责任	★★★
9.个人和团队：能够在多学科背景下的团队中承担个体、团队成员以及负责人的角色	★★★★
10.沟通：能够就复杂工程问题与业界同行及社会公众进行有效沟通和交流，包括撰写报告和设计文稿、陈述发言、清晰表达或回应指令。并具备一定的国际视野，能够在跨文化背景下进行沟通和交流	★★★★
11.项目管理：理解并掌握工程管理原理与经济决策方法，并能在多学科环境中应用	★★★★★
12.终身学习：具有自主学习和终身学习的意识，有不断学习和适应发展的能力	★★

Oracle数据库系统是美国Oracle公司(甲骨文)提供的以分布式数据库为核心的一组

软件产品，是目前最流行的客户/服务器(CLIENT/SERVER)或 B/S 体系结构的数据库之一，是当前最流行的大型关系数据库之一，拥有广泛的用户和大量的应用案例。Oracle 数据库管理与开发已经成为一门综合性高、实践性强、应用领域广的技术类课程。对于从事数据维护和计算机程序开发的人员，在当前互联网时代，掌握 Oracle 数据库管理与开发是非常必要的。本书主要特点如下：

(1) 强调 CDIO 理念。将工程认证的理念贯穿所有章节，每章都明确所述知识点与能力点的对应关系，方便读者有针对性的进行学习。

(2) 强调实践性。本书中每个重要的知识点都配备示例以及示例的实现过程，力求帮助读者在掌握知识的同时，能针对不同的需求进行实际问题的解决。同时在每一章后面，还附有练习和训练任务，在本书最后的附录中有练习的答案以及训练任务的详细完成过程。训练任务丰富、完整，完全可以作为课后实验项目。

(3) 强调项目实现。本书的每一个章节都在前一章节的基础上进行实现，循序渐进，从而达到对训练任务的迭代、升级。在最后的第 14 章，还帮助读者实现一个完整项目"小型商品销售系统"的开发，使读者可以从头至尾完整学习和掌握 Oracle 数据库的开发过程和实际应用项目开发过程。

(4) 辅助教学和学习资料全。本书配有实验指导教程、电子课件，项目的源代码等参考资料，读者可以从科学出版社网站上下载。

在使用本书进行教学时，建议采用过程化的考核方式，对每一章的训练任务完成情况进行记录，得到的成绩汇总平均后作为总成绩的 50%。期末再进行笔试，对书中的理论知识进行考核，得到的成绩作为总成绩的 40%。平时的签到和课堂提问成绩作为总成绩的 10%。

本书主编是赵卫东、刘永红，于曦，副主编是李立、鄢涛，编委是王仕平、蒋玲、陈天雄等。全书共有 14 章，其中赵卫东编写第 1~6 章，蒋玲编写第 7 章，刘永红编写第 8~10 章，李立编写第 11~12 章，于曦编写第 13 章，鄢涛编写第 14 章，王仕平和陈天雄对本书进行了统稿、修订以及部分程序设计。

由于作者水平有限，虽然对本书进行了反复审核与修订，但书中疏漏和不足之处在所难免，恳请广大读者及专家给予批评指正。

赵卫东
2017 年 8 月

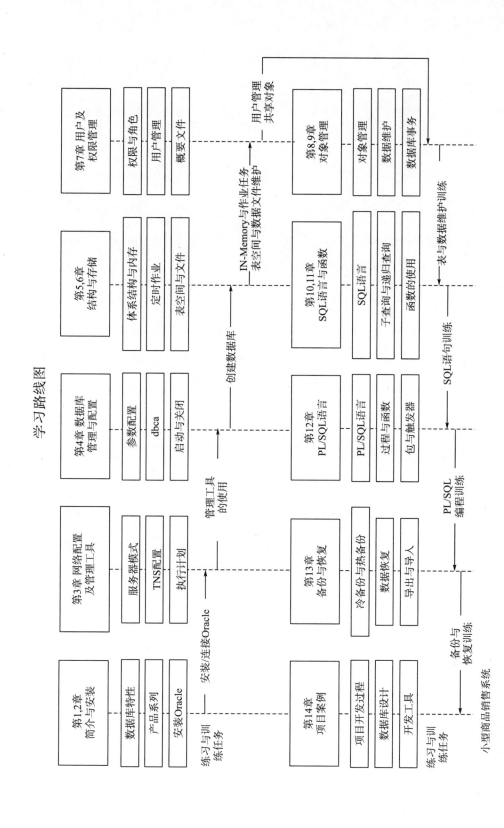

目　录

第1章　Oracle 12c 简介 ... 1
1.1　Oracle 12c 简介 ... 1
1.2　Oracle 12c 产品系列 ... 2
1.3　Oracle 12c 新特性 ... 2
　1.3.1　插接式数据库 PDB ... 2
　1.3.2　高可用性 ... 4
　1.3.3　XML DB ... 6
　1.3.4　In-Memory 数据库内存选件 ... 6
　1.3.5　Oracle JSON 文档存储 ... 7
　1.3.6　其他新特性 ... 7
练习 ... 7

第2章　Oracle 12c 的安装 ... 9
2.1　安装前配置 Linux 系统 ... 9
　2.1.1　配置 Linux 交换空间 ... 10
　2.1.2　创建 Oracle 用户和用户组 ... 10
　2.1.3　配置/etc/sysctl.conf 文件 ... 10
　2.1.4　其他配置 ... 11
2.2　安装 Oracle 12c ... 12
2.3　数据库连接测试 ... 25
2.4　Oracle 企业管理器 ... 28
2.5　安装后的检测 ... 30
　2.5.1　查看环境变量 ... 30
　2.5.2　查看目录及文件 ... 31
　2.5.3　查看 Oracle 进程 ... 32
　2.5.4　查看监听器状态 ... 32
2.6　设置开机启动 ... 34
练习 ... 36
训练任务 ... 37

第3章　网络配置及管理工具 ... 38
3.1　Oracle Net Services ... 39
3.2　服务器模式和数据库连接方式 ... 39

3.2.1	专用服务器模式	40
3.2.2	共享服务器模式	40
3.2.3	配置数据库支持共享模式	41
3.2.4	检测数据库的服务器模式	42
3.2.5	连接到不同的服务器模式	44
3.2.6	查看服务器连接进程	45

3.3 TNS 网络配置文件 …… 46
   3.3.1 lsnrctl 和 listener.ora …… 46
   3.3.2 监听器的动态注册 …… 47
   3.3.3 监听器的静态注册 …… 50
   3.3.4 tnsnames.ora …… 51
   3.3.5 sqlnet.ora …… 52

3.4 SQL*Plus …… 54
   3.4.1 SQL*Plus 连接数据库 …… 54
   3.4.2 SQL*Plus 命令列表 …… 57
   3.4.3 SQL*Plus 参数 …… 58
   3.4.4 SQL*Plus 替换变量 …… 59
   3.4.5 绑定变量 …… 60
   3.4.6 预设变量 …… 61
   3.4.7 PL/SQL 程序的运行 …… 61

3.5 Oracle SQL Developer …… 62
   3.5.1 SQL Developer 连接 Oracle …… 63
   3.5.2 Data Modeler …… 65

3.6 执行计划与 SQL 优化 …… 68
   3.6.1 授予查询执行计划的权限 …… 69
   3.6.2 分析和比较执行计划 …… 70
   3.6.3 统计信息与动态采样 …… 74
   3.6.4 SQL 语句的优化 …… 79
   3.6.5 自适应查询优化 …… 81

练习 …… 85
训练任务 …… 86

第 4 章　数据库管理与配置　88

4.1 常用的数据库配置查询方法 …… 89
4.2 使用 dbca 管理数据库实例 …… 90
   4.2.1 新建数据库实例 …… 90
   4.2.2 删除一个容器数据库 CDB …… 91

    4.3　在数据库实例之间切换 ··· 91
    4.4　配置插接式数据库 PDB ··· 92
        4.4.1　通过 SQL 语句创建插接式数据库 ··· 92
        4.4.2　通过 dbca 创建插接式数据库 ··· 93
        4.4.3　克隆插接式数据库 pdb ·· 96
        4.4.4　删除插接式数据库 pdb ·· 96
        4.4.5　插接式数据库的拔出与插入 ··· 97
    4.5　数据库的启动与关闭 ··· 97
        4.5.1　启动数据库 ·· 98
        4.5.2　启动异常处理 ·· 101
        4.5.3　关闭数据库 ·· 101
    4.6　数据库参数配置 ··· 102
    练习 ··· 103
    训练任务 ··· 104

第 5 章　Oracle 12c 数据库结构 ··· 105
    5.1　Oracle 12c 体系结构 ··· 106
        5.1.1　数据库物理存储结构 ··· 107
        5.1.2　逻辑存储结构 ·· 107
    5.2　Oracle 12c 内存结构 ··· 109
        5.2.1　基本内存结构 ·· 109
        5.2.2　PGA 概述 ·· 110
        5.2.3　SGA 概述 ·· 110
        5.2.4　In-Memory 列存储 ··· 112
    5.3　服务器进程 ··· 117
        5.3.1　后台进程 ·· 118
        5.3.2　定时执行作业任务 ··· 120
    练习 ··· 123
    训练任务 ··· 125

第 6 章　数据库存储管理 ·· 126
    6.1　表空间和数据文件的管理 ·· 127
    6.2　创建表空间 ··· 128
    6.3　查看表空间信息 ··· 131
    6.4　设置表空间 ··· 132
        6.4.1　修改表空间名称 ··· 132
        6.4.2　修改表空间大小 ··· 132
        6.4.3　切换表空间状态 ··· 133

| 6.5 删除表空间 | 134 |

6.6 控制文件的管理 ·· 135

6.7 重做日志文件与归档日志文件 ··· 137

    6.7.1 重做日志与归档日志的基本概念 ··· 137

    6.7.2 重做日志组管理 ··· 140

6.8 参数文件 ··· 142

    6.8.1 修改 spfile 参数值 ··· 143

    6.8.2 从 spfile 创建 pfile ·· 143

练习 ·· 144

训练任务 ··· 145

第 7 章 用户及权限管理 ··· 146

7.1 权限 ··· 147

    7.1.1 系统权限 ·· 147

    7.1.2 对象权限 ·· 150

7.2 角色 ··· 151

    7.2.1 系统预定义角色 ··· 152

    7.2.2 创建公共角色 ··· 152

    7.2.3 创建本地角色 ··· 155

    7.2.4 删除自定义角色 ·· 155

7.3 用户管理 ·· 156

    7.3.1 创建公共用户 ··· 156

    7.3.2 授予用户对象权限 ··· 159

    7.3.3 用户的其他常用操作 ·· 160

    7.3.4 监视用户 ·· 161

7.4 概要文件 ·· 161

    7.4.1 创建概要文件 ··· 162

    7.4.2 修改概要文件 ··· 164

    7.4.3 删除概要文件 ··· 164

练习 ·· 165

训练任务 ··· 166

第 8 章 数据库的对象管理 ··· 167

8.1 表 ··· 168

    8.1.1 数据类型 ·· 168

    8.1.2 创建表 ··· 169

    8.1.3 修改、删除表 ··· 173

    8.1.4 表的约束 ·· 176

8.2 分区表 176
8.2.1 分区类型 177
8.2.2 分区表的维护 182
8.3 索引 182
8.3.1 创建索引 183
8.3.2 修改、删除索引 187
8.4 簇表 187
8.4.1 簇的概念 187
8.4.2 创建簇表 188
8.4.3 查看簇信息 190
8.4.4 管理簇 191
8.5 视图 191
8.5.1 创建普通视图 192
8.5.2 操作普通视图 193
8.5.3 普通视图的更改与删除 194
8.5.4 创建物化视图 195
8.6 序列 196
8.6.1 创建序列 196
8.6.2 使用序列 197
8.6.3 修改、删除序列 198
8.6.4 自动序列 198
8.7 同义词 199
8.7.1 创建同义词 199
8.7.2 删除同义词 200
8.8 XML 和 Oracle 数据库 200
8.8.1 从关系数据生成 XML 200
8.8.2 XML DB 数据处理 202
练习 209
训练任务 210

第 9 章 表数据维护 211
9.1 使用 INSERT INTO 语句添加行 211
9.1.1 省略列的列表，默认值 212
9.1.2 为列指定空值 213
9.1.3 从一个表向另一个表复制行 214
9.2 使用 UPDATE 语句修改行 214
9.3 使用 DELETE 语句删除行 215

- 9.4 使用 MERGE 合并行 ……………………………………………………………… 215
- 9.5 数据库事务 ……………………………………………………………………… 216
  - 9.5.1 事务的提交和回滚 ………………………………………………………… 217
  - 9.5.2 事务的开始与结束 ………………………………………………………… 219
  - 9.5.3 保存点 ……………………………………………………………………… 219
  - 9.5.4 事务的 ACID 特性 ………………………………………………………… 220
  - 9.5.5 锁 …………………………………………………………………………… 226
- 练习 ……………………………………………………………………………………… 239
- 训练任务 ………………………………………………………………………………… 240

第 10 章　SQL 语言基础 ……………………………………………………………… 242
- 10.1 SQL 语言概述 …………………………………………………………………… 243
- 10.2 选择部分列 ……………………………………………………………………… 243
- 10.3 WHERE 子句 …………………………………………………………………… 244
- 10.4 列算术运算 ……………………………………………………………………… 246
- 10.5 禁止重复行 ……………………………………………………………………… 247
- 10.6 排序 ……………………………………………………………………………… 248
- 10.7 表别名及多表查询 ……………………………………………………………… 248
- 10.8 子查询 …………………………………………………………………………… 251
  - 10.8.1 单行子查询 ………………………………………………………………… 251
  - 10.8.2 多行子查询 ………………………………………………………………… 252
  - 10.8.3 Top N 查询 ………………………………………………………………… 253
  - 10.8.4 分页查询 …………………………………………………………………… 254
- 10.9 递归查询 ………………………………………………………………………… 256
- 练习 ……………………………………………………………………………………… 259
- 训练任务 ………………………………………………………………………………… 260

第 11 章　使用函数 ……………………………………………………………………… 261
- 11.1 单行函数 ………………………………………………………………………… 262
  - 11.1.1 字符处理函数 ……………………………………………………………… 262
  - 11.1.2 数值函数 …………………………………………………………………… 266
  - 11.1.3 类型转换函数 ……………………………………………………………… 269
  - 11.1.4 日期和时间函数 …………………………………………………………… 271
  - 11.1.5 正则表达式函数 …………………………………………………………… 272
- 11.2 分组查询及聚合函数 …………………………………………………………… 276
- 11.3 SQL 语句优化 …………………………………………………………………… 279
- 练习 ……………………………………………………………………………………… 283
- 训练任务 ………………………………………………………………………………… 284

第12章 PL/SQL 语言 ························· 286
12.1 PL/SQL 简介 ························· 287
12.1.1 PL/SQL 基本结构 ························· 288
12.1.2 变量和常量 ························· 289
12.1.3 可变数组 ························· 291
12.1.4 运算符 ························· 293
12.1.5 条件 ························· 294
12.1.6 循环 ························· 297
12.2 异常处理 ························· 299
12.2.1 预定义异常 ························· 299
12.2.2 自定义异常 ························· 300
12.2.3 引发应用程序异常 ························· 301
12.3 游标 ························· 303
12.3.1 游标的基本操作 ························· 303
12.3.2 游标 FOR 循环 ························· 306
12.3.3 引用游标 ························· 307
12.3.4 修改或删除游标结果集 ························· 308
12.4 存储过程 ························· 309
12.4.1 创建存储过程 ························· 310
12.4.2 调用存储过程 ························· 312
12.5 自定义函数 ························· 313
12.5.1 函数的创建与调用 ························· 313
12.5.2 函数参数的调用形式 ························· 314
12.6 删除过程和函数 ························· 315
12.7 块内存储过程和函数 ························· 315
12.8 过程与函数的比较 ························· 316
12.9 包 ························· 318
12.9.1 创建包 ························· 318
12.9.2 调用包 ························· 319
12.10 触发器 ························· 320
12.10.1 创建触发器 ························· 320
12.10.2 触发器的管理 ························· 324
12.10.3 行级触发器 ························· 325
12.10.4 系统级触发器 ························· 326
练习 ························· 328
训练任务 ························· 329

第 13 章 备份与恢复 ··· 331
13.1 备份与恢复概述 ·· 332
13.2 脱机备份与恢复 ·· 333
13.3 用户管理备份与恢复 ··· 334
13.4 RMAN 工具 ·· 339
13.4.1 备份集与镜像复制 ··· 339
13.4.2 启动 RMAN 并连接到数据库 ······································ 340
13.4.3 备份失效(Expired) ··· 341
13.4.4 备份过期(Obsolete) ·· 341
13.4.5 RMAN 备份和恢复命令 ··· 343
13.4.6 实用案例：完全恢复一个 PDB ···································· 347
13.4.7 实用案例：不完全恢复一个 PDB ································· 349
13.4.8 RMAN 批处理 ·· 351
13.5 闪回技术 Flashback ··· 352
13.5.1 Flashback Database ·· 353
13.5.2 Flashback Table ··· 354
13.5.3 回收站 ·· 355
13.6 数据导出与导入 ·· 356
13.6.1 Oracle 目录对象 ··· 357
13.6.2 数据导出 ··· 358
13.6.3 数据导入 ··· 359
练习 ·· 362
训练任务 ·· 363

第 14 章 小型商品销售系统 ··· 365
14.1 小型商品销售系统 E-R 模型 ··· 366
14.1.1 实体模型 ··· 366
14.1.2 实体联系模型 ··· 367
14.2 数据表的设计 ·· 368
14.3 用户创建与空间分配 ··· 370
14.4 创建表，约束和索引 ··· 372
14.5 创建触发器、序列和视图 ··· 375
14.6 创建程序包、函数和过程 ··· 378
14.7 数据库测试 ··· 380
14.8 应用程序开发 ·· 385
14.8.1 IDE 选择 ··· 385
14.8.2 程序目录结构和通用模块 ·· 385

  14.8.3 配置文件详述 ··· 386
  14.8.4 管理主界面与登录程序设计 ································· 389
  14.8.5 程序主要模块 ··· 390
附录 练习答案与训练任务的实现 ·· 404
 第 1 章 Oracle 12c 简介 ··· 404
 第 2 章 Oracle 12c 的安装 ··· 404
 第 3 章 网络配置及管理工具 ··· 406
 第 4 章 数据库管理与配置 ··· 410
 第 5 章 Oracle 12c 数据库结构 ··· 412
 第 6 章 数据库存储管理 ··· 415
 第 7 章 用户及权限管理 ··· 417
 第 8 章 数据库的对象管理 ··· 418
 第 9 章 表数据维护 ··· 421
 第 10 章 SQL 语言基础 ··· 422
 第 11 章 使用函数 ··· 423
 第 12 章 PL/SQL 语言 ··· 424
 第 13 章 备份与恢复 ··· 428

第1章 Oracle 12c 简介

本章目标

知识点	理解	掌握	应用
1.了解 Oracle 12c 的发展和特点	✓	✓	
2.了解 Oracle 12c 产品系列	✓	✓	
3.理解插接式数据库的特点	✓	✓	
4.了解 Oracle 12c 的新特性	✓		

知识能力点

能力点 \ 知识点	知识点 1	知识点 2	知识点 3	知识点 4
工程知识				
问题分析	✓	✓	✓	✓
设计/开发解决方案				
研究	✓	✓	✓	✓
使用现代工具				
工程与社会	✓	✓	✓	✓
环境和可持续发展	✓	✓	✓	✓
职业规范	✓		✓	✓
个人和团队	✓	✓	✓	✓
沟通				
项目管理				✓
终身学习			✓	✓

1.1 Oracle 12c 简介

Oracle 是当前最流行的大型关系数据库之一，拥有广泛的用户和大量的应用案例。2013 年 7 月，Oracle Database 12c 版本正式发布，首先发布的版本号是 12.1.0.1.0，支持包括 64 位 Windows、HP-UX、Solaris 和 Linux 等多种操作系统，本书将在 Linux 平台上安装和运行 Oracle 12c。

和甲骨文前几代数据库 Oracle 8i、9i、10g、11g 相比，Oracle 12c 命名上的"c"明确了这是一款针对云计算(Cloud)而设计的数据库。按照甲骨文公司披露的信息，Oracle 12c

增加了 500 多项新功能，其新特性主要涵盖了六个方面：云端数据库整合的全新多租户架构、数据自动优化、深度安全防护、面向数据库云的最大可用性、高效的数据库管理以及简化大数据分析。这些特性可以在高速度、高可扩展、高可靠性和高安全性的数据库平台之上，为客户提供一个全新的多租户架构，用户数据库向云端迁移后可提升企业应用的质量和应用性能，还能将数百个数据库作为一个整体进行管理，帮助企业在迈向云的过程中提高整体运营的灵活性和有效性。

1.2 Oracle 12c 产品系列

Oracle 12c 为适合不同规模的组织需要提供了多个量身定制的版本，并为满足特定的业务和 IT 需求提供了几个企业版专有选件。这三个版本是：标准版 1(SE1)，标准版(SE)和企业版(EE)。

1) 标准版 1(SE1)

标准版 1 为工作组、部门和 Web 应用程序提供空前的易用性、能力和性价比，运行在最多支持两个插槽的单一服务器上。

2) 标准版(SE)

标准版可运行在最多 4 个插槽的单一或集群服务器上使用。该版本包含 Oracle Real Application Clusters，这是一个标准特性，无需任何额外成本。

3) 企业版(EE)

可在无插槽限制的单一和集群服务器上使用。它为任务关键型事务应用程序、查询密集型数据仓库以及混合负载提供高效、可靠且安全的数据管理。

Oracle 12c 的所有版本均使用同一个代码库构建而成，彼此之间完全兼容。Oracle 12c 可用于多种操作系统，并且包含一组通用的应用程序开发工具和编程接口。客户可以从标准版 1 开始使用，随着业务的发展或根据需求的变化，可以轻松升级到标准版或企业版。升级过程非常简单：只是安装下一个版本的软件，无需对数据库或应用程序进行任何更改，便可在一个易于管理的环境中获得 Oracle 举世公认的性能、可伸缩性、可靠性和安全性。

1.3 Oracle 12c 新特性

1.3.1 插接式数据库 PDB

插接式数据库(Pluggable Database，PDB)是 Oracle 12c 最新最强的新特性之一，也称为多租户架构(Multitenant Architecture)。PDB 是可移植的模式、模式对象和非模式对象的集合，作为单独的数据库呈现到 Oracle Net 客户端。一个或者多个 PDB 合成为容器数据库(Container Database，CDB)。它们对用户和应用程序是完全透明的，并与之前的数据库版本完全兼容。通过 CDB 方式建立的私有数据库云架构可以使多个 PDB 共享服务器、操作系统和数据库,不用开很多虚拟机(虚拟机 DB 性能减半),仅在容器级别才需要内存和进程,同时又便于管理,比如统一备份,统一容灾,统一安全性管理等。

> 注意:在 Oracle 中,模式跟用户是一对一的关系,模式是数据库对象的集合,逻辑上这些对象分为模式对象和非模式对象,模式对象是用户直接访问的对象,如:表、索引、视图、存储过程、簇、序列和同义词等。非模式对象是用户依赖的对象,如用户、权限、表空间等。

容器(Container)可以是一个 PDB 或者 Root 容器(也称为 Root)。Root 容器是一个模式、模式对象和非模式对象的集合,所有的 PDB 都属于 Root。每个 CDB 都包含以下容器:

1)一个 Root

Root 包含 Oracle 的元数据和公用用户,例如 Oracle 提供的 PL/SQL 包的源代码。公用用户是每个容器中都可以使用的数据库用户。Root 容器的名称为 CDB$ROOT。

2)一个种子 PDB

种子 PDB 是系统提供的一个模板,可以用于 CDB 创建新的 PDB。种子 PDB 的名称为 PDB$SEED。用户不能添加或者修改 PDB$SEED 中的对象。

3)零个或者多个用户创建的 PDB。

PDB 由用户创建,一个 PDB 可以支持一个特定应用,例如人力资源或者销售。创建 CDB 时不会创建 PDB,可以基于业务需求添加 PDB。

图 1-1 显示了一个拥有 4 个容器的 CDB$ROOT、Seed、hrpdb 和 salespdb。hrpdb 和 salespdb 分别拥有自己的应用,并且由它自己的 PDB 管理员进行管理。一个公用用户在 CDB 中使用单个身份认证。公用用户 SYS 可以管理 Root 和每个 PDB。在物理层,该 CDB 拥有一个数据库实例和数据库文件,与非 CDB 一样。

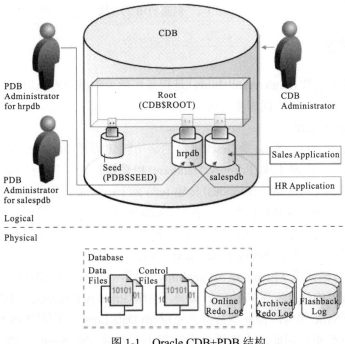

图 1-1　Oracle CDB+PDB 结构

在安装 Oracle 12c 时可以选择以 PDB 模式安装，Oracle 鼓励安装时使用 PDB 技术，它的好处包括降低成本、数据和代码分离、便于管理和监控，以及管理职责分离。与 CDB 和 PDB 的管理相关的任务可以用以下工具来执行：

1) sqlplus
2) dbca
3) 企业管理器云控制器 (Oracle Enterprise Manager Cloud Control)
4) Oracle SQL Developer
5) 服务器控制 (srvctl)

CDB 只起容器作用，包含很少或者不包含用户数据，用户数据应当保存在 PDB 中。只能在 CDB 中创建 PDB，而不能在 PDB 中创建 PDB。

图 1-2 描述了创建 PDB 的 5 种可选方式。

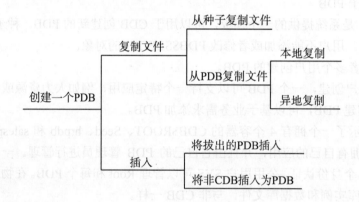

图 1-2 创建 PDB 的方式

1.3.2 高可用性

数据库的高可用性 (High Availability) 是指尽可能少的减少停机时间和减少因停机造成的数据损失，从而保证数据库服务的高度可用。

1) Active Data Guard

Active Data Guard Far Sync 是 Oracle 12c 的新功能 (也称为 Far Sync Standby)，Far Sync 功能的实现是通过在距离主库 (Primary Database) 相对较近的地点配置 Far Sync 实例，主库 (Primary Database) 同步 (Synchronous) 传输 redo 到 Far Sync 实例，然后 Far Sync 实例再将 redo 异步 (Asynchronous) 传输到终端备库 (Standby Database)。这样既可以保证零数据丢失又可以降低主库压力。Far Sync 实例只有密码文件，init 参数文件和控制文件，而没有数据文件。图 1-3 描述了 Active Data Guard 的工作过程。

如果 Redo 传输采用 Maximum Availability 模式，我们可以在距离生产中心 (Primary Database) 相对较近的地点配置 Far Sync 实例，主库 (Primary Database) 同步 (Synchronous) 传输 redo 到 Far Sync 实例，保证零数据丢失 (Zero Data Loss)，同时主库和 Far Sync 距离较近，网络延时很小，因此对主库性能影响很小。然后 Far Sync 实例再将 redo 异步

（Asynchronous）发送到终端备库（Standby Database）。

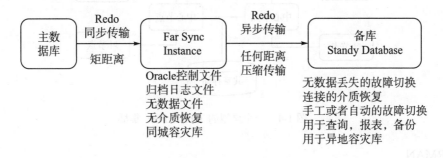

图 1-3 Active Data Guard 工作过程

如果 Redo 传输采用 Maximum Performance 模式，我们可以在距离生产中心（Primary Database）相对较近的地点配置 Far Sync 实例，主库（Primary Database）异步传输 redo 到 Far Sync 实例，然后 Far Sync 实例再负责传输 redo 到其他多个终端备库（Standby Database）。这样可以减少主库向多个终端备库（Standby Database）传输 redo 的压力（Offload）。

2）Transaction Guard

Oracle GoldenGate（OGG）可以帮助企业在传统数据库和云平台、大数据平台之间进行实时复制。新的 OGG 12c 支持更多的异构数据库和大数据平台，进一步提升可管理性和对混合云的支持。一个新的 Streams 迁移工具 Streams2OGG，帮助 Streams 用户迁移到 OGG 平台，利用 OGG 当前的集成捕获和冲突管理等新功能；支持 MS SQL Server 2012/2014 及 mysql 社区版；支持 MS SQL Server 2012/2014 及 mysql 社区版；支持 Socks5 通信协议，用户现在可以使用 Socks 兼容的协议进行数据传输，比如在没有 VPN 网络的内网和云平台之间进行数据传输；支持大数据：OGG For Java Adapter 可与 Oracle NoSQL、Hadoop、HDFS、HBase、Storm、Flume、Apache Kafka 等平台进行实时的数据集成，实现大数据平台的实时加载和分析；

3）Flex ASM 和 Flex Cluster

Oracle Real Application Cluster（RAC）是 Oracle 解决方案中的一个著名产品，Oracle RAC 允许在所有集群节点之间共享负载，采用 N-1 容错配置来应对节点故障，其中 N 是节点总数。Oracle 12c 包含"Flex ASM"和"Flex Cluster"两个属性，支持面向云计算的环境的各种苛刻需求。Oracle RAC 12c 引入以下了两个新概念。

中心节点：中心节点通过专用网络相互连接，并且可以直接访问共享存储。这些节点可以直接访问 Oracle 集群注册表（OCR）和表决磁盘（VD）。

叶节点：叶节点是轻型节点，彼此不互连，也不能像中心节点一样访问共享存储。每个叶节点与所连接的中心节点通信，并通过所连接的中心节点连接到集群。图 1-4 描绘了一个典型的 Oracle Flex 集群，包含 4 个叶节点和两个中心节点。其中只有中心节点可以直接访问 Oracle 集群注册表（OCR）和表决磁盘（VD）。但是应用可以通过叶节点访问数据库，而不必在叶节点上运行 ASM 实例。

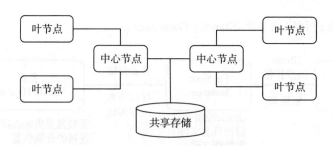

图1-4 一个典型的 Oracle Flex 集群

4）RMAN

RMAN 是 Oracle 的备份和恢复工具，在 Oracle 12c 中，这个工具已经有了大量的增强，其中包括：

（1）对本地数据库进行压缩备份，再进行异地恢复，可以让活动并行完成。

（2）可以在 RMAN 提示符下运行 SELECT 语句。

（3）不带 SQL 前缀运行大多数 SQL 命令。

（4）可以用命令 DESCRIBE 查询表结构。

（5）可以方便的仅恢复一个表或一组表。

1.3.3 XML DB

XML DB 从 Oracle 11g 开始就已经嵌入到 Oracle 产品中了，从而使 Oracle 可以直接访问 XML 数据，Oracle 12c 又在此基础上增强了 XML 引擎的访问能力。

（1）执行 Xquery 操作时，改善少量访问路径的效率。

（2）改为二进制 XML 格式和索引以优化 Xpath 和 Xquery 操作的执行。

（3）更多的 W3C 的查询语言功能，可以执行片断级和节点级更新。

（4）对普通物理分区方法的完全支持，例如管理员已经熟悉的对非 XML DB 数据的散列区和子分区。

（5）支持域索引使用散列和间隔分区。

（6）集成对调试和查看 Xquery 执行计划的支持，优化数据访问和吞吐量的处理。

1.3.4 In-Memory 数据库内存选件

In-Memory 特性使表的列存储在内存中，实现 4 个主要目标：

（1）显著增快 SQL 的全表扫描处理速度，全表扫描将增快 10~100 倍。

（2）显著增快复杂 SQL 的处理，在绝大多数场景中连接处理将变快 10 倍或者更多。聚集、排序和分组也将随之变快。

（3）显著增快事务处理，单行 DML 和批量 DML 都将运行地更快，主要来源于降低 10 倍的索引维护。

（4）100%的应用程序透明。所有的其他 Oracle 特性均可以与 In-Memory Option 一起工作，包括 Partitioning，Indexes，Text Indexes，没有明确的数据类型或者存储类型限制。

1.3.5　Oracle JSON 文档存储

Oracle 12c 还支持 Oracle JSON 文档存储，可以在 Oracle 关系型数据库中存储 JSON 文档数据，同时支持使用 SQL 或 REST 接口来对 JSON 数据进行查询。

1.3.6　其他新特性

1) PL/SQL 性能增强

类似在匿名块中定义过程，现在可以通过 WITH 语句在 SQL 中定义一个函数，采用这种方式可以提高 SQL 调用的性能。

2) 改善 Defaults

包括序列作为默认值、自增列、明确插入 NULL 时指定默认值、METADATA-ONLY，在 INSERT INTO 的 VALUES 子句中，DEFAULT 关键字代表默认值。在 Oracle 12c 中创建表的时候，可以对 NULL 列指定默认值。

3) 放宽多种数据类型长度限制

增加 VARCHAR2、NVARCHAR2 和 RAW 类型的长度到 32K，并且设置了初始化参数 MAX_SQL_STRING_SIZE 为 EXTENDED，这个功能不支持 CLUSTER 表和索引组织表。

4) 分页语句的实现

在 SELECT 语句中使用 FETCH FIRST/NEXT、OFFSET 以及 PERCENT，可以简便实现分页查询。

5) 行模式匹配

类似分析函数的功能，可以在行间进行匹配判断并进行计算。在 SQL 中新的模式匹配语句是"match_recognize"。

6) 临时 UNDO

将临时段的 UNDO 独立出来，放到 TEMP 表空间中，优点包括：减少 UNDO 产生的数量；减少 REDO 产生的数量；在 ACTIVE DATA GUARD 上允许对临时表进行 DML 操作。

练习

一、判断题

1. 容器(container)可以是一个 PDB 或者 Root 容器。　　　　　（　　）
2. Oracle 12c 只能在 Windows 平台上运行。　　　　　　　　　（　　）
3. 用户不能添加或者修改 PDB$SEED 中的对象。　　　　　　　（　　）
4. 可以在 PDB 中创建 PDB。　　　　　　　　　　　　　　　　（　　）
5. Oracle 12c 表中可以为空的列不能设置默认值。　　　　　　　（　　）
6. In-Memory 使用内存列存储来实现表的快速访问。　　　　　　（　　）

二、填空题

1. Oracle 12c 中的字母 c 表示_____。

2. Oracle 12c 有三个版本，分别是标准版 1，标准版和_____版。

3. Root 容器的名称为_____。

4. 种子 PDB 的名称为_____。

5. high availability 是指数据库的_____性。

6. 数据库的高可用性是指尽可能减少_____时间和减少因停机造成的数据损失，从而保证数据库服务的高度可用。

三、简答题

1. Oracle 12c 有哪些版本，各有何特点？

2. PDB 多租户架构有什么好处。

3. 简述 Active Data Guard 工作过程。

4. 简述 In-Memory Option 数据库内存选件的功能。

第 2 章　Oracle 12c 的安装

本章目标

知识点	理解	掌握	应用
1.安装前配置 Linux	✓		
2.安装 Oracle 12c	✓	✓	✓
3.安装后的数据库连接测试		✓	✓
4.安装后的检测		✓	✓

训练任务

- 安装 Oracle 12c，连接到数据库并且观察运行状态。

知识能力点

能力点 \ 知识点	知识点 1	知识点 2	知识点 3	知识点 4
工程知识	✓	✓		
问题分析	✓	✓		
设计/开发解决方案		✓		
研究		✓	✓	✓
使用现代工具			✓	✓
工程与社会			✓	✓
环境和可持续发展	✓			
职业规范		✓		
个人和团队	✓			
沟通	✓	✓		
项目管理		✓		
终身学习				✓

2.1　安装前配置 Linux 系统

Oracle 12c 只支持 64 位的 Linux 系统，不支持 32 位 Linux 平台。本书使用的 Linux 版本是 CentOS release 6.5（Final），内核版本号是 2.6.32-431.el6.x86_64。Oracle 12c 对系统内存的最低要求为 1G，推荐 2G 或更大的内存。Oracle 12c 大约需要 6G 的磁盘空间，建议在学习时使用 50G 以上的硬盘空间。

2.1.1 配置 Linux 交换空间

Oracle 需要使用操作系统部的交换空间,建议的交换空间大小与内存有关,如果内存是 1GB 到 2GB,交换空间是内存的 1.5 倍,如果内存是 2GB 到 16GB,交换空间等于内存大小,如果内存超过 16GB,交换空间等于 16GB。Linux 系统的交换空间通常都比较小,不能满足 Oracle 12c 的要求,在 Oracle 的安装和运行时可能出现警告或异常。下面是一个增加 Linux 交换空间的示例,将原有交换空间 5GB(5119KB),增加到了 9GB(9119KB)。

【示例 2-1】 增加 Linux 的交换空间

本例中使用目录/opt/swap 来作为新增交换空间的目录。

```
#free -m
Swap:         5119        0        5119
#dd if=/dev/zero of=/opt/swap bs=1024
#mkswap /opt/swap
#swapon /opt/swap
#free -m
Swap:         9119        0        9119
```

上述操作第一次 free -m 命令看到当前的交换空间大小是 5119KB,经过增加后,最后再用 free -m 命令可以看见交换空间已经增到了 9119KB。

2.1.2 创建 Oracle 用户和用户组

在安装 Oracle 之前,需要在 Linux 中创建安装和管理 Oracle 的用户以及用户组,以便以新用户身份安装。安装后,新用户也自然拥有数据库的管理权限。

【示例 2-2】 创建用户 oracle 和用户组 dba, oinstall

```
#groupadd dba
#groupadd oinstall
#useradd -g oinstall -G dba oracle
#id oracle
uid=503(oracle)gid=503(oinstall)组=503(oinstall),504(dba)
#passwd oracle
```

创建了 oracle 用户以及相关用户组后,需要设置 oracle 用户密码。创建过程产生的用户号和组号不一定是 503、504,要以实际产生的号为准。注意 dba 组号要在下面的 sysctl.conf 文件中使用。

2.1.3 配置/etc/sysctl.conf 文件

安装前还需要配置/etc/sysctl.conf 文件,设置一些系统参数。

【示例2-3】 配置/etc/sysctl.conf文件

```
# Controls the maximum shared segment size, in bytes
kernel.shmmax = 68719476736
# Controls the maximum number of shared memory segments, in pages
kernel.shmall = 4294967296
kernel.sem =250 32000 100 128
kernel.shmmni = 4096
net.core.wmem_default = 262144
net.core.wmem_max = 1048576
net.core.rmem_default = 262144
net.core.rmem_max = 4194304
fs.file-max=6815744
fs.aio-max-nr=1048576
ip_local_port_range=9000 65535
#id oracle
#uid=503(oracle)gid=503(oinstall)组=503(oinstall),504(dba)
vm.hugetlb_shm_group = 504
```

注意，参数修改完成后，运行下面的命令才能使修改生效：

```
# sysctl -p
```

2.1.4 其他配置

为了在安装界面中显示为中文，应该将/etc/sysconfig/i18n文件中的LANG="en_US.UTF-8"换成LANG="zh_CN.UTF-8"，然后重启系统。另外，还可能需要安装一些其他依赖库。

```
# yum install compat-libcap1-1.10
# yum install compat-libstdc++-33-3.2.3
# yum install ksh
# yum install libaio-devel-0.3.107
# yum install gcc-c++
```

由于服务器的IP地址可能会变动，最好设置固定的主机名hostname，在Oracle监听器内部用hostname代替IP地址，从而方便数据库的访问。下面两个步骤将hostname名称设置为oracle-server。

（1）在/etc/hosts文件中增加一行：127.0.0.1。

（2）在/etc/sysconfig/network文件设置HOSTNAME=oracle-server。

最后，因为Oracle的监听服务要用到1521端口，所以安装前必须确保1521端口没有被其他程序占用。

2.2 安装 Oracle 12c

可以在 Oracle 官方网站下载 Oracle 12c 的 Linux 安装文件，一共有两个：linuxamd64_12102_database_1of2.zip 和 linuxamd64_12102_database_2of2.zip。

以创建的 Oracle 用户登录后，将这两个安装文件都解压到同一个目录，然后运行"./runInstaller"就开始安装了。

```
$ unzip linuxamd64_12102_database_1of2.zip
$ unzip linuxamd64_12102_database_2of2.zip
$ cd database
$ ./runInstaller
正在启动 Oracle Universal Installer...
检查临时空间：必须大于 500 MB。    实际为 30295 MB    通过
检查交换空间：必须大于 150 MB。    实际为 9119 MB    通过
检查监视器：监视器配置至少必须显示 256 种颜色。    实际为 16777216    通过
准备从以下地址启动 Oracle Universal Installer/tmp/OraInstall2015-
02-13_08-12-33AM。请稍候...
```

安装过程及安装选项如图 2-1，图 2-13 所示：

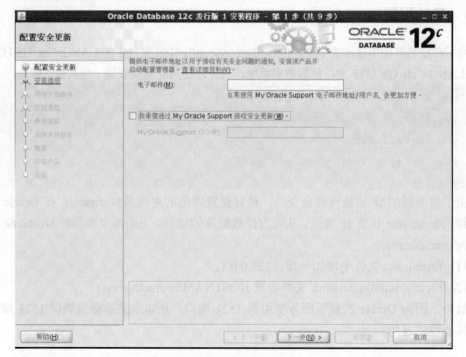

(a)

第 2 章　Oracle 12c 的安装　　13

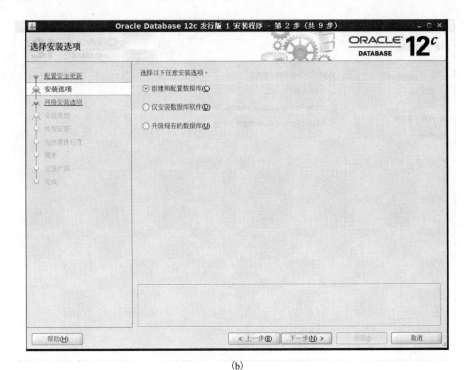

(b)

图 2-1　Oracle 12c 安装过程 1

图 2-2 选择服务器类，这是要进行自定义安装，有更多选项，同时也可以安装 Oracle 提供的示例数据库，如果要快捷安装，可以选择"桌面类"。

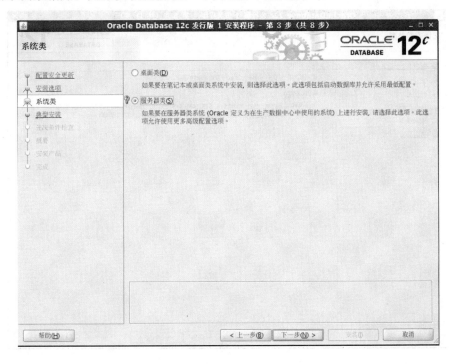

(a)

(b)

图 2-2 Oracle 12c 安装过程 2

选择高级安装和产品语言:

(a)

(b)

图 2-3 Oracle 12c 安装过程 3

选择安装版本和安装目录的位置:

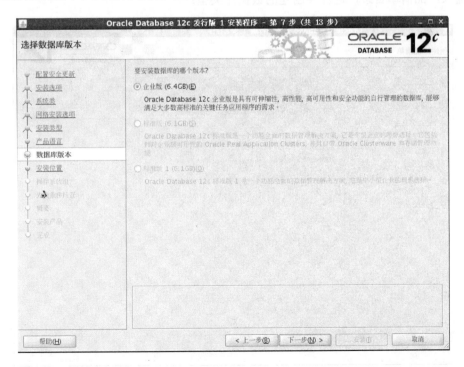

(a)

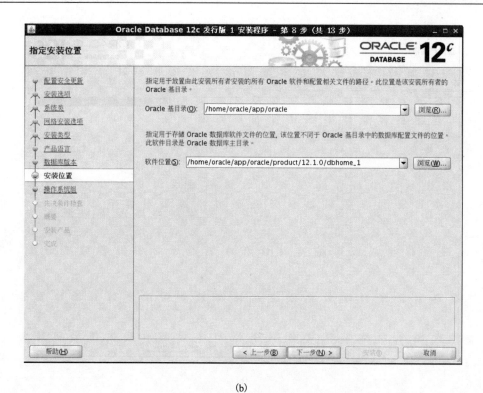

(b)

图 2-4　Oracle 12c 安装过程 4

输入产品清单目录，选择要创建的数据库类型：

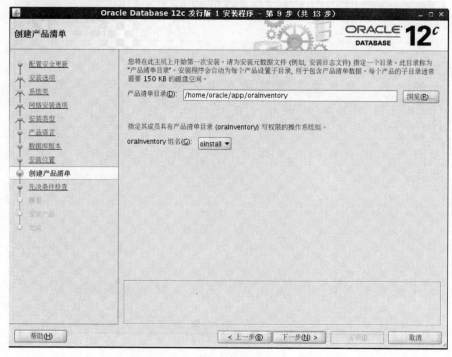

(a)

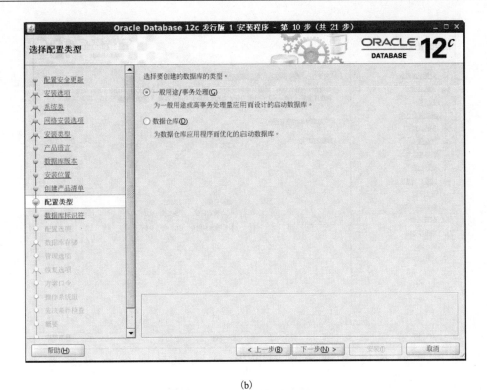

(b)

图 2-5　Oracle 12c 安装过程 5

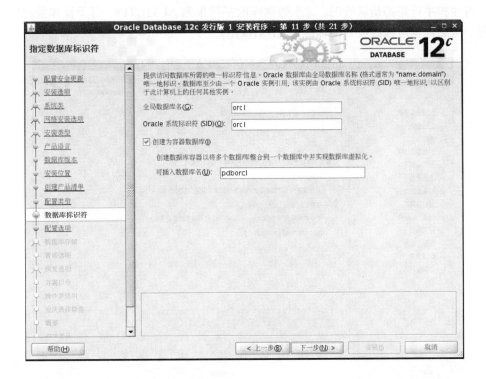

(a)

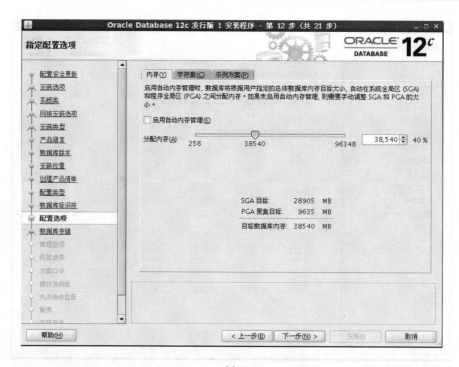

(b)

图 2-6　Oracle 12c 安装过程 6

内存和字符集都是缺省的，这里数据库的字符集为 AL32UTF8，等于操作系统的语言设置，当然，也可以改成"简体中文 ZHS16GBK"。

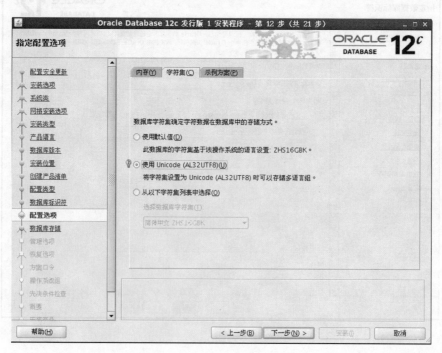

(a)

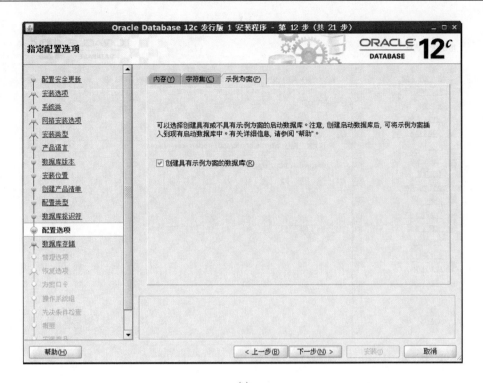

(b)

图 2-7　Oracle 12c 安装过程 7

选择启用恢复：

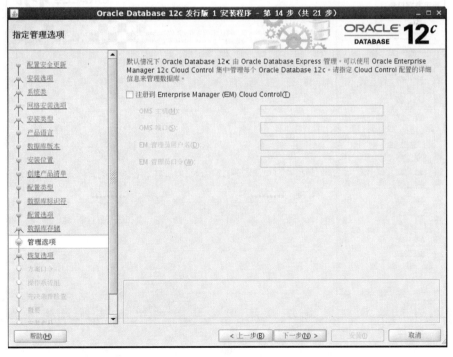

(a)

(b)

图 2-8　Oracle 12c 安装过程 8

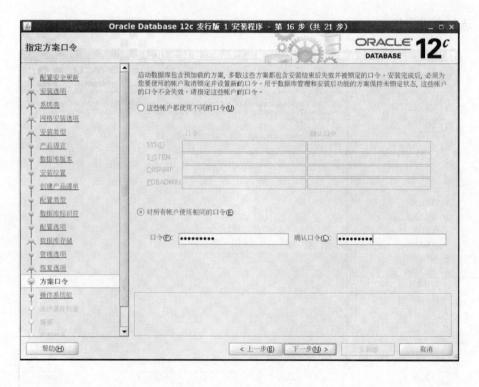

(a)

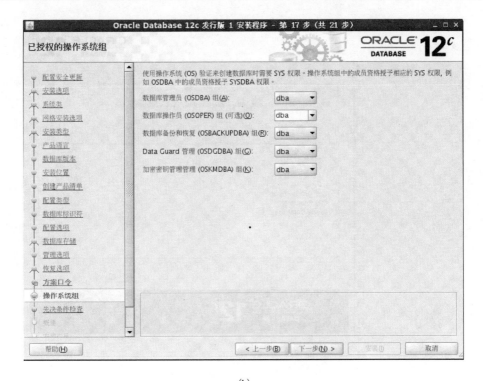

(b)

图 2-9　Oracle 12c 安装过程 9

(a)

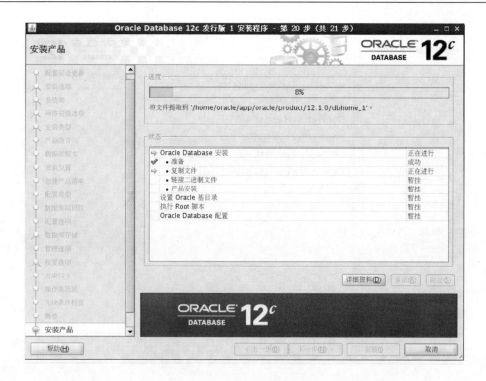

(b)

图 2-10　Oracle 12c 安装过程 10

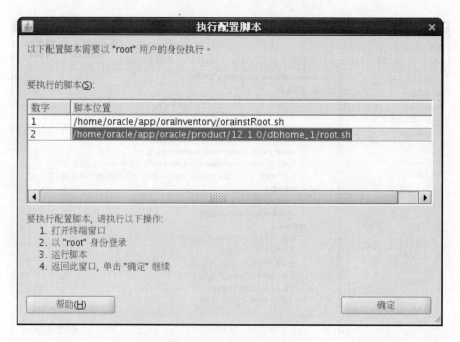

图 2-11　Oracle 12c 安装过程 11

在看见图 2-11 的提示后，需要在 Shell 终端中执行以下两个命令：

```
# /home/oracle/app/oraInventory/orainstRoot.sh
更改权限/home/oracle/app/oraInventory.
添加组的读取和写入权限。
删除全局的读取，写入和执行权限。
更改组名/home/oracle/app/oraInventory 到 oinstall.
脚本的执行已完成。
# /home/oracle/app/oracle/product/12.1.0/dbhome_1/root.sh
Performing root user operation.
The following environment variables are set as:
    ORACLE_OWNER= oracle
    ORACLE_HOME=  /home/oracle/app/oracle/product/12.1.0/dbhome_1
Enter the full pathname of the local bin directory: [/usr/local/bin]:
   Copying dbhome to /usr/local/bin ...
   Copying oraenv to /usr/local/bin ...
   Copying coraenv to /usr/local/bin ...
Creating /etc/oratab file...
Entries will be added to the /etc/oratab file as needed by
Database Configuration Assistant when a database is created
Finished running generic part of root script.
Now product-specific root actions will be performed.
```

执行完成后，点击如图 2-11 的"确定"按钮，安装继续进行，直到安装成功，如图 2-12 所示。

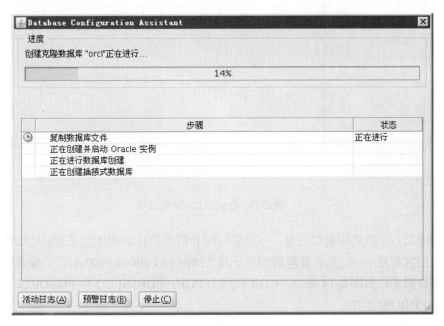

(a)

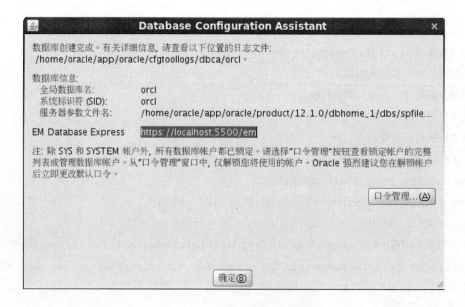

(b)

图 2-12 Oracle 12c 安装过程 12

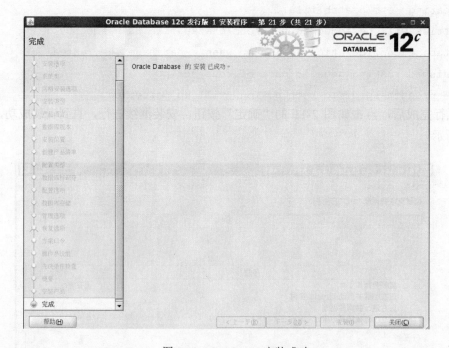

图 2-13 Oracle 12c 安装成功

安装成功后,在关闭窗口之前,一定要仔细分析并记住界面上的信息。比如界面上的数据库实例名称是 orcl,企业管理器的地址是"https://localhost:5500/em"。安装完成后一共有 3 个数据库,如图 2-14 所示,CDB 和两个 PDB:PDB$SEED 和 PDBORCL,示例方案存储在 PDBORCL 中:

第 2 章　Oracle 12c 的安装

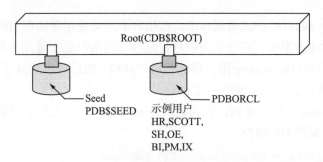

图 2-14　全局数据库实例 orcl 的 CDB 和 PDB

2.3　数据库连接测试

安装成功后可以通过一些工具连接并登录到数据库进行测试了，Oracle 主要的管理工具有 Sqlplus、Sqldeveloper 和企业管理器（Oracle Enterprise Manager）。

【示例 2-4】 sqlplus 测试连接

本示例使用安装 Oracle 时候创建的 oracle 用户登录，在 Linux 命令行下运行 sqlplus 命令。

```
$ sqlplus / as sysdba
SQL*Plus: Release 12.1.0.2.0 Production on 星期一 2月 13 09:54:38 2017
Copyright (c)1982, 2014, Oracle.  All rights reserved.
连接到:
Oracle Database 12c Enterprise Edition Release 12.1.0.2.0 - 64bit Production
With the Partitioning, OLAP, Advanced Analytics and Real Application Testing
options
SQL> SELECT con_id, name FROM v$database;
  CON_ID    NAME
---------- ----------
     0     ORCL
SQL> exit
从 Oracle Database 12c Enterprise Edition Release 12.1.0.2.0 - 64bit
Production
With the Partitioning, OLAP, Advanced Analytics and Real Application Testing
options 断开
$
```

本例的测试过程是这样的：用 sqlplus 命令以操作系统认证方式登录 Oracle，然后通过 SELECT 查询语句查询容器数据库的 ID 和 NAME，最后执行 exit 退出 Sqlplus 返回操作系统。

由于在安装的时候选择了安装样例数据库，所以现在可以查询一下样例数据的信息，样例数据存储在 pdborcl 接插式数据库中，按用户不同又分别存储在 HR、SCOTT、OE、IX、BI、PM 和 SH 方案中。在安装完成后，这些方案用户处在过期并且锁定的状态，不能直接登录，我们可以用 system 用户将这些用户解锁，然后可以登录了。下面以查看 HR 用户的表信息为例，介绍一下操作过程。

首先，用 system 用户登录到插接式数据库 pdborcl，并给 HR 用户解锁。

【示例 2-5】 解锁 HR 用户

```
$ sqlplus system/***@localhost:1521/pdborcl
SQL> SELECT username, user_id, account_status FROM dba_users;
USERNAME         USER_ID     ACCOUNT_STATUS
-------------    --------    ---------------------------------
BI               108         EXPIRED & LOCKED
PM               105         EXPIRED & LOCKED
IX               106         EXPIRED & LOCKED
SH               107         EXPIRED & LOCKED
OE               104         EXPIRED & LOCKED
HR               103         EXPIRED & LOCKED
SCOTT            109         EXPIRED & LOCKED
SQL> ALTER USER hr ACCOUNT UNLOCK;
用户已更改。
SQL> SELECT username, user_id, account_status FROM dba_users WHERE username='HR';
USERNAME         USER_ID     ACCOUNT_STATUS
-------------    --------    ---------------------------------
HR               103         EXPIRED
SQL> ALTER USER hr IDENTIFIED BY 123;
SQL> SELECT username, user_id, account_status FROM dba_users WHERE username='HR';
USERNAME         USER_ID     ACCOUNT_STATUS
-------------    --------    ---------------------------------
HR               103         OPEN
SQL> exit
```

注意：登录命令"sqlplus system/***@localhost:1521/pdborcl"中的***表示 system 用户的密码，这是安装数据库时候设置的，应该用真实的密码代替。本书后面的样例也这样约定。

上面的命令中，通过查询 dba_users 知道有哪些用户以及每个用户的状态，一开始 HR 的状态是 EXPIRED&LOCKED，表示用户已经过期并且被锁定了，通过 unlock 把 HR 解除锁定，再通过改密码解除过期。这几步操作之后，HR 用户就可以登录了。要启用其他用户也可以做同样的操作。比如下面的操作设置所有 Oracle 样例用户的密码为 123，并且解除锁定。

【示例 2-6】给所有样例用户设定密码并解锁

```
SQL> ALTER USER "HR" ACCOUNT UNLOCK ;
SQL> ALTER USER "HR" IDENTIFIED BY 123 ;
SQL> ALTER USER "SCOTT" ACCOUNT UNLOCK ;
SQL> ALTER USER "SCOTT" IDENTIFIED BY 123 ;
SQL> ALTER USER "SH" ACCOUNT UNLOCK ;
SQL> ALTER USER "SH" IDENTIFIED BY 123 ;
SQL> ALTER USER "BI" ACCOUNT UNLOCK ;
SQL> ALTER USER "BI" IDENTIFIED BY 123 ;
SQL> ALTER USER "OE" ACCOUNT UNLOCK ;
SQL> ALTER USER "OE" IDENTIFIED BY 123 ;
SQL> ALTER USER "IX" ACCOUNT UNLOCK ;
SQL> ALTER USER "IX" IDENTIFIED BY 123 ;
SQL> ALTER USER "PM" ACCOUNT UNLOCK ;
SQL> ALTER USER "PM" IDENTIFIED BY 123 ;
```

接下来，以 HR 用户登录，查询 HR 用户的 regions 表的信息。

【示例 2-7】用 HR 用户登录，查询样例数据

```
$ sqlplus hr/123@localhost:1521/pdborcl
SQL> SELECT table_name FROM user_tables;
TABLE_NAME
-----------------------------------------------------------
REGIONS
COUNTRIES
LOCATIONS
DEPARTMENTS
JOBS
EMPLOYEES
JOB_HISTORY
已选择 7 行。
SQL> SELECT * FROM regions WHERE region_id=1;
 REGION_ID REGION_NAME
---------- -------------------------
         1 Europe
```

上述操作通过 user_tables 查询用户能访问的所有表。在 Oracle 中有一些以 user_ 开头的视图，表示用户可以访问的对象视图，除了 user_tables 外，还有 user_triggers、user_queues、user_sequences、user_indexes、user_views 以及 user_procedures 等。

2.4　Oracle 企业管理器

Oracle 企业管理器是基于浏览器的 Oracle 管理工具，直观、方便。在安装完成后的界面中有企业管理器地址的提示：https://localhost:5500/em，打开这个地址，输入用户名 sys 或者 system，以及安装时候设置的密码就可以登录到管理器了。图 2-15 和图 2-16 分别是企业管理器的登录页面和主页面。

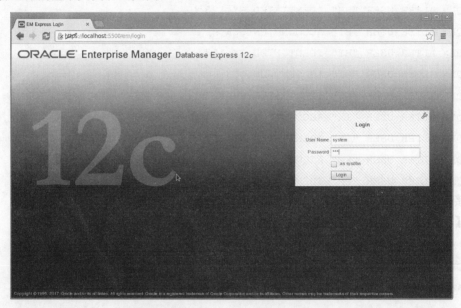

图 2-15　Oracle 12c 企业管理器的登录页面

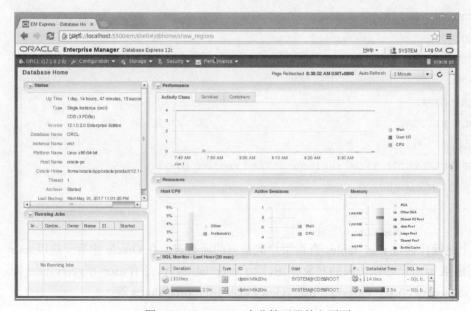

图 2-16　Oracle 12c 企业管理器的主页面

> 注意：Oracle 12c 的企业管理器与 11g 的功能和界面都完全不同了。Oracle 11g 的企业管理器的默认端口号是 1158，而 Oracle 12c 的是 5500。

可以允许以 http 或者 https 登录到 Oracle 企业管理器，并且设置登录到哪个数据库。通过 http 或者 https 的端口号指定是否允许登录，如果端口号为 0 表示禁止登录。http 和 https 可以同时指定，同时有效。使用 http 的好处是浏览器兼容性强，访问方便，但缺省安装的时候没有开放这个访问，需要重新设置。

【示例 2-8】设置登录到 CDB 的企业管理器的端口号

本示例为容器数据库 CDB 分配两个企业管理器端口号：https 的 5500 和 http 的 5502。其中的 EXEC 表示执行 Oracle 的存储过程。

```
$ sqlplus / as sysdba
SQL> EXEC DBMS_XDB_CONFIG.SETHTTPSPORT(5500);
PL/SQL procedure successfully completed.
SQL> EXEC DBMS_XDB_CONFIG.SETHTTPPORT(5502);
PL/SQL procedure successfully completed.
SQL>
```

【示例 2-9】设置登录到 PDB 的企业管理器的端口号

本示例为插接式数据库 pdborcl 分配两个企业管理器端口号：https 的 5501 和 http 的 5505，注意，通过命令"ALTER SESSION SET CONTAINER=pdborcl；"从 CDB 切换到数据库 pdborcl。

```
$ sqlplus / as sysdba
SQL> ALTER SESSION SET CONTAINER=pdborcl;
Session altered.
SQL> EXEC DBMS_XDB_CONFIG.SETHTTPSPORT(5500);
PL/SQL procedure successfully completed.
SQL> EXEC DBMS_XDB_CONFIG.SETHTTPPORT(5502);
PL/SQL procedure successfully completed.
SQL>
```

如果端口号是 0，则表示取消这个端口的监听，网站就无法访问了，比如：

```
SQL> EXEC DBMS_XDB_CONFIG.SETHTTPPORT(0);
```

【示例 2-10】查看已经分配的 HTTPS 端口号

如果数据库的 CDB 中分配了两个企业管理器端口号：https 的 5500 和 http 的 5502，CDB 包含的插接式数据库也有两个企业管理器端口号：https 的 5501 和 http 的 5505，那么可以通过以下命令查询出来。

```
$ sqlplus / as sysdba
SQL> SELECT dbms_xdb_config.gethttpsport FROM DUAL;
GETHTTPSPORT
------------
5500
```

```
SQL> SELECT dbms_xdb_config.gethttpport FROM DUAL;
GETHTTPPORT
-----------
5502
```
从 CDB 切换会话到 pdborcl：
```
SQL> ALTER SESSION SET CONTAINER =pdborcl;
Session altered.
SQL> SELECT dbms_xdb_config.gethttpsport FROM DUAL;
GETHTTPSPORT
------------
5501
SQL> SELECT dbms_xdb_config.gethttpport FROM DUAL;
GETHTTPPORT
-----------
5505
```
也可以通过 Oracle 的操作系统命令"lsnrctl status"查看端口占用情况。

2.5 安装后的检测

安装成功后，应该查看环境变量、安装路径中的文件及目录分配、数据库进程和监听器状态，做到心中有数，方便今后的数据库维护。

2.5.1 查看环境变量

Oracle 的环境变量极其重要，必须有正确的值才能正常运行和访问。在安装成功后，Oracle 环境变量有：ORACLE_OWNER、ORACLE_HOME、ORACLE_SID 以及 NLS_LANG。环境变量存储在/etc/profile 文件中，如果没有变量，或者变量的值不正确，可以进行修改。下面是安装成功后/etc/profile 文件的部分内容：

```
export ORACLE_OWNER=oracle
export ORACLE_HOME=/home/oracle/app/oracle/product/12.1.0/dbhome_1
export ORACLE_SID=orcl
export PATH=$ORACLE_HOME/bin:/usr/bin:$PATH
export NLS_LANG='SIMPLIFIED CHINESE_CHINA.AL32UTF8'
```

其中，ORACLE_OWNER 表示 Oracle 进程的所有者用户，ORACLE_HOME 是 Oracle 的安装目录，ORACLE_SID 是 Oracle 的默认实例名称，如果在本机中安装了多个实例，ORACLE_SID 的值就非常重要了，它表示客户端默认登录的实例。到目前为止，只有一个数据库实例 orcl。PATH 变量是设置搜索路径，可以让 Linux 找到 Oracle 命令。

尤其值得关注的是 NLS_LANG，它表示访问数据库的客户端的国家字符集，这个值决定 Linux 客户端上使用的消息语言，日期，数字格式以及实际数据的字符编码，这个值必须和安装时选择的数据库字符集相同，如果不相同，终端上显示汉字的数据或者提示的时候，可能为"乱码"，简言之，就是要求客户端字符集与数据库端字符集相同才会避免乱码。

NLS_LANG 由 3 部分组成，整体格式是：<language>_<territory>.<client character set>，其中：

  language：显示 Oracle 消息，校验，日期命名

  territory：指定默认日期、数字、货币等格式

  client character set：指定客户端将使用的字符集，这是最重要的配置，必须和服务器端的字符集完全相同。查看服务器端字符集的命令是：

```
SQL> SELECT * FROM V$NLS_PARAMETERS;
PARAMETER                     VALUE                         CON_ID
----------------------------- ----------------------------- ----------
NLS_LANGUAGE                  SIMPLIFIED CHINESE            0
NLS_TERRITORY                 CHINA                         0
NLS_CURRENCY                  ￥                            0
NLS_ISO_CURRENCY              CHINA                         0
NLS_NUMERIC_CHARACTERS        .,                            0
NLS_CALENDAR                  GREGORIAN                     0
NLS_DATE_FORMAT               DD-MON-RR                     0
NLS_DATE_LANGUAGE             SIMPLIFIED CHINESE            0
NLS_CHARACTERSET              AL32UTF8                      0
NLS_SORT                      BINARY                        0
NLS_TIME_FORMAT               HH.MI.SSXFF AM                0
NLS_TIMESTAMP_FORMAT          DD-MON-RR HH.MI.SSXFF AM      0
NLS_TIME_TZ_FORMAT            HH.MI.SSXFF AM TZR            0
NLS_TIMESTAMP_TZ_FORMATTZR    DD-MON-RR HH.MI.SSXFF AM      0
NLS_DUAL_CURRENCY             ￥                            0
NLS_NCHAR_CHARACTERSET        AL16UTF16                     0
NLS_COMP                      BINARY                        0
NLS_LENGTH_SEMANTICS          BYTE                          0
NLS_NCHAR_CONV_EXCP           FALSE                         0
```

上述查询结果中，NLS_LANGUAGE、NLS_TERRITORY 和 NLS_CHARACTERSET 这 3 行组合起来就是$NLS_LANG。

2.5.2 查看目录及文件

Oracle 安装完成之后，有必要查看相关文件以及存储目录。

1)Oracle 主目录

Oracle 的主目录由环境变量$ORACLE_HOME 决定，多数文件都放在主目录下。Oracle 的数据库启动日志文件和监听器启动日志文件也存储在此。

2)执行程序目录

执行程序目录是$ORACLE_HOME/bin，在这个目录中存储了 sqlplus，dbca，lsnrctl 等所有的可执行文件

3)网络配置文件目录

网络配置文件目录是$ORACLE_HOME/network/admin，这个目录中包件 listener.ora，tnsnames.ora，sqlnet.ora 等网络配置文件。只有正确配置网络配置文件，才能成功访问 Oracle 的各个数据库。

4)初始化文件目录

初始化文件存储的目录是$ORACLE_HOME/dbs，在这个目录里面，存放了 Oracle 实例启动时候的参数值。这些文件有：init.ora，spfile$ORACLE_SID.ora。

5)数据文件目录

数据文件目录是 app/oracle/oradata/$ORACLE_SID，默认安装的情况下，数据文件目录没有放在$ORACLE_HOME 目录下面，每个数据库实例单独一个子目录。

6)控制文件目录

默认安装的情况下，控制文件目录和数据文件目录相同。

7)重做日志文件目录

默认安装的情况下，重做日志文件目录和数据文件目录相同。

8)自动备份目录

自动备份及恢复目录是 app/oracle/fast_recovery_area。默认情况下这个目录中存储 Oracle 的归档日志文件，控制文件的备份，在线重做日志文件组的克隆文件。

2.5.3 查看 Oracle 进程

Oracle 在 Linux 中的进程名称是以"ora_"开头的，可以通过 ps 命令查看。

```
$ ps -ef | grep ora_
oracle    6808     1  0 May31 ?        00:00:07 ora_pmon_orcl
oracle    6810     1  0 May31 ?        00:00:28 ora_psp0_orcl
oracle    6812     1  3 May31 ?        00:46:51 ora_vktm_orcl
oracle    6816     1  0 May31 ?        00:00:07 ora_gen0_orcl
oracle    6818     1  0 May31 ?        00:00:05 ora_mman_orcl
...
```

2.5.4 查看监听器状态

Oracle 安装成功后，除了启动 Oracle 进程之外，还会启动监听进程，监听进程的作用是提供外部应用访问 Oracle 服务的接口。查看监听器状态的命令是 lsnrctl status。

```
$ lsnrctl status
LSNRCTL for Linux: Version 12.1.0.2.0 - Production on 15-2月-2017 18:41:32

Copyright (c)1991, 2014, Oracle.  All rights reserved.
正在连接到
(DESCRIPTION=(ADDRESS=(PROTOCOL=TCP)(HOST=deep02)(PORT=1521)))
LISTENER 的 STATUS
------------------------
别名                       LISTENER
版本                       TNSLSNR for Linux: Version 12.1.0.2.0 - Production
启动日期                    15-2月-2017 17:15:06
正常运行时间                 0 天 1 小时 26 分 25 秒
跟踪级别                    off
安全性                      ON: Local OS Authentication
SNMP                       OFF
监听程序参数文件
/home/oracle/app/oracle/product/12.1.0/dbhome_1/network/admin/listener.ora
监听程序日志文件
/home/oracle/app/oracle/diag/tnslsnr/deep02/listener/alert/log.xml
监听端点概要...
  (DESCRIPTION=(ADDRESS=(PROTOCOL=tcp)(HOST=deep02)(PORT=1521)))
  (DESCRIPTION=(ADDRESS=(PROTOCOL=ipc)(KEY=extproc)))

(DESCRIPTION=(ADDRESS=(PROTOCOL=tcps)(HOST=deep02)(PORT=5500))(Security=(my_wallet_directory=/home/oracle/app/oracle/admin/orcl/xdb_wallet))(Presentation=HTTP)(Session=RAW))
服务摘要..
服务 "PLSExtProc" 包含 1 个实例。
  实例 "PLSExtProc", 状态 UNKNOWN, 包含此服务的 1 个处理程序...
服务 "orcl" 包含 2 个实例。
  实例 "orcl", 状态 UNKNOWN, 包含此服务的 1 个处理程序...
  实例 "orcl", 状态 READY, 包含此服务的 1 个处理程序...
服务 "orclXDB" 包含 1 个实例。
  实例 "orcl", 状态 READY, 包含此服务的 1 个处理程序...
服务 "pdborcl" 包含 1 个实例。
  实例 "orcl", 状态 READY, 包含此服务的 1 个处理程序...
命令执行成功
```

2.6 设置开机启动

设置开机启动数据库实例，如果有多个实例，文件/etc/oratab 就应该有多行。该文件要被 dbstart 和 dbshut 命令调用，每一行的结尾有 Y/N 两个选项，Y 表示 dbstart 和 dbshut 命令要启动或者停止该行实例，N 则表示不影响该实例。以下操作需要以 root 身份进行。

```
# vi /etc/oratab
orcl:/home/oracle/app/oracle/product/12.1.0/dbhome_1:N
orcl2:/home/oracle/app/oracle/product/12.1.0/dbhome_1:N
```

改为：

```
orcl:/home/oracle/app/oracle/product/12.1.0/dbhome_1:Y
orcl2:/home/oracle/app/oracle/product/12.1.0/dbhome_1:Y

# cd /etc/init.d
# vi dbora
```

/etc/init.d/dbora 文件内容如下：

```
#!/bin/bash
# oracle: Start/Stop Oracle Database 11g R2/12c
#
# chkconfig: 345 90 10
# description: The Oracle Database Server is an RDBMS created by Oracle Corporation
#
# processname: oracle

. /etc/rc.d/init.d/functions

LOCKFILE=/var/lock/subsys/oracle
ORACLE_HOME=/home/oracle/app/oracle/product/12.1.0/dbhome_1
ORACLE_USER=oracle
case "$1" in
'start')
   if [ -f $LOCKFILE ]; then
      echo $0 already running.
      exit 1
   fi
   echo -n $"Starting Oracle Database: "
```

```
    su - $ORACLE_USER -c "$ORACLE_HOME/bin/lsnrctl start"
    su - $ORACLE_USER -c "$ORACLE_HOME/bin/dbstart $ORACLE_HOME"
    touch $LOCKFILE
    ;;
'stop')
    if [ ! -f $LOCKFILE ]; then
        echo $0 already stopping.
        exit 1
    fi
    echo -n $"Stopping Oracle Database: "
    su - $ORACLE_USER -c "$ORACLE_HOME/bin/lsnrctl stop"
    su - $ORACLE_USER -c "$ORACLE_HOME/bin/dbshut $ORACLE_HOME"
    rm -f $LOCKFILE
    ;;
'restart')
    $0 stop
    $0 start
    ;;
'status')
    if [ -f $LOCKFILE ]; then
        echo $0 started.
    else
        echo $0 stopped.
    fi
    ;;
*)
    echo "Usage: $0 [start|stop|status]"
    exit 1
esac
exit 0
```

最后设置开机自动启动：

```
# chgrp dba dbora
# chmod 750 dbora
# ln -s /etc/init.d/dbora /etc/rc.d/rc0.d/K01dbora
# ln -s /etc/init.d/dbora /etc/rc.d/rc3.d/S99dbora
# ln -s /etc/init.d/dbora /etc/rc.d/rc5.d/S99dbora
```

手工启动或者停止 Oracle 的命令是：

service dbora [start / stop / restart / status]，比如：

```
# service dbora start
# service dbora stop
# service dbora restart
# service dbora status
```

所有工作完成后重启 Linux，Oracle 12c 可自动启动，启动过程信息保存在启动日志文件$ORACLE_HOME/startup.log 中，可以查看启动过程是否正常。

> 注意：通过 Linux 命令 service dbora start 启动 oracle 只能启动外层的根容器数据库 CDB，不能自动启动 CDB 内部的插接式数据库 PDBs，如果要启动 PDBs，需要在 CDB 中创建一个启动触发器。

练习

一、判断题

1. Oracle 12c 只支持 32 位和 64 位的 Linux 系统。   （   ）
2. Oracle 的监听服务默认的端口号是 1521   （   ）
3. 登录方式"sqlplus / as sysdba"是操作系统认证方式。   （   ）
4. 监听进程的作用是提供内部应用访问 Oracle 服务的接口。   （   ）
5. Oracle 企业管理器是基于浏览器的 Oracle 管理工具。   （   ）

二、单项选择题

1. Oracle 的主目录环境变量是：（   ）
   A. $ORACLE_HOME   B. $ORACLE_SID
   C. $ORACLE_OWNER   D. $NLS_LANG
2. Oracle 客户端的国家字符集的环境变量是：（   ）
   A. $ORACLE_HOME   B. $ORACLE_SID
   C. $ORACLE_OWNER   D. $NLS_LANG
3. 连接 Oracle 数据库的命令是：（   ）
   A. sqlplus   B. lsnrctl   C. dbstart   D. dbshut
4. 运行监听器的命令是：（   ）
   A. sqlplus   B. lsnrctl   C. dbstart   D. dbshut
5. 通过 linux 命令 service 停止 Oracle 服务的命令是：（   ）
   A. service dbora start   B. service dbora restart
   C. service dbora stop   D. service dbora status

三、填空题

1. user_tables 视图的作用是_____。
2. 在安装 Oracle 之前，需要创建 Oracle 的用户和_____，以便有独立的管理权限。
3. Oracle 12c 的根数据库容器名称是_____。
4. Oracle 12c 的种子数据库名称是_____。

5. 在 Oracle 中有一些以_____开头的视图，表示用户能访问的对象。

6. Oracle 在 Linux 中的进程名称是以_____开头的。

四、简答题

1. 在安装 Oracle 12c 前应该具备什么条件，安装过程中应注意什么问题？
2. Oracle 的环境变量主要有哪些，有何作用？

训练任务

1 在 Linux 环境下安装 Oracle 12c

培养能力	工程知识、问题分析、研究、职业规范		
掌握程度	★★★★	难度	高
结束条件	安装 Oracle 12c 成功，并能在本地连接 Oracle，特别是在 Linux 运行级别 3 模式下(即排图形界面模式下)安装 Oracle 12c		
训练内容： 本训练的目的是能在 Linux 环境中安装 Oracle 12c			

2 连接/检测 Oracle 12c

培养能力	工程知识、问题分析、研究、项目管理		
掌握程度	★★★★★	难度	中
结束条件	连接并访问 Oracle 12c 成功		
训练内容： 本训练的目的是能够在本地/异地连接并访问 Oracle 12c，查看环境变量、安装目录及文件、Oracle 进程以及监听器状态			

第 3 章 网络配置及管理工具

本章目标

知识点	理解	掌握	应用
1.数据库服务器模式和数据库连接方式	✓	✓	
2.配置 TNS 网络配置文件	✓	✓	
3.掌握 SQL*Plus 的使用		✓	✓
4.掌握 SQL Developer 的使用		✓	✓
5.能够理解和分析 SQL 执行计划	✓	✓	
6.能够进行 SQL 语句的优化指导			

训练任务

- 熟悉 SQL*Plus 和 SQL Developer 两个工具的使用。
- 配置服务器同时工作在专用服务器模式和共享服务器模式。
- 分析 SQL 执行计划,执行 SQL 语句的优化指导。

知识能力点

能力点 \ 知识点	知识点 1	知识点 2	知识点 3	知识点 4	知识点 5	知识点 6
工程知识	✓		✓	✓	✓	✓
问题分析	✓	✓			✓	
设计/开发解决方案	✓	✓				
研究	✓	✓	✓	✓	✓	✓
使用现代工具			✓	✓	✓	✓
工程与社会	✓				✓	
环境和可持续发展	✓				✓	
职业规范			✓	✓		✓
个人和团队	✓					
沟通			✓	✓	✓		✓
项目管理		✓	✓	✓	✓	✓
终身学习	✓				✓	

3.1 Oracle Net Services

无论是应用程序还是管理工具要连接到数据库,都必须建立客户端与数据库服务器之间的连接,Oracle 提供的 Oracle Net Service 组件为 Oracle 环境提供了一个安全、可扩展和易于使用的高可用性网络基础架构。它降低了网络配置和管理的复杂性,实现了性能最大化,并且提高了网络安全性和诊断功能。Oracle Net Services 的特性有:

1) 连接性

Oracle Net Services 支持从客户端应用到 Oracle 数据库服务器的网络会话。建立网络会话之后,Oracle Net 将充当客户端应用和数据库服务器之间的数据信使。它负责建立和维护客户端应用和数据库服务器之间的连接,以及在它们之间交换消息。

2) 可管理性

它包括位置透明性、集中配置和管理、快速安装和配置。位置透明性服务使数据库客户端可以识别目标数据库服务器;为了实现此目标,提供了几种命名方法:Oracle 网络目录命名、本地命名(tnsnames.ora)、主机命名和外部命名。

3) 集中配置和管理

让大型网络环境中的管理员可以轻松访问中央信息库(即符合 LDAP 的目标服务器,如 Oracle Internet Directory),从而指定和修改网络配置。

4) 快速安装和配置

Oracle 数据库服务器和客户端的网络组件已针对大多数环境进行了预配置。使用各种命名方法对 Oracle 数据库服务进行解析。因此,客户端和服务器可以在安装后立即连接。

5) 性能和可扩展性

像数据库驻留连接池(连接池)、共享服务器(会话多路复用)以及可扩展的事件模型(轮询)这样的特性可提高性能和可扩展性。

6) 网络安全性

Oracle Net Services 使用防火墙访问控制和协议访问控制的特性来实现数据库访问控制。

7) 可诊断性

Trace Assistant 诊断和性能分析工具会在出现问题时提供有关问题的起源和上下文的详细信息。

3.2 服务器模式和数据库连接方式

Oracle 服务器可以工作在专用服务器模式(Dedicated Server)和共享服务器模式(Shared Server)下或者混合模式下,相对应的,Oracle 支持两种数据库连接方式,一种是专用连接方式,另一种是共享连接方式。区别是专用连接方式是一个用户会话对应一个数据库服务器进程,而共享服务器连接方式是多个用户不定向轮流使用一个服务器进程。

3.2.1 专用服务器模式

Oracle 默认的连接是专用连接方式,一个客户端会话(session)对应一个服务器进程,优点是减少竞争,对于较长事务很有用,但是会耗费更多的 SGA 资源。图 3-1 是专用连接方式。

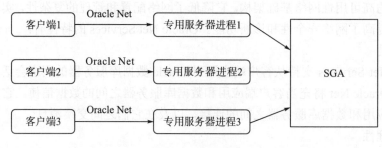

图 3-1 专用连接模式

当用户登录时,总会创建一个进程来为会话提供专门的服务。每当一个新会话建立,监听程序就会创建新的专用服务器来提供专门服务,会话与专门服务器一一对应。用户的客户进程会通过某种网络通道与专门服务器直接通信,来响应提交的 SQL、PL/SQL 调用。

3.2.2 共享服务器模式

在连接用户数比较多,事务比较小,但会话多的情况下,可以采用共享服务器模式方式连接数据库,这种方式中可供使用的连接数是一定的,通过资源调度器(Dispatchers)来动态管理会话与实例建立连接,这些连接供所有的会话共享,可以有效的减少资源负载。如图 3-2 所示:

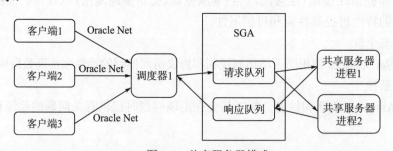

图 3-2 共享服务器模式

共享服务器进程机制需要在客户进程与共享服务器通信之间建立一种调度机制(调度程序)来保障调度的有效性。用户提出连接请求,监听器接收到连接请求后,会从可调用的调度程序中选择一个,将连接端口等信息返回给用户,调度程序在相应的端口等待,客户进程将与该调度程序进行连接,与调度程序连接后,将用户请求转入 SGA 的请求队列中,第一个空闲的共享服务器会得到这个请求,并进行处理。待处理完成,共享服务器会将结果放在响应的队列中。调度程序一直在监听响应队列,一旦发现有结果了,就会把结果传给用户。

共享服务器模式有很多优点，例如：

（1）如果连接用户数比较多，事务都比较小，则共享服务器可以高效的配置资源并提高效率。

（2）基于节省服务器内存开销的考虑，共享服务器可以在有效的资源下更加高效，例如：负载均衡等特性要求采用共享服务器。

（3）减少操作系统进程数量，减少操作系统上下文切换的资源消耗，提高效率。

（4）有效地限制并发度，将系统消耗限制在一个合理的范围。

（5）共享服务器模式也有一些缺点。

（6）在客户进程与共享服务器之间多了一层调度器。执行效率有所下降，所以不太适合实时性要求高的应用。

3.2.3 配置数据库支持共享模式

专用服务器模式是 Oracle 的默认启动选项，不需要额外的设置。如果要启用共享服务器模式，就必须要设置系统参数，共享模式需要设置的系统参数有：

1）DISPATCHERS

这个参数是共享服务器最重要的参数，它是一个格式字符串，如果该字符串为空，表示数据库不工作在共享模式下，不允许共享连接。如果不为空，它的基本格式是：(PROTOCOL=TCP)(HOST=HOST_NAME)(DISPATCHERS=DISPATCHERS_NUM)(SERVICE=SERVICE_NAMES)，里面有 4 个子参数，其中只有 PROTOCOL 参数是必须的，表示 TCP 网络协议。HOST 表示在该地址的情况下使用共享模式，如果没有该参数，则在所有地址下都使用共享模式。子参数 DISPATCHERS 表示调度器的数量，如果没有该参数，默认调度器数量为 1，SERVICE 子参数表示哪些数据库使用共享模式，SERVICE_NAMES 是用逗号分隔的 SERVICE_NAME 或者 listener.ora 中静态注册的 GLOBAL_DBNAME，如果没有 service 子参数，则表示所有数据库都开启共享模式。比如命令：ALTER SYSTEM SET DISPATCHERS="(PROTOCOL=TCP)(DISPATCHERS=3)(SERVICE=pdborcl，pdbtest)"，将设置 pdborcl，pdbtest 数据库工作在共享模式下。

2）SHARED_SERVERS

初始化启动的服务器进程数和保留的最小服务器进程数；如果使用共享连接，这个参数必须大于 0。

3）MAX_SHARED_SERVERS

最大并发的共享服务器进程数

4）SHARED_SERVER_SESSIONS

可并发处理的最大共享用户会话数，超过此数值的用户会话将使用专用连接。

5）MAX_DISPATCHERS

可同时并发的共享连接的最大的调度器数量，此参数目前可忽略；

6）CIRCUITS

指定应用于入站和出站的网络会话的虚拟回路总数。

【示例 3-1】 设置服务器开启共享服务器模式，调度器数量为 3

```
$ sqlplus / as sysdba
SQL> ALTER SYSTEM SET dispatchers="(PROTOCOL=TCP)(dispatchers=3)";
系统已更改。
SQL> ALTER SYSTEM SET max_dispatchers=5;
系统已更改。
SQL> ALTER SYSTEM SET shared_servers = 1;
系统已更改。
SQL> ALTER SYSTEM SET max_shared_servers=20;
系统已更改。
SQL> ALTER SYSTEM SET shared_server_sessions=200;
系统已更改。
SQL> show parameter shared_server
NAME                           TYPE        VALUE
------------------------------ ----------- -----------
max_shared_servers             integer     20
shared_server_sessions         integer     200
shared_servers                 integer     1
SQL>
```

本例中：shared_servers 的值为 1(大于 0)，表示启动了共享模式，如果共享会话数量超过了 shared_server_sessions 的时候，新建共享连接会失败，但仍然可以进行专用连接。

在设计调度器的数量 dispatchers 的时候，要看客户端 session 的预计数量，比如数据库当前有 500 个 TCP/IP session，每个调度器管理 50 个 session，那么就需要 10 个（即 500/50）调度器。参数设置如下：DISPATCHERS="(PROTOCOL=TCP)(DISPATCHERS=10)"，也可将调度器设置为连接池：DISPATCHERS="(PROTOCOL=TCP)(DISPATCHERS=1)(POOL=ON)(CONNECTIONS=500)(SESSIONS=1000)"，表示启动连接池，一个调度器的连接数最大为 500 个，sessions 数目最大为 1000 个。

> 注意：只能在 CDB 容器数据库下设置数据库服务器模式，不能在 PDB 插接式数据库下设置。修改后的设置不能立即生效，需要强制性地杀掉现在的 DISPATCH 进程，或者重启数据库。

3.2.4 检测数据库的服务器模式

Oracle 提供了一个重要的管理程序 lsnrctl，它用来管理网络监听器（listener），查看监听器的状态。通过查看监听器的状态，也就能查看数据库的工作进程的状态，可以通过 lsnrctl 查看数据库的工作模式：

【示例 3-2】 通过 lsnrctl service 命令查看数据库的完整工作模式。

```
$ lsnrctl service
LSNRCTL for Linux: Version 12.1.0.2.0 - Production on 01-MAR-2017 23:59:03
Copyright (c)1991, 2014, Oracle. All rights reserved.
Connecting to
(DESCRIPTION=(ADDRESS=(PROTOCOL=TCP)(HOST=localhost)(PORT=1521)))
Services Summary...
Service "orcl" has 2 instance(s).
  Instance "orcl", status UNKNOWN, has 1 handler(s)for this service...
    Handler(s):
      "DEDICATED" established: 0 refused: 0
        LOCAL SERVER
  Instance "orcl", status READY, has 4 handler(s)for this service...
    Handler(s):
      "DEDICATED" established: 0 refused: 0 state: ready
        LOCAL SERVER
      "D002" established: 0 refused: 0 current: 0 max: 1022 state: ready
        DISPATCHER <machine: oracle-pc, pid: 8776>
        (ADDRESS=(PROTOCOL=tcp)(HOST=oracle-pc)(PORT=61946))
      "D001" established: 0 refused: 0 current: 0 max: 1022 state: ready
        DISPATCHER <machine: oracle-pc, pid: 8774>
        (ADDRESS=(PROTOCOL=tcp)(HOST=oracle-pc)(PORT=39113))
      "D000" established: 0 refused: 0 current: 0 max: 1022 state: ready
        DISPATCHER <machine: oracle-pc, pid: 8771>
        (ADDRESS=(PROTOCOL=tcp)(HOST=oracle-pc)(PORT=47446))
Service "pdborcl" has 1 instance(s).
  Instance "orcl", status READY, has 4 handler(s)for this service...
    Handler(s):
      "DEDICATED" established: 0 refused: 0 state: ready
        LOCAL SERVER
      "D002" established: 0 refused: 0 current: 0 max: 1022 state: ready
        DISPATCHER <machine: oracle-pc, pid: 8776>
        (ADDRESS=(PROTOCOL=tcp)(HOST=oracle-pc)(PORT=61946))
      "D001" established: 0 refused: 0 current: 0 max: 1022 state: ready
        DISPATCHER <machine: oracle-pc, pid: 8774>
        (ADDRESS=(PROTOCOL=tcp)(HOST=oracle-pc)(PORT=39113))
      "D000" established: 0 refused: 0 current: 0 max: 1022 state: ready
```

```
          DISPATCHER <machine: oracle-pc, pid: 8771>
          (ADDRESS=(PROTOCOL=tcp)(HOST=oracle-pc)(PORT=47446))
The command completed successfully
```

结果分析：orcl 实例(Instance "orcl")中包含了两个 SERVICE_NAME，即 Service "orcl"和 Service " pdborcl "。这两个服务名中 orcl 是 CDB，pdborcl 是 PDB。两个数据库都运行在 DEDICATED 和 SHARED 双模式下。其中 D000，D001，D002 是共享模式下的 3 个调度器，这 3 个调度器是由"DISPATCHERS=3"决定的。

注意：除了通过 lsnrctl service 查看完整的工作模式之外，还可以通过 lsnrctl status 命令查看数据库的简略的工作模式。

3.2.5 连接到不同的服务器模式

数据库的专用模式和共享模式设定好之后，就可以分别连接到这两个不同的模式了。在 sqlplus 命令中，在 SERVICE_NAME 参数后加上":dedicated"或者":shared"表示连接到专用模式或者共享模式，也可以省略。如果省略，Oracle 12c 的缺省模式是共享模式优先，即如果不指定连接模式，并且服务器又工作在共享模式下，则自动连接到共享模式。

【示例 3-3】指定以专用模式连接数据库

本例指定以专用模式连接到 pdborcl 数据库，然后通过 v$session 查询是否以专用模式连接到数据库。

```
$ sqlplus hr/***@localhost/pdborcl:dedicated
SQL> SELECT server FROM v$session WHERE  SID=(SELECT distinct SID FROM v$mystat);
SERVER
---------
DEDICATED
```

【示例 3-4】指定以共享模式连接数据库

本例指定以共享模式连接到 pdborcl 数据库，然后通过 v$session 查询是否以共享模式连接到数据库。

```
$ sqlplus hr/***@localhost/pdborcl:shared
SQL> SELECT server FROM v$session WHERE  SID=(SELECT DISTINCT SID FROM v$mystat);
SERVER
---------
SHARED
```

下面再次通过 lsnrctl service 命令查看数据库的完整工作状态，在运行这个命令之前，可以在不关闭以前的连接的情况下，在更多的终端中运行上面两个示例，从而打开更多的专用连接和共享连接。我们就会看到下面的结果：

```
$ lsnrctl service
…
Service "pdborcl" has 1 instance(s).
  Instance "orcl", status READY, has 4 handler(s)for this service...
    Handler(s):
      "DEDICATED" established: 6 refused: 0 state: ready
         LOCAL SERVER
      "D002" established: 4 refused: 0 current: 4 max: 1022 state: ready
         DISPATCHER <machine: oracle-pc, pid: 8776>
         (ADDRESS=(PROTOCOL=tcp)(HOST=oracle-pc)(PORT=61946))
      "D001" established: 3 refused: 0 current: 3 max: 1022 state: ready
         DISPATCHER <machine: oracle-pc, pid: 8774>
         (ADDRESS=(PROTOCOL=tcp)(HOST=oracle-pc)(PORT=39113))
      "D000" established: 5 refused: 0 current: 4 max: 1022 state: ready
         DISPATCHER <machine: oracle-pc, pid: 8771>
         (ADDRESS=(PROTOCOL=tcp)(HOST=oracle-pc)(PORT=47446))
The command completed successfully
```

从输出的结果可以看到,""DEDICATED" established: 6"表示有 6 个专用连接,""D000" established:5" 表示 D000 调度器中有 5 个共享连接,同理,D001 中有 3 个共享连接,D002 中有 4 个共享连接。

3.2.6 查看服务器连接进程

在专用连接模式下,每一次连接都会在操作系统中创建一个进程,进程的名称是 oracle$ORACLE_SID。其中$ORACLE_SID 是表示数据库实例的系统变量。

【示例 3-5】查看专用连接进程

```
$ ps -ef | grep oracleorcl
oracle   10102  7318  0 00: 59 ?        00: 00: 01 oracleorcl
(DESCRIPTION=(LOCAL=YES)(ADDRESS=(PROTOCOL=beq)))
oracle   10416     1  0 01: 03 ?        00: 00: 00 oracleorcl (LOCAL=NO)
oracle   10457     1  0 01: 03 ?        00: 00: 00 oracleorcl (LOCAL=NO)
```

从命令的输出结果看,一共有 3 个专用连接进程,对应 3 个专用连接,实际情况是有多少专用连接,就会有多少连接进程,其中"LOCAL=YES"这样的专用进程表示不通过网络连接的本地进程,在 lsnrctl 没有运行的情况下仍然可以连接。比如命令"sqlplus / as sysdba"。而"LOCAL=NO"这样的专用进程表示通过网络连接创建的异地进程,必须在 lsnrctl 运行的情况下才能连接成功。比如命令"sqlplus hr/***@localhost/pdborcl:dedicated"。

在专用连接模式下,由于每增加一个连接,就会增加一个系统进程,这对服务器的资源消耗很大,当然,其优势是一旦建立了连接,由于是直接连接到专用服务器,访问效率就比共享连接高。

与专用连接模式不同，在共享连接模式下，不会每增加一个连接就新增一个操作系统进程，共享模式下的连接进程的数量是固定的，等于调度器的数量（dispatchers）。共享连接进程的名称样式是"ora_d 三位数字_$ORACLE_SID"。

【示例 3-6】查看共享连接进程

```
$ ps -ef | grep ora_d[0-9].*[_orcl$]
oracle    10086     1  0 00:59 ?        00:00:00 ora_d000_orcl
oracle    10088     1  0 00:59 ?        00:00:00 ora_d001_orcl
oracle    10090     1  0 00:59 ?        00:00:00 ora_d002_orcl
```

可以看出，共享进程数量不会因为连接数量的增加而增加，这也是共享连接的优势所在，当然，这种优势是以牺牲一定的效率换取的。

3.3 TNS 网络配置文件

TNS 是指透明网络底层（Transparence Network Substrate），它是 Oracle Net 的一部分，是专门用来管理和配置 Oracle 数据库和客户端连接的协议，在大多数情况下客户端和数据库要通讯，必须配置 TNS。TNS 的配置文件包括服务器端文件和客户端文件两部分。服务器端文件有 listener.ora、sqlnet.ora 和 tnsnames.ora；客户端文件有 tnsnames.ora，sqlnet.ora。这些文件固定存放在"$ORACLE_HOME/network/admin"目录下，是文本文件，可以通过手工修改这些文件进行网络配置，也可以通过一些命令来配置，这些命令是 netmgr，netca。

3.3.1 lsnrctl 和 listener.ora

Oracle 的监听进程的作用是接受远程对数据库的接入申请并转交给 Oracle 的服务器进程。所以如果不是使用远程连接，Listener 进程就不是必需的，同样，关闭 Listener 进程并不会影响已经存在的数据库连接。Oracle 的监听控制程序是 lsnrctl，启动后的进程名称是 tnslsnr，该程序的启动参数就是来自参数文件 listener.ora 的内容，注意 listener.ora 文件的内容只为 lsnrctl 进程使用，数据库的 PMON 进程不会使用该文件，PMON 进程使用 tnsnames.ora 配置文件，PMON 进程名称是 ora_pmon_$ORACLE_SID。

【示例 3-7】listener.ora 典型内容

```
#lsnrctl 程序使用的监听器的名称，不是 PMON 使用的名称
LISTENER =
  (DESCRIPTION_LIST =
    (DESCRIPTION =
      (ADDRESS = (PROTOCOL = TCP)(HOST = localhost)(PORT = 1521))
      (ADDRESS = (PROTOCOL = IPC)(KEY = EXTPROC1521))
    )
```

```
  )
  #静态注册
  SID_LIST_LISTENER =
  (SID_LIST =
    (SID_DESC =
    (GLOBAL_DBNAME = pdborcl)
    (SID_NAME = orcl)
    )
    (SID_DESC =
    (GLOBAL_DBNAME = orcl)
    (SID_NAME = orcl)
    )
  )
```

Listener.ora 中 LISTENER 部分是 lsnrctl 程序默认的监听器的名称。LISTENER 中的 DESCRIPTION_LIST 表示目录列表，包括 DESCRIPTION 节点，DESCRIPTION 节点由多个地址（ADDRESS）节组成。ADDRESS 节点有两种基本协议（PROTOCOL）类型：TCP 和 IPC。

TCP 协议是基本的远程通信协议，它的参数有 HOST 和 PORT，HOST 可以是主机名或者 IP 地址，PORT 是监听的端口号，默认值是 1521，在设置端口的时候最好使用 Linux 命令"netstat -an"检查该端口是否已经被占用了。

IPC 是进程间通信协议，Oracle 使用该协议调用第三方程序库，IPC 协议需要指定 IPC 的键值 KEY。

LISTENER 中可以配置多个 TCP，设置不同的主机地址或者端口号，以适应各种网络环境。比如，一般专业的服务器上有多块网卡（区分为内外网），每块网卡对应不同 IP 地址，我们可以设置多个监听地址和监听端口，将内外网区分开来。

Listener.ora 中的 SID_LIST_LISTENER 部分是静态注册配置。其中 SID_LIST 是目录列表，它包含多个静态注册 SID_DESC 节，每个 SID_DESC 节包括：

GLOBAL_DBNAME：全局数据库名，这个名称就是对外提供的服务名，外部应用可以根据这个名称访问数据库。通过"SELECT * FROM global_name；"查询得出。

ORACLE_HOME：Oracle 系统的根目录。

SID_NAME：数据库的系统标识符 SID。

3.3.2 监听器的动态注册

监听器的动态注册是数据库实例启动的时候 PMON 进程根据文件 init.ora 中的 instance_name，service_names 两个参数将实例和服务动态注册到 listener 中。动态注册默认只注册到默认的监听器上（名称是 LISTENER、端口是 1521、协议是 TCP）。

命令 lsnrctl start 用于启动监听器，lsnrctl status 用于查询服务的注册结果。服务的动态注册成功与 Oracle 服务器主进程的启动和监听程序的启动顺序有关。那么什么时候才能注册成功呢？如果监听器 lsnrctl 先启动（lsnrctl start），Oracle 之后启动（startup），可以立刻运态注册成功，如果 Oracle 先启动，lsnrctl 之后启动，不会立即注册动态监听，几分钟之后才能注册成功。这是因为数据库实例启动的时候会启动 PMON 进程，启动的时候如果监听器在运行，就与这个监听器通信并传递如服务名和实例的负载等参数，如果监听器没有启动，PMON 会定期地尝试连接监听来注册实例。可以通过 lsnrctl status 查看注册结果：

数据库未动态注册成功的时候：

```
$ lsnrctl status
...
Listening Endpoints Summary...
  (DESCRIPTION=(ADDRESS=(PROTOCOL=tcp)(HOST=oracle-pc)(PORT=1521)))
  (DESCRIPTION=(ADDRESS=(PROTOCOL=ipc)(KEY=EXTPROC1521)))
The listener supports no services
The command completed successfully
```

动态注册成功的时候：

```
$ lsnrctl status
...
Listening Endpoints Summary...
  (DESCRIPTION=(ADDRESS=(PROTOCOL=tcp)(HOST=oracle-pc)(PORT=1521)))
  (DESCRIPTION=(ADDRESS=(PROTOCOL=ipc)(KEY=EXTPROC1521)))
...
Services Summary...
Service "orcl" has 1 instance(s).
  Instance "orcl", status READY, has 4 handler(s)for this service...
Service "pdborcl" has 1 instance(s).
  Instance "orcl", status READY, has 4 handler(s)for this service...
The command completed successfully
```

在 lsnrctl status 的查询结果中"Service 服务名 has n instance(s)"表示服务名有 n 个连接。服务名是 Oracle 自动取名的，取名的方式是：后台进程 PMON 自动在监听器中注册初始化参数 SERVICE_NAMES 中定义的服务名，SERVICE_NAMES 默认为 DB_NAME+DOMAIN_NAME。客户端 tns 配置文件（tnsnames.ora）中 SERVICE_NAME 的名称必须是 SERVICE_NAMES 或其中的一个 NAME。

如果需要向非默认监听注册，则需要配置 local_listener 参数！比如，我们希望将 1521 端口修改为 1522，又希望能够动态注册所有的数据库以便外部应用异地访问。我们就必须修改 local_listener 参数。local_listener 参数是一个字符串，它的默认值是空值。

【示例 3-8】直接设置 local_listener

本例设置自定义 local_listener 的值为 1522：

```
$ sqlplus / as sysdba
SQL> ALTER SYSTEM SET
```

```
    LOCAL_LISTENER='(DESCRIPTION_LIST=(DESCRIPTION=(ADDRESS=(PROTOCOL =
TCP)(HOST = localhost)(PORT = 1522))(ADDRESS = (PROTOCOL = IPC)(KEY =
EXTPROC1521))))';
    SQL> ALTER SYSTEM REGISTER;
```

除了用"ALTER SYSTEM"命令之外,Oracle 允许将监听的信息添加到 tnsnames.ora 文件中。下面是 tnsnames.ora 文件的部分内容:

```
LISTENER1522 =
  (DESCRIPTION_LIST =
    (DESCRIPTION =
      (ADDRESS = (PROTOCOL = TCP)(HOST = localhost)(PORT = 1522))
      (ADDRESS = (PROTOCOL = IPC)(KEY = EXTPROC1521))
    )
  )
```

然后以 sys 用户登录后运行:

```
$ sqlplus / as sysdba
SQL> ALTER SYSTEM SET LOCAL_LISTENER= LISTENER1522;
SQL> ALTER SYSTEM REGISTER;
```

无论是直接修改或者是间接修改了 local_listener,都只是改变了数据库的 PMON 进程的动态注册方式,lsnrctl 进程都不知道这些改变,所以还必须同时改变 listener.ora 文件的配置,将 listener.ora 文件的 LISTENER 配置节修改为相应的内容,然后重启 lsnrctl:

```
LISTENER =
  (DESCRIPTION_LIST =
    (DESCRIPTION =
      (ADDRESS = (PROTOCOL = TCP)(HOST = localhost)(PORT = 1522))
      (ADDRESS = (PROTOCOL = IPC)(KEY = EXTPROC1521))
    )
  )
```

动态注册和静态注册可以同时并存,这就提供了连接数据库的多种方式:

```
$ lsnrctl status
...
Listening Endpoints Summary...
  (DESCRIPTION=(ADDRESS=(PROTOCOL=tcp)(HOST=oracle-pc)(PORT=1521)))
  (DESCRIPTION=(ADDRESS=(PROTOCOL=ipc)(KEY=EXTPROC1521)))
...
Services Summary...
Service "orcl" has 2 instance(s).
  Instance "orcl", status UNKNOWN, has 1 handler(s)for this service...
  Instance "orcl", status READY, has 4 handler(s)for this service...
```

```
  Service "pdborcl" has 2 instance(s).
    Instance "orcl", status UNKNOWN, has 1 handler(s) for this service...
    Instance "orcl", status READY, has 4 handler(s) for this service...
The command completed successfully
```

查询结果中,"status READY"表示动态注册的服务。"status UNKNOWN"表示静态注册,这是由于静态注册并不知道服务器是否启动,所以状态为UNKNOWN。

3.3.3 监听器的静态注册

监听器的注册就是将数据库作为一个服务注册到监听程序。客户端不需要知道数据库名和实例名,只需要知道该数据库对外提供的服务名就可以申请连接到数据库。这个服务名可能与实例名一样,也有可能不一样。如果数据库名称没有注册,外部应用则不能访问到数据库。

Oracle 提供了静态注册的方式,可以比较容易地注册数据库,这个方式只需要在listener.ora 文件中定义静态注册配置节。默认的静态注册节点的名称是SID_LIST_LISTENER,该名称的格式是 SID_LIST_监听器名称,由于默认的监听器名称是 LISTENER,所以默认的静态注册节点的名称就是 SID_LIST_LISTENER。

要正确设置静态注册,必须正确设置 LISTENER 以及 SID_LIST_LISTENER 节点的值,如果正确设置了静态注册节点,并且启动 lsnrctl 之后,那么静态注册会立即生效,这也是静态设置的优点。与动态注册相比,它的缺点是不能全部设置每个数据库,必须将数据库逐个地编写在 SID_DESC 中。

【示例 3-9】查看数据库静态注册状态

本例启动监听后看到的监听状态是"未知:status UNKNOWN",这表示静态注册,因为只有客户端请求连接到实例时候才会去检查实例是否存在。

```
$ lsnrctl
LSNRCTL> start
Listening Endpoints Summary...
  (DESCRIPTION=(ADDRESS=(PROTOCOL=tcp)(HOST=oracle-pc)(PORT=1521)))
  (DESCRIPTION=(ADDRESS=(PROTOCOL=ipc)(KEY=EXTPROC1521)))
Services Summary...
Service "orcl" has 1 instance(s).
  Instance "orcl", status UNKNOWN, has 1 handler(s) for this service...
Service "pdborcl" has 1 instance(s).
  Instance "orcl", status UNKNOWN, has 1 handler(s) for this service...
The command completed successfully
LSNRCTL> status
...
```

> 注意：监听器 start 之后，如果立即运行 status 只能看到静态注册信息，不能看到动态注册信息。等一段时间之后，才能看到动态注册信息。监听器启动日志文件是 $ORACLE_HOME/listener.log。

3.3.4　tnsnames.ora

tnsnames.ora 文件是客户端的网络服务名配置文件，用于存放客户端配置的可连接实例的参数，所以每个客户都应该有一个 tnsnames.ora 文件。数据库注册成功后，用户就可以根据服务名连接数据库了，比如 sqlplus hr/***@localhost:1521/pdborcl:dedicated，其中 pdborcl 就是服务名。通过配置文件 tnsnames.ora 可以更快捷地连接到数据库，tnsnames.ora 的样例是：

```
PDBORCL_S =
  (DESCRIPTION =
    (ADDRESS_LIST =
      (ADDRESS = (PROTOCOL = TCP)(HOST = localhost)(PORT = 1521))
    )
    (CONNECT_DATA =
      (SERVER = SHARED)
      (SERVICE_NAME = pdborcl)
    )
  )
PDBORCL_D =
  (DESCRIPTION =
    (ADDRESS_LIST =
      (ADDRESS = (PROTOCOL = TCP)(HOST = localhost)(PORT = 1521))
    )
    (CONNECT_DATA =
      (SERVER = DEDICATED)
      (SERVICE_NAME = pdborcl)
    )
  )
ORCL =
  (DESCRIPTION =
    (ADDRESS_LIST =
      (ADDRESS = (PROTOCOL = TCP)(HOST = localhost)(PORT = 1521))
    )
    (CONNECT_DATA =
      (SERVER = DEDICATED)
```

```
            (SERVICE_NAME = orcl)
        )
    )
```

上例提供了三个 TNS 名称(PDBORCL_S，PDBORCL_D，ORCL)，用于快捷连接，比如命令"sqlplus hr/***@pdborcl_d"也一样可以登录成功。很明显这种方式可以屏蔽服务器的地址和服务名称，登录方式更友好。

另外，为了验证 tnsnames.ora 配置的正确性，Oracle 提供了命令 tnsping 来检验 TNS 名称是否设置正确。

【示例 3-10】 tnsping 测试 tns 名称的连接

本示例测试 tns 名称 pdborcl_d 是否可以解析。输出"OK (0 msec)"表示 tnsping 成功连通，用时 0 毫秒。

```
$ tnsping pdborcl_d
...
Used TNSNAMES adapter to resolve the alias
Attempting to contact (DESCRIPTION = (ADDRESS_LIST = (ADDRESS = (PROTOCOL = TCP)(HOST = localhost)(PORT = 1521)))(CONNECT_DATA = (SERVER = DEDICATED)(SERVICE_NAME = pdborcl)))
OK (0 msec)
$ sqlplus hr/***@pdborcl_d
SQL>
```

3.3.5 sqlnet.ora

sqlnet.ora 文件内容也是由一些参数组成的，有服务端参数和客户端参数，用于配置连接服务端 Oracle 的相关参数，比较常用的参数有：

1) NAMES.DEFAULT_DOMAIN(客户端参数)

NAMES.DEFAULT_DOMAIN 是域名(domain)的定义，在用 sqlplus 访问数据库时，会在 TNS 别名后面自动加上".domain"，比如设置：NAMES.DEFAULT_DOMAIN = com.cn 后，在客户端执行命令："sqlplus username/password@local_dev"时，会在 tnsnames.ora 中寻找 local_dev.com.cn 的 TNS 名称。相当于在没有定义 NAMES.DEFAULT_DOMAIN 时执行：

```
sqlplus username/password@local_dev.com.cn
```

2) NAMES.DIRECTORY_PATH(客户端参数)

NAMES.DIRECTORY_PATH 定义了在客户端连接数据库时，采用什么样的匹配方式连接地址。比如：NAMES.DIRECTORY_PATH= (TNSNAMES, ONAMES, HOSTNAME)。那么在客户端执行 sqlplus username/password@local_dev 连接数据库时，首先采用 tnsnames.ora 的 TNS 别名配置连接数据库，如果没有连接上，再采用 ONAMES 进行解析，最后采用主机名进行解析。ONAMES 表示 Oracle 使用自己的名称服务器(Oracle Name Server)来解析，目前 Oracle 建议使用轻量目录访问协议 LDAP 来取代 ONAMES，

HOSTNAME 表示使用 host 文件、DNS 和 NIS 等来解析；

3）SQLNET.AUTHENTICATION_SERVICES（客户端参数）

SQLNET.AUTHENTICATION_SERVICES 定义登录数据库的认证方式（数据库身份认证，还是操作系统身份认证），NONE 表示数据库方式认证，ALL（Linux）或者 NTS（Windows）表示操作系统方式认证。SQLNET.AUTHENTICATION_SERVICES 在 Linux 环境和 Windows 环境下的值有一点点不同：

◆ 在 Linux 环境下：

如果没有定义 SQLNET.AUTHENTICATION_SERVICES 就表示可以双重认证：数据库身份认证和操作系统身份认证。等同于：SQLNET.AUTHENTICATION_SERVICES =（NONE，ALL）。

SQLNET.AUTHENTICATION_SERVICES=（NONE），只能是数据库身份认证。

SQLNET.AUTHENTICATION_SERVICES=（ALL），只能是操作系统身份认证。

SQLNET.AUTHENTICATION_SERVICES=（NONE，ALL），可以是双重认证：数据库身份认证和操作系统身份认证。

◆ 在 Windows 环境下：

没有定义 SQLNET.AUTHENTICATION_SERVICES：不能基于操作系统验证，等同于 SQLNET.AUTHENTICATION_SERVICES =（NONE）。

SQLNET.AUTHENTICATION_SERVICES =（NTS），只能基于操作系统验证。

SQLNET.AUTHENTICATION_SERVICES =（NONE），只能基于数据库身份验证。

SQLNET.AUTHENTICATION_SERVICES =（NONE，NTS），可以是双重认证：数据库身份认证和操作系统身份认证。

4）tcp.validnode_checking=yes（服务器端参数）

表示启用客户端的 IP 检查，非法的 IP 将被拒绝访问 Oracle。这个参数与 tcp.invited_nodes 和 tcp.excluded_nodes 参数联合使用。

5）tcp.invited_nodes=（IP1，IP2，IP3...）（服务器端参数）

表示允许哪些 IP 访问 Oracle。

6）tcp.excluded_nodes=（IP1，IP2，IP3...）（服务器端参数）

表示拒绝哪些 IP 访问 Oracle。

注意：服务器端参数修改这三个参数后，需要重新启动 lsnrctl。

【示例 3-11】sqlnet.ora 文件内容样例

这个样例不允许 IP 地址 192.168.0.101 登录：

```
NAMES.DIRECTORY_PATH=(TNSNAMES, EZCONNECT)
#NAMES.DEFAULT_DOMAIN=com.cn
SQLNET.AUTHENTICATION_SERVICES=(NONE, ALL)
tcp.validnode_checking=yes
tcp.excluded_nodes=(192.168.0.101)
```

3.4 SQL*Plus

SQL*Plus 是与 Oracle 进行交互的客户端工具。在 SQL*Plus 中，可以运行 SQL*Plus 命令、SQL 语句以及 PL/SQL 程序。在操作系统中运行命令"sqlplus"可以打开 SQL*Plus 窗口。

sqlplus 是基本的，最常用的工具，主要功能有：

(1) 数据库的维护，如启动、关闭等，这一般在服务器端操作。
(2) 执行 SQL 语句和 PL/SQL 语句块。
(3) 捕获 PL/SQL 程序的错误。
(4) 执行 SQL 脚本。
(5) 数据的导出，报表。
(6) 应用程序开发、测试 SQL 语句或者 PL/SQL 程序。
(7) 生成新的 SQL 脚本。
(8) 定义变量。
(9) 供应用程序调用，如安装程序中进行脚本的安装。
(10) 用户管理及权限维护等。
(11) 网络配置。
(12) 数据恢复。

3.4.1 SQL*Plus 连接数据库

sqlplus 命令连接数据库的语法是：

```
sqlplus {<用户名>[/<密码>][@<连接标识符>] | / }
        [AS {SYSDBA | SYSOPER | SYSASM | SYSBACKUP | SYSDG | SYSKM}]
```

其中"连接标识符"的格式是：@[<net_service_name>|[//] Host[: Port] /<service_name>]，"连接标识符"可以有两种方式：一种是网络服务名称(Net Service Name)，另一种是简易连接字符串(Easy Connect)，简易连接字符串的格式是<主机名>[：端口号]/<服务名>。

如果没有指定连接标识符，sqlplus 就连接默认的数据库，默认数据库由环境变量 $ORACLE_SID 指定。Net Service Name 通常在 tnsnames.ora 文件中定义。

连接标识符后面的可选项有 AS SYSDBA、AS SYSOPER、AS SYSASM、AS SYSBACKUP、AS SYSDG 以及 AS SYSKM。

默认情况下，使用命令 sqlplus 命令连接后，就会打开一个 SQL Plus 窗口，该窗口是一个行编辑器，在"SQL>"提示符后面可以输入 SQL 语句或者 sqlplus 命令。

【示例 3-12】SYS 用户登录并设置查询格式参数

本示例不需要输入用户名和密码，以操作系统认证方式登录数据库，但只能在服务器本机登录，不能在远端登录。登录后查询视图 v$instance，在查询之前用 desc 命令描述 v$instance 的结构，列出所有字段。

```
$ sqlplus / as sysdba
SQL*Plus: Release 12.1.0.2.0 Production on 星期一 2月 20 11:26:55 2017
Copyright (c) 1982, 2014, Oracle.  All rights reserved.
连接到:
Oracle Database 12c Enterprise Edition Release 12.1.0.2.0 - 64bit Production
With the Partitioning, OLAP, Advanced Analytics and Real Application Testing options
SQL> desc v$instance;
 名称                     是否为空?      类型
 ----------------------- -----------   ------------------
 INSTANCE_NUMBER                       NUMBER
 INSTANCE_NAME                         VARCHAR2(16)
 HOST_NAME                             VARCHAR2(64)
 VERSION                               VARCHAR2(17)
 STARTUP_TIME                          DATE
 STATUS                                VARCHAR2(12)
 PARALLEL                              VARCHAR2(3)
 THREAD#                               NUMBER
 ARCHIVER                              VARCHAR2(7)
 LOG_SWITCH_WAIT                       VARCHAR2(15)
 LOGINS                                VARCHAR2(10)
 SHUTDOWN_PENDING                      VARCHAR2(3)
 DATABASE_STATUS                       VARCHAR2(17)
 INSTANCE_ROLE                         VARCHAR2(18)
 ACTIVE_STATE                          VARCHAR2(9)
 BLOCKED                               VARCHAR2(3)
 CON_ID                                NUMBER
 INSTANCE_MODE                         VARCHAR2(11)
 EDITION                               VARCHAR2(7)
 FAMILY                                VARCHAR2(80)
```

列出 v$instance 的所有字段后，选择部分字段查询：

```
SQL> SELECT instance_number, host_name, version, status
  2  FROM v$instance;
INSTANCE_NUMBER HOST_NAME
--------------- ---------------------------------------------------------
VERSION           STATUS
----------------- ------------
              1 deep02
```

```
12.1.0.2.0        OPEN
```

可以发现上面的查询语句输出结果有点"乱",这是因为 SQL*Plus 的行尺寸太小了,通过 show linesize 可以看出,默认的 linesize 只有 80 个字符。我们可以设置更大一点的值,例如设置为 1024,其输出结果如下:

```
SQL> show linesize
linesize 80
SQL> set linesize 1024
SQL> SELECT instance_number, host_name, version, status
  2 FROM v$instance;
INSTANCE_NUMBER HOST_NAME             VERSION       STATUS
--------------- ---------------------------------------------------------------- ----------------- ------------
              1 deep02                12.1.0.2.0    OPEN
```

设置了 linesize 后的结果仍然不太友好,原因是部分列太长,我们可以通过设置列宽来解决这个问题,下面的"col host_name format a20"命令是设置 host_name 列的宽度为 20 个字符,最后可以看到,输出结果比较整齐了。

```
SQL> col host_name format a20
SQL> SELECT instance_number, host_name, version, status
  2 FROM v$instance;
INSTANCE_NUMBER HOST_NAME            VERSION          STATUS
--------------- -------------------- ---------------- ------------
              1 deep02               12.1.0.2.0       OPEN
```

SQL Plus 任务完成后,可以输入 exit 命令返回到操作系统:

```
SQL> exit
从 Oracle Database 12c Enterprise Edition Release 12.1.0.2.0 - 64bit Production
With the Partitioning, OLAP, Advanced Analytics and Real Application Testing options 断开
$
```

【示例 3-13】 使用 Easy Connect 方式登录

Oracle 样例数据库中有一个用户叫人力资源用户 HR(Humen Resource),现在以 Easy Connect 方式登录 HR,并查看 HR 用户的 Regions。在安装 Oracle 时我们选择了将样例数据库安装为 pdborcl,所以应该这样登录:

```
$ sqlplus hr/***@localhost:1521/pdborcl
SQL> SELECT * FROM regions;
 REGION_ID REGION_NAME
---------- -------------------------
```

```
1          Europe
2          Americas
3          Asia
4          Middle East and Africa
```

本例可以在任何能连接到服务器的计算机上执行,只需要将 localhost 换成服务器的地址即可,并且:1521 也可以省略。

3.4.2　SQL*Plus 命令列表

除了可以在 SQL*Plus 窗口中输入和执行 SQL 语句之外,还可以输入 SQL Plus 特有的命令对 SQL*Plus 的输入输出进行配置,对数据库进行管理。

表 3-1 是 SQL Plus 的常用命令列表,表中给出了常用的 SQL*Plus 命令以及基本描述。

表 3-1　SQL*Plus 常用命令列表

命令	基本描述
@	运行指定脚本中的 SQL 语句
@@	与@相似,只是运行前一个脚本相同目录中的文件
/	执行 SQL 命令或 PL/SQL 块
accept	读取输入的一行,并把它存储在指定的用户变量中
append	向缓冲区中的当前行尾部添加指定的文本
archive log	启动(start)、停止(stop)归档,查看(list)归档
attribute	类似于 column 命令,作用对象为 Object Type
break	对输出结果进行分组显示,如隐藏重复列的值、组分页等
btitle	在报告页底部设置一个标题及格式
change	在缓冲区中的当前行中进行文本替换
clear	清除/重置 break 或者 column 的格式
column	定义列的别名,输出格式
compute	计算或者显示汇总行
connect	连接到数据库
copy	将查询结构复制到本地或者远端数据库表中
define	定义用户变量
del	删除一行或者多行
describe	显示指定表、视图、过程或者函数的定义
disconnect	挂起连接,记录当前用户,但并不退出 SQL Plus
edit	打开外部文本编辑器,编辑指定文件或者缓冲区中的内容
execute	执行一条 PL/SQL 语句

续表

命令	基本描述
exit	退出 SQL Plus，返回操作系统
get	把操作系统文件装载到缓冲区
help	访问 SQL Plus 帮助系统
host	在 SQL Plus 环境中运行操作系统命令
input	在当前行后面添加一行或者多行文本
list	显示缓冲区的一行或者多行
password	修改口令
pause	暂停显示文本，直到按回车键
print	显示指定变量的值
prompt	发送指定信息到屏幕
quit	与 exit 相同
recover	执行数据库的介质恢复
remark	在脚本中标注注释信息的开始
repfooter	显示、替换或者定义报告底部的页脚格式
repheader	显示、替换或者定义报告顶部的页眉格式
run	显示并运行当前缓冲区中的 SQL 命令或 PL/SQL 块
save	将当前缓冲区中的内容保存为脚本
set	设置系统变量的值
show	显示 SQL Plus 系统变量的值或者当前的 SQL Plus 环境
shutdown	关闭当前运行的数据库实例或者 pdb
spool	将查询结果保存到文件中
start	运行指定脚本中的 SQL Plus 语句
startup	启动一个 Oracle 实例
store	将当前 SQL Plus 环境的属性保存为脚本文件
timing	定义时钟，记录一段时间内的时间数据
ttitle	显示、替换或者定义指定报告顶部的标题格式
undefine	删除一个或者多个由 define 定义的用户变量
variable	声明一个变量，可以在 PL/SQL 程序中使用

3.4.3 SQL*Plus 参数

可以在 SQL*Plus 窗口中通过 show 查看一些参数的值，可以通过 set 设置一些参数的值。在使用 show 时，可以使用"show 参数名"查看某个参数的值，也可以通过"show all"

查看所有参数的名称和参数值。

【示例 3-14】设置 TIME 参数

TIME 参数为 ON 时，命令提示符上会显示时间，这有助于观察当前时间。

```
SQL> show time
time OFF
SQL> set time on
12: 03: 57 SQL> show time
time ON
```

【示例 3-15】设置 TIMING 参数：

TIMING 参数为 ON 时，会显示命令执行时间长度。

```
SQL> set timing on
SQL> SELECT count(*)FROM jobs;
  COUNT(*)
----------
  19
已用时间：  00: 00: 00.01
```

3.4.4　SQL*Plus 替换变量

在 SQL*Plus 中可以定义一些替换变量，存储临时的数值。对于字符串或者日期类型的替换变量，在使用时必须加上单引号。替换变量可以用于 WHERE 子句、ORDER BY 子句以及表的名称的替换。替换变量的定义方式有两种，分别是 define 和 accept。

Oracle 自身定义了一些替换变量，都是以下划线开头的，比如_DATE 等，只需要运行命令 define 就能查询出它们的值。

```
SQL> define
DEFINE _DATE          = "2017-02-21 00: 52: 38" (CHAR)
DEFINE _CONNECT_IDENTIFIER = "localhost/pdborcl" (CHAR)
DEFINE _USER          = "HR" (CHAR)
DEFINE _PRIVILEGE     = "" (CHAR)
DEFINE _SQLPLUS_RELEASE = "1201000200" (CHAR)
DEFINE _EDITOR        = "ed" (CHAR)
DEFINE _O_VERSION     = "Oracle Database 12c Enterprise Edition Release 12.1.0.2.0 - 64bit Production
With the Partitioning, OLAP, Advanced Analytics and Real Application Testing options" (CHAR)
DEFINE _O_RELEASE     = "1201000200" (CHAR)
```

【示例 3-16】通过替换变量 JOB 查询 jobs 表。

本示例第一次是精确查询出 job_id 等于 MK_MAN 的工作，第二次是模糊查询出

job_id 以 MK 开头的所有工作。

```
SQL> define job='MK_MAN'
SQL> SELECT * FROM jobs WHERE  job_id='&job';
old   1: SELECT * FROM jobs WHERE  job_id='&job'
new   1: SELECT * FROM jobs WHERE  job_id='MK_MAN'

JOB_ID      JOB_TITLE                            MIN_SALARY  MAX_SALARY
----------  -----------------------------------  ----------  ----------
MK_MAN      Marketing Manager                          9000       15000


SQL> define job='MK%'
SQL> SELECT * FROM jobs WHERE  job_id like '&job';
old   1: SELECT * FROM jobs WHERE  job_id like '&job'
new   1: SELECT * FROM jobs WHERE  job_id like 'MK%'


JOB_ID      JOB_TITLE                            MIN_SALARY  MAX_SALARY
----------  -----------------------------------  ----------  ----------
MK_MAN      Marketing Manager                          9000       15000
MK_REP      Marketing Representative                   4000        9000
SQL>
```

【示例 3-17】使用 accept 定义替换变量

使用 accept 可以定义替换变量的类型和提示字符串。本例也是定义 job 替换变量，它的类型是 char，在输入值时有一个提示字符串"job_id："。

```
SQL> accept job char prompt 'job_id: '
job_id: MK_MAN
SQL> SELECT * FROM jobs WHERE  job_id like '&job';
old   1: SELECT * FROM jobs WHERE  job_id like '&job'
new   1: SELECT * FROM jobs WHERE  job_id like 'MK_MAN'


JOB_ID      JOB_TITLE                            MIN_SALARY  MAX_SALARY
----------  -----------------------------------  ----------  ----------
MK_MAN      Marketing Manager                          9000       15000
```

3.4.5 绑定变量

Oracle SQL*Plus 还允许另一种类型的变量：绑定变量(bind variable)。这种变量的值可以被方便地修改，可以通过 print 语句输出。

【示例 3-18】绑定变量的使用

本例定义绑定变量 region_id，然后多次给它赋不同的值。

```
SQL> var region_id number;
SQL> EXEC:region_id:= 1;
PL/SQL procedure successfully completed.
SQL> SELECT * FROM regions WHERE  region_id =: region_id;
 REGION_ID REGION_NAME
---------- -------------------------
  1          Europe
SQL> EXEC:region_id:= 2;
PL/SQL procedure successfully completed.
SQL> SELECT * FROM regions WHERE  region_id =: region_id;
 REGION_ID REGION_NAME
---------- -------------------------
  2          Americas
SQL> print region_id
    2
```

> 注意：绑定变量与替换变量不同，绑定变量不是简单替换，而是真正的变量，上例的查询语句：SELECT * FROM regions WHERE region_id =:region_id;执行了两次，由于完全相同，会被看作是同一条语句，Oracle 不会重复编译，因此它的效率比使用替换变量高。

3.4.6 预设变量

有些值我们可能经常要改变，如果每次启动 sqlplus 去重新设定就不是很方便。Oracle 在启动 sqlplus 之后会自动调用 glogin.sql 文件，我们可以修改这个文件，这样就不用每次都去设置了。glogin.sql 文件在目录$ORACLE_HOME/sqlplus/admin/中。下面是这个文件的典型设置：

```
set sqlprompt "_user'@'_connect_identifier> "
set linesize 1000 ;
set pagesize 100;
set serveroutput on;
ALTER SESSION SET NLS_DATE_FORMAT='YYYY-MM-DD HH24: MI: SS';
```

这样设置以后，提示符变更为用户@连接标识符，每行的宽度为 1000，每 100 条记录一页，允许显示 dbms_output 的输出，日期格式为年-月-日 时：分：秒。

3.4.7 PL/SQL 程序的运行

PL/SQL 和 SQL 语句不同，它是指 SQL 的编程语言。PL/SQL 是由多个语句块组成"程序块"，一次性批量执行。PL/SQL 程序块结构是：

```
DECLARE
    变量定义
BEGIN
    语句
END;
/
```

每个变量定义和每条语句都必须以分号结束，变量定义部分是可选的。在 PL/SQL 程序中，可以使用程序包 dbms_output 输出变量的值，在输出值之前还必须要设置 serveroutput 参数值为 on。

【示例 3-19】 编辑并运行 PL/SQL 程序

本例定义了一个变量 a，然后将 1+2 的运算结果赋值给 a，最后打印输出变量 a 的值：

```
SQL> set serveroutput on
SQL> DECLARE
2   a NUMBER;
3   BEGIN
4     a: = 1+2;
5     DBMS_OUTPUT.put_line(a);
6   END;
7   /
3
PL/SQL procedure successfully completed.
SQL> //
3
```

上例运行的每条程序语句前面有一个行号，行号本身不属于程序语句的部分，是 SQL*Plus 自动添加的，上例中有两个命令符号，一个是/，它表示运行程序，另一个是//，它表示重复运行刚刚运行过的程序。

如果 PL/SQL 程序比较复杂，行数多，在 sqlplus 窗口中编写代码非常不方便，这时候可以在 sqlplus 外面用其他工具先将程序代码保存为一个文本文件，然后在 sqlplus 中运行这个文件即可。比如，将上面的代码放在文件 a.sql 中，然后运行@a.sql，一样可以达到同样的目的。

```
SQL> @a.sql
3
```

3.5 Oracle SQL Developer

Oracle SQL Developer 是一个免费的 GUI 图形界面的管理和开发工具，可以提高工作效率并简化数据库开发任务。SQL Developer 可以在没有安装数据库的客户端上运行，支持 Windows、Linux 和 Mac OS X 系统。

可以在 Oracle 官方网站上下载，直接使用，无需安装。Oracle SQL Developer 是基于 Java 的应用程序，如果客户端没有安装 Java，就需要下载带自带有 Java 的 SQL Developer，如果客户端已经安装了 Java，就可以下载不带 Java 的版本。图 3-3 是 SQL Developer 的运行界面。

3.5.1 SQL Developer 连接 Oracle

SQL Developer 连接 Oracle 的方式比较多，常用的有两种，分别是基本方式和 TNS 方式。

【示例 3-20】基本方式连接 Oracle

如图 3-4 所示，以基本方式连接 Oracle，需要输入连接名。连接名与数据库无关，便于记忆即可，还需要输入用户名和口令，需要输入主机名、端口号、服务名或者数据库的 SID 名称。服务名和 SID 名称是安装 Oracle 时输入的名称，SID 只能是 CDB 的名称，服务名可以是 PDB 名称。一旦登录成功，连接名就会保存起来，下次直接通过这个连接名就可以立即连接到 Oracle。

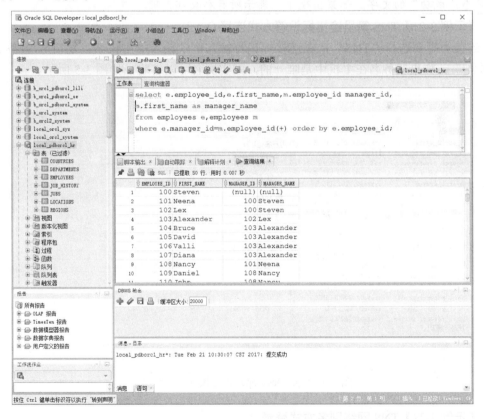

图 3-3　SQL Developer 运行界面

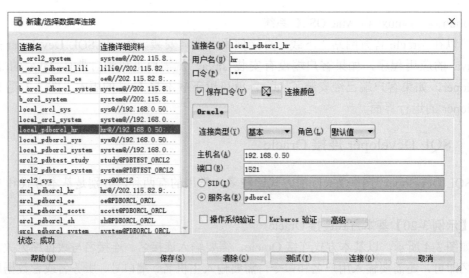

图 3-4　SQL Developer 基本方式连接

> 注意：在所有参数输入完成后，最好单击"测试"按钮，看看界面左下角是否出现"状态：成功"的文字。如果出现了，表示所有参数都正确。

【示例 3-21】TNS 简易连接标识符方式连接 Oracle

连接类型要选择"TNS"，在连接标识符输入框中输入"主机名：端口号/服务名[/实例名]"。如果服务名在主机中是唯一的，则实例名可以省略，如图 3-5 所示。

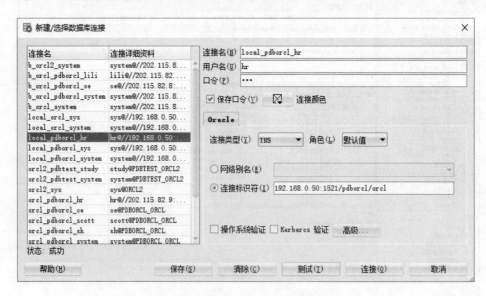

图 3-5　SQL Developer 以连接标识符简易连接

【示例 3-22】TNS 网络别名方式登录。

上面两种方式都需要明确输入主机名、端口号、服务名。Oracle 也支持 TNS 网络别名方式登录，所以 TNS 网络别名就是将连接标识符按 TNS 命名规则将主机名、端口号、服务名存储在自定义的文本文件中，同时取一个名称。在登录时就可以直接采用该名称，从而在 SQL Developer 界面中屏蔽掉具体的服务信息。本列所用的网络别名是 LOCAL_ORCL_PDBORCL。

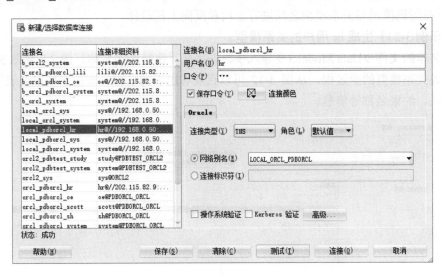

图 3-6　使用网络别名方式连接

在本例中，自定义的 TNS 文本文件的名称和存储目录都是任意指定的，为了让 SQL Developer 能够识别到这个文件，必须设置 Windows 环境变量 TNS_ADMIN 的值为该文件所在目录，Oracle SQL Developer 会访问这个目录中的所有文件，所以在具体存储时，存储 TNS 文件的目录最好只存储 TNS 文件。本例中 TNS 文本文件的内容是：

```
LOCAL_ORCL_PDBORCL=
  (DESCRIPTION =
    (ADDRESS = (PROTOCOL = TCP)(HOST = 192.168.0.50)(PORT = 1521))
    (CONNECT_DATA =
      (SERVER = DEDICATED)
      (SERVICE_NAME = pdborcl)
      (INSTANCE_NAME = orcl)
    )
  )
```

3.5.2　Data Modeler

Oracle SQL Developer 中有一个功能模块 Data Modeler，它的作用是提供广泛的数据建模能力，功能有：

(1) 捕捉业务规则和信息。
(2) 创建流程模型、逻辑模型、关系模型、物理模型和多维模型。
(3) 从各种源导入数据模型，如 Oracle Designer、VAR 文件、Erwin 等。

(4) 应用设计规则来检查和实现设计的完整性和一致性。
(5) 将元数据信息存储在 XML 文件中。
(6) 同步关系模型与数据字典。
(7) 在逻辑模型与关系模型之间进行正向和反向工程设计。

【示例 3-23】生成 hr 用户的关系模型。

从菜单的文件→Data Modeler→导入→数据字典，选择连接名称。图 3-7 到图 3-12 演示了生成关系模型的全过程，从图 3-12 可以直观看出表的名称、表的字段以及表间的主外键关联、约束名称等信息。

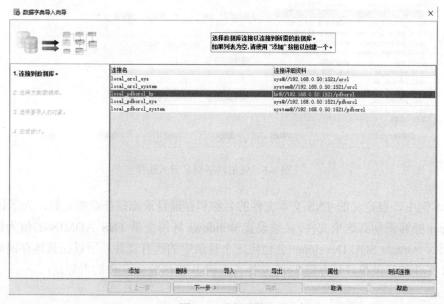

图 3-7　选择连接名称

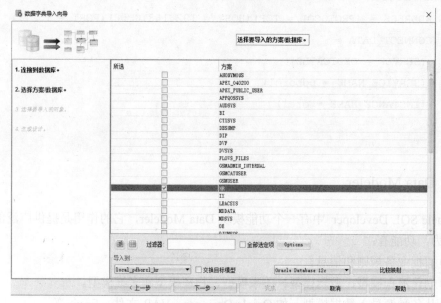

图 3-8　选择方案(用户)

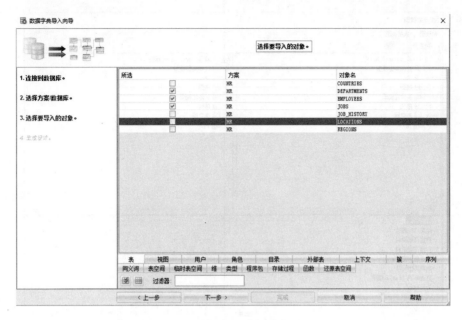

图 3-9 选择对象

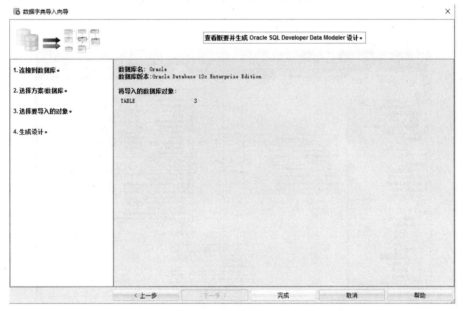

图 3-10 选择"完成"按钮

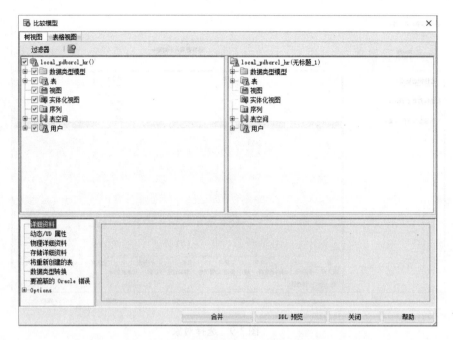

图 3-11 在比较模型中选择"合并"

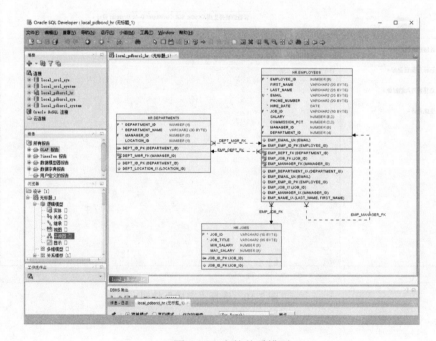

图 3-12 表的关系模型

3.6 执行计划与 SQL 优化

执行计划是数据库根据 SQL 语句和相关表的统计信息作出的一个最有效的执行计

划和查询方案,目的是以最快的方式查询出结果。查询方案是由查询优化器自动分析产生的。Oracle 的优化器有两种优化方式,即基于规则的优化方式(Rule-Based Optimization,RBO)和基于成本的优化方式(Cost-Based Optimization,CBO),Oracle 12c 推荐用 CBO 的方式。

CBO 方式根据表的统计信息分析来决定 SQL 语句的执行计划,比如:如果一条 SQL 语句从 10 万条记录的表中查 1 条记录,那查询优化器会选择"索引查找"方式,如果该表只剩下 5000 条记录了,那查询优化器就会改变方案,采用"全表扫描"方式。

表的统计信息包括列的离散度、列的直方图、索引的可用性以及索引上的聚簇因子。当这些信息是真实完整的情况下,CBO 优化器通常都可以制定出最优的执行计划。

3.6.1 授予查询执行计划的权限

普通用户不允许查询执行计划,必须有 plustrace 角色才可以。Oracle 的插接式数据库本身并没有默认创建 plustrace 角色,所以需要首先在 pdborcl 数据库中创建角色 plustrace,方法是用 sys 登录到 PDB 数据库,然后运行$ORACLE_HOME/sqlplus/admin/plustrce.sql 脚本文件,最后通过"GRANT plustrace to 用户名;"命令将 plustrace 赋予用户。

【示例 3-24】给 hr 授予 plustrace 角色权限。

```
$ sqlplus sys/***@localhost/pdborcl as sysdba
SQL> @$ORACLE_HOME/sqlplus/admin/plustrce.sql
SQL> create role plustrace;
Role created.
SQL>
SQL> GRANT SELECT ON v_$sesstat TO plustrace;
Grant succeeded.
SQL> GRANT SELECT ON v_$statname TO plustrace;
Grant succeeded.
SQL> GRANT SELECT ON v_$mystat TO plustrace;
Grant succeeded.
SQL> GRANT plustrace TO dba WITH ADMIN OPTION;
Grant succeeded.
SQL> set echo off
SQL>
SQL> GRANT plustrace TO hr;
Grant succeeded.

SQL>GRANT SELECT ON v_$sql TO scott;
SQL>GRANT SELECT ON v_$sql_plan TO scott;
SQL>GRANT SELECT ON v_$sql_plan_statistics_all TO scott;
```

```
SQL>GRANT SELECT ON v_$session TO scott;
SQL>GRANT SELECT ON v_$parameter TO scott;
```

3.6.2 分析和比较执行计划

hr 用户被授予了 plustrace 后，就可以查看 SQL 语句的执行计划了。在 sqlplus 中查询执行计划是通过设置 autotrace 参数的值来进行的。set autotrace 的选项和各种组合情况有：

SET AUTOT[RACE] {OFF | ON | TRACE[ONLY]} [EXP[LAIN]] [STAT[ISTICS]]

set autotrace off：缺省值，将不生成 autotrace 报告。

set autotrace on：输出查询结果，显示执行计划和统计信息。

set autotrace traceonly：不输出查询结果，只显示执行计划和统计信息。

set autotrace traceonly explain：不输出查询结果，只显示执行计划。

set autotrace traceonly statistics：不输出查询结果，只显示统计信息。

set autotrace on explain：输出查询结果和显示执行计划。

set autotrace on statistics：输出查询结果和显示统计信息。

【示例 3-25】查询执行计划

hr 用户有一个员工表 employees，有三个字段属性：employee_id、first_name 和 manager_id，分别表示员工号、姓名、员工的管理员号，其中员工的管理员号也代表一个员工，管理员的信息也在员工表中。本例除了要查询每个员工的这三个属性外，还要查询出管理员的 first_name，把这个管理员的 first_name 命名为 manager_name。

```
SQL> SET AUTOTRACE ON
SQL> SELECT e.EMPLOYEE_ID, e.FIRST_NAME, e.MANAGER_ID,
  2        (SELECT    M.FIRST_NAME    FROM    employees    m    WHERE
m.EMPLOYEE_ID=e.MANAGER_ID)
  3 AS MANAGER_NAME FROM hr.employees e ORDER BY e.EMPLOYEE_ID;
EMPLOYEE_ID FIRST_NAME           MANAGER_ID MANAGER_NAME
----------- -------------------- ---------- --------------------
        100 Steven
        101 Neena                       100 Steven
        102 Lex                         100 Steven
        103 Alexander                   102 Lex
        104 Bruce                       103 Alexander
...
107 rows selected.
Execution Plan
----------------------------------------------------------
Plan hash value: 4265739349
```

```
---------------------------------------------------------------------
| Id | Operation                    | Name          | Rows | Bytes | Cost(%CPU)| Time     |
---------------------------------------------------------------------
|  0 | SELECT STATEMENT             |               |  107 |  1605 |   21 (0)| 00:00:01 |
|  1 | TABLE ACCESS BY INDEX ROWID  | EMPLOYEES     |    1 |    11 |    1 (0)| 00:00:01 |
|* 2 |  INDEX UNIQUE SCAN           | EMP_EMP_ID_PK |    1 |       |    0 (0)| 00:00:01 |
|  3 | TABLE ACCESS BY INDEX ROWID  | EMPLOYEES     |  107 |  1605 |    3 (0)| 00:00:01 |
|  4 |  INDEX FULL SCAN             | EMP_EMP_ID_PK |  107 |       |    1 (0)| 00:00:01 |
---------------------------------------------------------------------

Predicate Information (IDENTIFIED BY operation id):
---------------------------------------------------------
   2 - access("M"."EMPLOYEE_ID"=:B1)
Statistics
---------------------------------------------------------
        1  recursive calls
        0  db block gets
       45  consistent gets
        0  physical reads
        0  redo size
     4431  bytes sent via SQL*Net to client
      629  bytes received via SQL*Net FROM client
        9  SQL*Net roundtrips to/from client
        0  sorts (memory)
        0  sorts (disk)
      107  rows processed
```

从本例输出的执行计划来看，执行计划有三个部分，分别是：

Execution Plan（执行计划）：执行计划的输出是一个表，属性有 ID、Operation、Name、Rows、Bytes、Cost（%CPU）和 Time。

ID：序号，带*号的是谓词操作。ID 号不是执行顺序，执行顺序是从下到上，从右向左。

Operation：当前操作的内容。

Rows：当前操作的 Cardinality，即当前操作的返回结果集行数。

Cost（CPU）：成本，用于说明 SQL 执行的代价。

Time：Oracle 估计当前操作的时间。

Predicate Information（谓词信息）：谓词就是 WHERE 子句的查询条件。其中"：B1"是绑定变量，其实就是 SQL 语句中的 e.MANAGER_ID。

access：表示这个谓词条件的值将会影响数据的访问路径，选择索引时才会是 access。

filter：表示谓词条件的值不会影响数据的访问路径，只起过滤的作用。

statistics（统计信息）：统计信息的属性有：

- recursive calls：递归调用次数。
- db block gets：从 buffer cache 中读取的 block 的数量。
- consistent gets：从 buffer cache 中读取的 undo 数据的 block 数量。
- physical reads：从磁盘读取的 block 的数量。
- redo size：DML 生成的 redo 的大小。
- sorts（memory）：在内存执行的排序量。
- sorts（disk）：在磁盘上执行的排序量。

physical reads 和 consistent gets 通常是我们最关心的，如果 physical reads 很高，说明要从磁盘请求大量的数据到 Buffer Cache 里，通常意味着系统里存在大量全表扫描的 SQL 语句，这会影响到数据库的性能，因此尽量避免语句做全表扫描，对于全表扫描的 SQL 语句，建议增加相关的索引，优化 SQL 语句来解决。physical reads、db block gets 和 consistent gets 这三个参数之间有一个换算公式：

数据缓冲区的使用命中率 =
1-(physical reads/(db block gets + consistent gets))。

上例中 SQL 语句的 consistent gets=45，cost 成本=21，它并不是最有效率的语句，原因是用了子查询语句作为一个字段属性，使得每输出 employees 的一行都要再次查询一次该表，更好的办法是用多表外连接方式查询。优化后的语句如下：

【示例 3-26】更优的查询语句

```
SQL> SELECT e.EMPLOYEE_ID, e.FIRST_NAME, e.MANAGER_ID, m.FIRST_NAME
  2  AS MANAGER_NAME FROM employees e, employees m WHERE
  3  e.MANAGER_ID=m.EMPLOYEE_ID(+)ORDER BY e.EMPLOYEE_ID;
...
107 rows selected.
Execution Plan
----------------------------------------------------------
Plan hash value: 2948531109
----------------------------------------------------------
| Id | Operation          | Name | Rows | Bytes | Cost (%CPU)| Time     |
----------------------------------------------------------
|  0 | SELECT STATEMENT   |      |  107 |  2782 |    6  (17)| 00:00:01 |
```

```
|   1 |  SORT ORDER BY          |              |   107 |  2782 |     6
(17)| 00: 00: 01 |
|*  2 |   HASH JOIN OUTER       |              |   107 |  2782 |     5
(0) | 00: 00: 01 |
|   3 |    TABLE ACCESS FULL    | EMPLOYEES    |   107 |  1605 |     3
(0) | 00: 00: 01 |
|   4 |    VIEW                 | index$_join$_002 | 107 | 1177 |   2
(0) | 00: 00: 01 |
|*  5 |     HASH JOIN           |              |       |       |
|     |
|   6 |      INDEX FAST FULL SCAN| EMP_EMP_ID_PK |  107 | 1177 |    1
(0) | 00: 00: 01 |
|   7 |      INDEX FAST FULL SCAN| EMP_NAME_IX   |  107 | 1177 |    1
(0) | 00: 00: 01 |
-------------------------------------------------------------------

Predicate Information (IDENTIFIED BY operation id):
-------------------------------------------------------------------

   2 - access("E"."MANAGER_ID"="M"."EMPLOYEE_ID"(+))
   5 - access(ROWID=ROWID)
Statistics
-------------------------------------------------------------------
          0  recursive calls
          0  db block gets
         15  consistent gets
          0  physical reads
          0  redo size
       4431  bytes sent via SQL*Net to client
        629  bytes received via SQL*Net from client
          9  SQL*Net roundtrips to/from client
          1  sorts (memory)
          0  sorts (disk)
        107  rows processed
```

> 注意：执行计划中的代价 cost 要影响执行时间，但 cost 不是唯一影响执行时间的因素，因此，要综合判断执行计划中的所有输出值才能判断一个执行计划的好坏。

【示例 3-27】显示执行计划的另一种办法

本例先运行"explain plan for ..."SQL 语句，再运行语句"SELECT * FROM table(dbms_xplan.display)；"也可以显示执行计划。

```
SQL> explain plan for SELECT salary FROM hr.employees WHERE
  2  first_name like 'Pat';
Explained.
SQL> SELECT * FROM table(dbms_xplan.display);
PLAN_TABLE_OUTPUT
Plan hash value: 612698390
---------------------------------------------------------------------------
| Id  | Operation                           | Name         | Rows | Bytes | Cost (%CPU)| Time     |
---------------------------------------------------------------------------
|   0 | SELECT STATEMENT                    |              |   1  |   11  |   2   (0)| 00:00:01 |
|   1 |  TABLE ACCESS BY INDEX ROWID BATCHED| EMPLOYEES    |   1  |   11  |   2   (0)| 00:00:01 |
|*  2 |   INDEX SKIP SCAN                   | EMP_NAME_IX  |   1  |       |   1   (0)| 00:00:01 |
---------------------------------------------------------------------------
Predicate Information (IDENTIFIED BY operation id):
---------------------------------------------------
   2 - access("FIRST_NAME"='Pat')
       filter("FIRST_NAME"='Pat')
15 rows selected.
```

explain plan 并不执行当前的 SQL 语句，而是根据数据字典中记录的统计信息获取最佳的执行计划并加载到表 plan_table。由于统计信息和执行环境的变化，explain plan 与实际的执行计划可能会有差异。对于运行时间较长的 SQL 语句，不需要等到结果输出即可提前获得该 SQL 的执行计划，对于在生产环境中调试的情况，这样会减轻数据库负荷。

3.6.3 统计信息与动态采样

统计信息(Statistic)的收集对 Oracle 是非常重要的。它会提前收集数据库中对象的详细信息，并存储在相应的数据字典里。根据这些统计信息，查询优化器可以对每个 SQL 选择最好的执行计划。

动态采样(Dynamic Sampling)是对统计信息收集的一种补充。在 CBO(基于代价的优化器模式)条件下，如果某个表没有统计信息，即在段(表、索引、分区)没有分析的情况下，为了使 CBO 优化器得到足够的信息以保证做出正确的执行计划，Oracle 会对表进行动态采样。动态采样技术可以通过直接从需要分析的对象上收集数据块(采样)来获得 CBO 需要的统计信息。Oracle 的初始化参数 optimizer_dynamic_sampling 决定了如何进行动态采样。

统计信息关乎 SQL 的执行计划是否正确。Oracle 的初始化参数 statistics_level 控制收集统计信息的级别，有三个值：BASIC（收集基本的统计信息），TYPICAL（收集大部分统计信息），这是数据库的默认设置，ALL（收集全部统计信息）。统计信息包含：

行统计信息(user_tables)：行数(NUM_ROWS)、块数(BLOCKS)、行平均长度(AVG_ROW_LEN)；

列统计信息(user_tab_columns)：列中唯一值的数量(NUM_DISTINCT)，NULL 值的数量(NUM_NULLS)，数据分布直方图(HISTOGRAM)；

索引统计(user_index)：叶块数量(LEAF_BLOCKS)、等级(BLEVEL)、聚簇因子(CLUSTERING_FACTOR)。

Oracle 提供了一个程序包 dbms_stats 用于收集统计信息，dbms_stats 中包含的主要函数有：

gather_database_stats：分析数据库的所有对象。

gather_schema_stats：分析用户所有的对象(包括表、索引、簇)。

gather_table_stats：分析表。

gather_index_stats：分析索引。

delete_database_stats：删除数据库统计信息。

delete_schema_stats：删除用户方案统计信息。

delete_table_stats：删除表统计信息。

delete_index_stats：删除索引统计信息。

delete_column_stats：删除列统计信息。

DBMS_STATS.GATHER_TABLE_STATS 的语法如下：

```
DBMS_STATS.GATHER_TABLE_STATS (
    ownname            VARCHAR2,
    tabname            VARCHAR2,
    partname           VARCHAR2 DEFAULT NULL,
    estimate_percent   NUMBER,
    block_sample       BOOLEAN  DEFAULT FALSE,
    method_opt         VARCHAR2,
    degree             NUMBER,
    granularity        VARCHAR2,
    cascade            BOOLEAN,
    stattab            VARCHAR2 DEFAULT NULL,
    statid             VARCHAR2 DEFAULT NULL,
    statown            VARCHAR2 DEFAULT NULL,
    no_invalidate      BOOLEAN,
    stattype           VARCHAR2 DEFAULT 'DATA',
    force              BOOLEAN  DEFAULT FALSE);
```

参数说明：

ownname：要分析表的拥有者。

tabname：要分析的表名。

partname：分区的名字，只对分区表或分区索引有用。

estimate_percent：采样行的百分比，取值范围[0.000001，100]，null 为全部分析，不采样。常量：DBMS_STATS.AUTO_SAMPLE_SIZE 是默认值，由 Oracle 决定最佳取采样值。

block_sample：是否用块采样代替行采样。

method_opt：决定直方图(histograms)信息是怎样被统计的。method_opt 的取值如下：

for all columns：统计所有列的 histograms。

for all indexed columns：统计所有 indexed 列的 histograms。

for all hidden columns：统计隐藏列的 histograms。

for columns <list> SIZE <N> | REPEAT | AUTO | SKEWONLY：统计指定列的 histograms。N 的取值范围[1，254]，如果 N=1 表示不做直方图分析；REPEAT 上次统计过的 histograms；AUTO 由 Oracle 决定 N 的大小；SKEWONLY 只会根据 column 的数据分布情况决定是否收集 histogram。

degree：决定并行度，默认值为 null。

granularity：统计收集器的颗粒度，只当表是分区表的时候才用。

cascade：是否收集索引的信息，默认为 false。

stattab：指定要存储统计信息的表。

statid：统计信息 ID，当 stattab 相同的时候，区别不同的统计信息，是 stattab 的补充。

statown：存储统计信息表的拥有者。

如果 stattab、statid 和 statown 三个参数都不指定，统计信息会直接更新到数据字典中。

no_invalidate：如果设置为 TRUE，不会使依赖的游标无效，如果设置为 FALSE，将使依赖的游标立即失效。

force：强制统计，即使表锁住了也收集统计信息。

【示例 3-28】表的统计信息收集前后的对比

本示例首先创建一个测试表 emp_test，该表来源是 hr 用户的 employees 表，emp_test 的数据是 employees 的 4 倍。

```
SQL> CREATE TABLE hr.emp_test as SELECT * FROM hr.employees;
Table created.
SQL> INSERT INTO hr.emp_test SELECT * FROM employees;
107 rows created.
SQL> INSERT INTO hr.emp_test SELECT * FROM employees;
107 rows created.
SQL> INSERT INTO hr.emp_test SELECT * FROM employees;
107 rows created.
SQL> SELECT COUNT(*)FROM hr.emp_test WHERE  employee_id=110;
  COUNT(*)
```

```
         ----------
              4
SQL> explain plan for SELECT * FROM hr.emp_test
  2   WHERE  employee_id=110;
Explained.
SQL> SELECT * FROM TABLE(dbms_xplan.display);
Plan hash value: 3124080142
---------------------------------------------------------------
| Id  | Operation          | Name      | Rows  | Bytes | Cost (%CPU)| Time     |
---------------------------------------------------------------
|   0 | SELECT STATEMENT   |           |     1 |    69 |     3   (0)| 00:00:01 |
|*  1 |  TABLE ACCESS FULL | EMP_TEST  |     1 |    69 |     3   (0)| 00:00:01 |
---------------------------------------------------------------

Predicate Information (IDENTIFIED BY operation id):
---------------------------------------------------

   1 - filter("EMPLOYEE_ID"=110)
13 rows selected.
```

可以看出,该执行计划的输出行 Rows 为 1,实际情况应该是 4 行,所以这个执行计划是不准确的,原因就在于没有收集表的统计信息。下面进行统计信息的收集,然后再观察查询计划的变化。

```
SQL> EXEC DBMS_STATS.GATHER_TABLE_STATS('HR', 'EMP_TEST');
PL/SQL procedure successfully completed.


SQL> explain plan for SELECT * FROM hr.emp_test
  2 WHERE  employee_id=110;
Explained.
SQL> SELECT * FROM TABLE(dbms_xplan.display);
PLAN_TABLE_OUTPUT
Plan hash value: 3124080142
---------------------------------------------------------------
| Id  | Operation          | Name      | Rows  | Bytes | Cost (%CPU)| Time     |
---------------------------------------------------------------
|   0 | SELECT STATEMENT   |           |     4 |   276 |     6   (0)| 00:00:01 |
|*  1 |  TABLE ACCESS FULL | EMP_TEST  |     4 |   276 |     6   (0)| 00:00:01 |
---------------------------------------------------------------

Predicate Information (IDENTIFIED BY operation id):
---------------------------------------------------
```

```
           1 - filter("EMPLOYEE_ID"=110)
13 rows selected.
```

可以看出,该执行计划的输出行为 4 行,实际情况也是 4 行,所以这个执行计划是准确的,原因就在于收集了表的统计信息。在实际应用中,如果有了这样的计划,就会达到最快的查询速度。

统计信息也可以被删除,删除统计信息后,再观察查询计划,执行计划仍然是 4 行。如下操作所示:

```
SQL> EXEC DBMS_STATS.DELETE_TABLE_STATS('HR', 'EMP_TEST');
PL/SQL procedure successfully completed.
SQL> explain plan for SELECT * FROM hr.emp_test
  2 WHERE  employee_id=110;

Explained.

SQL> SELECT * FROM TABLE(dbms_xplan.display);
Plan hash value: 3124080142

---------------------------------------------------------------------
| Id | Operation          | Name     | Rows | Bytes | Cost (%CPU)| Time     |
---------------------------------------------------------------------
|  0 | SELECT STATEMENT   |          |    4 |   532 |     6   (0)| 00:00:01 |
|* 1 |  TABLE ACCESS FULL | EMP_TEST |    4 |   532 |     6   (0)| 00:00:01 |
---------------------------------------------------------------------

Predicate Information (IDENTIFIED BY operation id):
---------------------------------------------------

   1 - filter("EMPLOYEE_ID"=110)

Note
-----
   - dynamic statistics used: dynamic sampling (level=2)
17 rows selected.
SQL>
```

可以看出,该执行计划的输出行 Rows 仍为 4 行,实际情况也是 4 行,所以这个执行计划是准确的,但这时表 emp_test 的统计信息已经被删除了,那为什么执行计划没有回到最初的时候呢?

这是因为 Oracle 有动态采样机制,它会对没有统计过的表进行自动采样,尽可能实现最优的查询计划。注意,输出的结果最后有一行注意(Note):"dynamic statistics used:dynamic sampling(level=2)",其中 level=2 表示初始化参数 optimizer_dynamic_sampling

的值为 2。如果 optimizer_dynamic_sampling 的值为 0 就会禁用动态采样。

我们有时会希望查询表是否被分析过，Oracle 提供了 user_tables 视图，不仅可以查询用户的所有表，也能通过 LAST_ANALYZED 属性查询这些统计信息的分析时间。

【示例 3-29】查询表何时被分析过

本例以 HR 登录，查看每个表最后被分析的时间 LAST_ANALYZED，LAST_ANALYZED 为空表示该表未被分析过。

```
SQL> col table_name format a30
SQL> SELECT table_name, last_analyzed FROM user_tables;
TABLE_NAME                     LAST_ANALYZED
------------------------------ -----------------
EMP_TEST
JOB_HISTORY                    2014-07-07 06:56:25
EMPLOYEES                      2017-02-22 23:43:52
JOBS                           2014-07-07 06:56:25
DEPARTMENTS                    2014-07-07 06:56:25
LOCATIONS                      2014-07-07 06:56:26
REGIONS                        2014-07-07 06:56:26
COUNTRIES                      2014-07-07 06:56:24
10 rows selected.
```

3.6.4 SQL 语句的优化

有时候我们并不能直观找到一些复杂的 SQL 语句的优化方法，这时可以借助 Oracle SQL Developer 工具的 SQL 语句优化指导功能。通常，普通的用户没有这个权限，需要添加 SELECT_CATALOG_ROLE、SELECT ANY DICTIONARY、ADVISOR 以及 ADMINISTER SQL TUNING SET 权限才可以进行优化指导。

【示例 3-30】让户 HR 可以进行 SQL 语句的优化指导。

首先以 system 身份登录到 pdborcl，然后授权。

```
$ sqlplus system/***@localhost/pdborcl
SQL> GRANT select_catalog_role TO hr;
Grant succeeded.
SQL> GRANT SELECT ANY DICTIONARY TO hr;
Grant succeeded.
SQL> GRANT ADVISOR TO hr;
Grant succeeded.
SQL> GRANT ADMINISTER SQL TUNING SET TO hr;
Grant succeeded.
```

只要有上述权限，一个用户在 SQL Developer 中很容易获取一条 SQL 语句的优化建议,我们现在假定要查询一个 EMPLOYEE_ID=100 的员工,为了看到测试效果,我们这样写 SQL 语句:"SELECT * FROM employees WHERE EMPLOYEE_ID+1=100+1;",故意在等号左右两边都加 1,由于 EMPLOYEE_ID 是主键,是唯一索引,这一点改动就会让查询变成全表搜索,效率大大降低!执行效果如图 3-13 所示。从图 3-13 中可以看到,SQL 优化指导给出了三条优化信息:

查找结果:谓词 "EMPLOYEES"."EMPLOYEE_ID"+1=101(在执行计划的行 ID 1 处使用)包含索引列 "EMPLOYEE_ID" 的表达式。此表达式使优化程序无法选择表 "HR"."EMPLOYEES" 的索引。

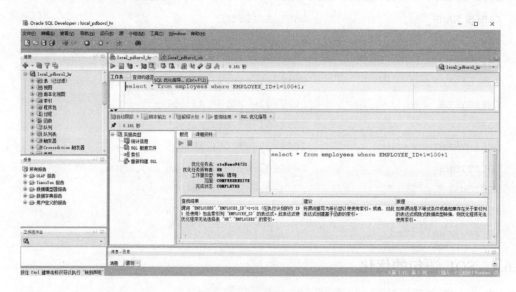

图 3-13 SQL 语句优化指导

建议:将谓词重写为等价型以便使用索引,或者对此表达式创建基于函数的索引。

原理:如果谓词是不等式条件,或者存在关于索引列的表达式或隐式数据类型转换,则优化程序无法使用索引。

根据这个优化指导,我们不难得到最优的 SQL 语句:"SELECT * FROM employees WHERE EMPLOYEE_ID=100;",将修改后的 SQL 语句再次执行优化指导,就看不到任何优化建议了,表示已经是最优的了,如图 3-14 所示。

在实际编写 SQL 语句时要把使用优化指导工具和人工分析结合起来,有时候优化指导工具没有提示任何优化建议,但并不表示 SQL 语句就是效率最高的,SQL 语句是否最优的判断最后还要手动确认。

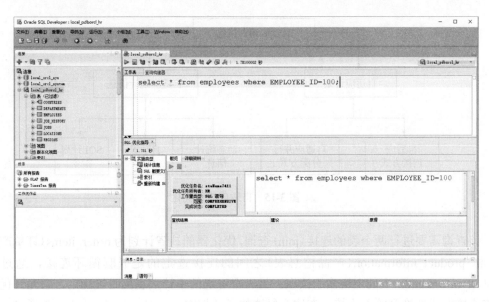

图 3-14　SQL 语句无优化指导

3.6.5　自适应查询优化

Oracle 12c 数据库中最大的变化是自适应查询优化（Adaptive Query Optimization）。优化器在某些条件下会选择自适应计划。例如，当查询包括连接和复杂谓词时，使准确地估计基数变得很困难。自适应计划使得优化器能够把执行一个语句的计划推迟到执行的时候才确定。优化器在它所选择的计划（缺省计划）中植入统计收集器，从而在运行的时候，它能够判断基数估算与实际看到的行数之间是否有很大的偏差。如果有显著的区别，那么这个计划或者计划的一部分就会被自动调整，以避免不理想的性能。自适应查询优化包括两个方面的作用：

1）自适应计划（Adaptive Plans）

自适应计划改善一个查询的初次执行计划，避免第一次查询就采用到最坏的执行计划。

2）自适应统计信息（Adaptive Statistics）

自适应统计信息为后续的执行提供了额外的信息。如图 3-15 所示。图 3-15 中自适应连接方式是指通过为执行计划中的某些分支预先确定多个子计划，优化器能够实时调整连接方式。我们假定查询 OE 用户的两个表：

```
SQL> SELECT p.product_name FROM oe.order_items o,
  2 oe.product_information p WHERE  o.unit_price=15
  3 AND o.quantity>1 AND p.product_id=o.product_id;
```

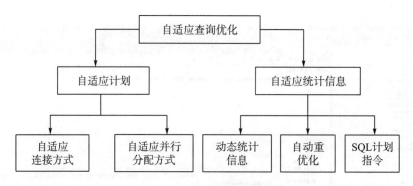

图 3-15　自适应查询优化

该查询需要进行两个表的连接(join)查询,优化器的缺省计划为 order_items(订单产品表)和 product_information(产品信息表)之间的连接选定的是嵌套循环连接,通过对 product_information 表的主键索引读取。另一个可选的子计划也同时被确定,它允许优化器将连接方式切换到哈希连接,在这个候选的子计划中,product_information 是通过全表扫描来读取的,如图 3-16 所示。

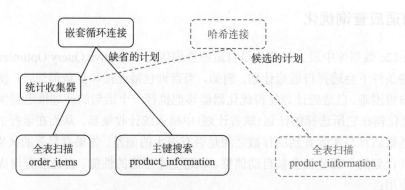

图 3-16　缺省的计划是嵌套循环连接(order_items 行数小于阈值时)

【示例 3-31】查看缺省的嵌套循环连接(nested loops join)

```
SQL> ALTER SESSION SET optimizer_dynamic_sampling=0;
SQL> explain plan for SELECT p.product_name
  2    FROM oe.order_items o, oe.product_information p
  3    WHERE  o.unit_price=15 AND o.quantity>1 AND
  4    p.product_id=o.product_id;
已解释。
SQL> SELECT * FROM TABLE(dbms_xplan.display());
```

上述示例中临时设置了参数动态采样参数 optimizer_dynamic_sampling 为 0,表示暂时不使用动态采样,这样可以稳定得到不采样的查询计划。查询结果如图 3-17 所示。

第 3 章　网络配置及管理工具　　83

```
Plan hash value: 1255158658

| Id  | Operation                     | Name                | Rows | Bytes | Cost (%CPU)| Time     |
|   0 | SELECT STATEMENT              |                     |    4 |   128 |     7   (0)| 00:00:01 |
|   1 |  NESTED LOOPS                 |                     |    4 |   128 |     7   (0)| 00:00:01 |
|   2 |   NESTED LOOPS                |                     |    4 |   128 |     7   (0)| 00:00:01 |
|*  3 |    TABLE ACCESS FULL          | ORDER_ITEMS         |    4 |    48 |     3   (0)| 00:00:01 |
|*  4 |    INDEX UNIQUE SCAN          | PRODUCT_INFORMATION_PK |  1 |       |     0   (0)| 00:00:01 |
|   5 |   TABLE ACCESS BY INDEX ROWID | PRODUCT_INFORMATION |    1 |    20 |     1   (0)| 00:00:01 |

Predicate Information (identified by operation id):
---------------------------------------------------

   3 - filter("O"."UNIT_PRICE"=15 AND "O"."QUANTITY">1)
   4 - access("P"."PRODUCT_ID"="O"."PRODUCT_ID")

Note
-----
   - this is an adaptive plan

已选择 22 行。
```

图 3-17　缺省的自适应计划结果

在 SQL 语句第一次执行时，统计收集器收集了关于这次执行的信息，并且将一部分进入到子计划的数据行缓存起来。优化器会确定要收集哪些统计信息，以及如何根据统计的不同值来确定计划。它会算出一个"拐点"，两个计划选项在这个值是一样好的。例如，如果当 order_items 表的扫描产生的行数少于 10 行，则嵌套循环连接是最佳，当 order_items 表的扫描产生的行数多于 10 行，则哈希连接是最佳。

在图 3-18 中，因为来自 order_items 表的行数大于优化器最初的估计，优化器最终决定使用哈希连接，并禁用统计收集器。

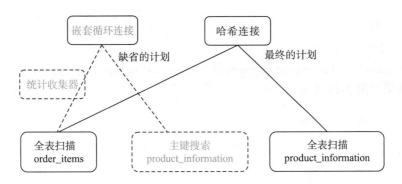

图 3-18　最终的计划是哈希连接(order_items 行数大于阈值)

为了看到自适应计划中所有的操作，包括统计收集器的位置，必须使用 DBMS_XPLAN.display_cursor() 函数，该函数与 DBMS_XPLAN.display() 不同，DBMS_XPLAN.display() 函数显示的是优化器选定的初始执行计划，而 DBMS_XPLAN.display_cursor() 显示的是实际的执行计划，如果再指定额外的格式参数'+adaptive'，就可以显示实际的完整的执行计划。在这个模式下，Id 栏会出现一个额外的(-)记号，指明在计划中未被采用(非激活)的操作。

【示例 3-32】查看最终的哈希连接(hash join)

```
SQL> ALTER SESSION SET optimizer_dynamic_sampling=2;
SQL> SELECT /*+ gather_plan_statistics */ p.product_name
  2  FROM oe.order_items o, oe.product_information p
```

```
  3  WHERE  o.unit_price=15 AND o.quantity>1 AND
  4  p.product_id=o.product_id;
PRODUCT_NAME
----------------------------------------------------
Screws <B.28.S>
Screws <B.28.S>
Screws <B.28.S>
Screws <B.28.S>
Screws <B.28.S>
Screws <B.28.S>
Screws <B.28.S>
Screws <B.28.S>
Screws <B.28.S>
Screws <B.28.S>
Screws <B.28.S>
Screws <B.28.S>
Screws <B.28.S>

已选择 13 行。
SQL> SELECT * FROM
  2  TABLE(dbms_xplan.display_cursor(format=>'+adaptive'));
```

执行结果如图 3-19 所示：

```
Plan hash value: 1553478007

---------------------------------------------------------------------------------------------------
|   Id  | Operation                            | Name                    | Rows | Bytes | Cost (%CPU)| Time     |
---------------------------------------------------------------------------------------------------
|    0  | SELECT STATEMENT                     |                         |      |       |   7 (100)|          |
|  * 1  |  HASH JOIN                           |                         |   4  |  128  |   7   (0)| 00:00:01 |
|  - 2  |   NESTED LOOPS                       |                         |   4  |  128  |   7   (0)| 00:00:01 |
|  - 3  |    NESTED LOOPS                      |                         |   4  |  128  |   7   (0)| 00:00:01 |
|  - 4  |     STATISTICS COLLECTOR             |                         |      |       |          |          |
|  * 5  |      TABLE ACCESS FULL               | ORDER_ITEMS             |   4  |   48  |   3   (0)| 00:00:01 |
|  -* 6 |     INDEX UNIQUE SCAN                | PRODUCT_INFORMATION_PK  |   1  |       |   0   (0)|          |
|    7  |    TABLE ACCESS BY INDEX ROWID       | PRODUCT_INFORMATION     |   1  |   20  |   1   (0)| 00:00:01 |
|    8  |   TABLE ACCESS FULL                  | PRODUCT_INFORMATION     |   1  |   20  |   1   (0)| 00:00:01 |
---------------------------------------------------------------------------------------------------

Predicate Information (identified by operation id):

   1 - access("P"."PRODUCT_ID"="O"."PRODUCT_ID")
   5 - filter(("O"."UNIT_PRICE"=15 AND "O"."QUANTITY">1))
   6 - access("P"."PRODUCT_ID"="O"."PRODUCT_ID")

Note
-----
   - this is an adaptive plan (rows marked '-' are inactive)
```

图 3-19 最终自适应计划 HASH JOIN

在上面的操作中将 OPTIMIZER_DYNAMIC_SAMPLING 参数设置为了默认值为 2，表示采样 block 数大于 32 并且没有被分析的表就进行少量的动态采样（Dynamic Sampling）。在 Oracle 12c 中，初始化参数 OPTIMIZER_DYNAMIC_SAMPLING 引入了新的取样级别 11。11 级使得优化器能够自动为任何 SQL 语句使用动态统计信息，即使所有基本的表统计信息都已经存在。优化器做出使用动态统计的决定，是基于所用谓词的复杂性、已经存在的基础统计信息，以及预期的 SQL 语句总执行时间。

练习

一、判断题

1. Oracle 不能同时工作在专用服务器模式和共享服务器模式下。（　）
2. Oracle 可以仅指定部分 PDB 工作在共享服务器模式。（　）
3. 共享服务器模式是多用户可以轮流使用一个服务器进程。（　）
4. 系统参数 SHARED_SERVERS=1 时表示启用共享服务器模式。（　）
5. TNS 的配置文件 listener.ora 是客户端配置文件。（　）
6. listener 的动态注册和静态注册是可以同时并存的。（　）
7. 启动 lsnrctl 的时候，静态注册会立即生效。（　）
8. 只含有绑定变量的 SQL 会被看作是同一条语句，Oracle 不会重复编译，因此它的效率比使用替换变量高。（　）
9. 命令"//"表示重复运行刚刚运行过的 PL/SQL 程序。（　）
10. 功能模块 Data Modeler 的作用是提供广泛的数据建模能力。（　）

二、单项选择题

1. sqlplus 中指定共享连接模式的选项是（　　）

A. shared   B. :shared
C. dedicated   D. :dedicated

2. sqlplus 中指定专用连接模式的选项是（　　）

A. shared   B. :shared
C. dedicated   D. :dedicated

3. lsnrctl 的哪个子命令表示查看监听器的服务信息（　　）

A. start   B. stop
C. status   D. services

4. Oracle 监听器的启动参数文件名称是（　　）

A. listener.ora   B. tnsnames.ora
C. sqlnet.ora   D. shrept.lst

5. Oracle 客户端的网络服务名配置文件是（　　）

A. listener.ora   B. tnsnames.ora
C. sqlnet.ora   D. shrept.lst

三、填空题

1. Oracle 的连接模式可以工作在专用_____服务器和_____服务器下或者混合模式下。
2. 专有连接方式是一个用户会话对应_____个数据库服务器进程。
3. sqlplus 替换变量的定义方式有两种，分别是_____和 accept。

4. Oracle 的优化器有两种方式，即基于_____的优化方式和基于_____的优化方式。
5. sqlplus 命令 column 的作用是_____。
6. sqlplus 命令 show 的作用是_____。
7. sqlplus 命令设置 time 参数的意义是_____。
8. sqlplus 命令设置 timing 参数的意义是 _____。
9. 执行计划是数据库根据 SQL 语句和相关表的统计信息作出的一个查询方案，目的是为了以_____的方式查询出结果。
10. Oracle 的动态采样机制是对_____表进行自动采样，尽可能实现最优的查询计划。
11. sqlnet.ora 文件中 tcp.invited_nodes 表示_____。
12. sqlnet.ora 文件中 tcp.excluded_nodes 表示_____。

四、简答题

1. Oracle Net Services 有哪些特性？
2. 共享服务器模式和专用服务器模式各有何优点和缺点？
3. 怎样设置服务器处于共享服务器模式？
4. 监听器的注册有什么作用？静态注册和动态注册有何区别？
5. 统计信息的作用是什么？动态采样的作用又是什么？
6. 执行计划中包括哪些信息，各有何意义？
7. 如何使用 SQL Developer 的优化指导功能优化 SQL 语句？

训练任务

1. 熟悉 SQL*Plus 和 SQL Developer 两个工具的使用

培养能力	工程知识、研究、职业规范、项目管理		
掌握程度	★★★★★	难度	中
结束条件	熟悉使用 sqlplus 和 sqldeveloper 两个工具软件		

训练内容：
本训练的目的是熟练使用 sqlplus 和 sqldeveloper 两个工具以及相关命令集。要求分别使用 sqlplus 和 sqldeveloper 两个工具：
(1) 使用两种方式连接到 Oracle 服务器：简易连接和 TNS 名称连接。
(2) 运行一些 SQL 语句，并在 SQL 语句中使用绑定变量，观察语句执行时间。
首先使用 sqlplus 以简单方式登录 pdborcl，连接字符串是 "localhost:1521/pdborcl"。登录用户是 hr，登录后查询某个部门的部门总人数和平均工资。SQL 语句中定义了绑定变量 dn，表示需要查询的一个部门。编写其实现代码

2. 配置服务器同时工作在专用服务器模式和共享服务器模式

培养能力	研究、职业规范、项目管理、设计/开发解决方案		
掌握程度	★★★★	难度	高
结束条件	使用 lsnrctl service 命令查询到两种服务器模式		
训练内容： 本训练的目的是配置服务器同时工作在专用服务器模式和共享服务器模式下，设置共享服务器模式参数。使用 lsnrctl 工具控制和监视服务器的工作模式			

3. 分析 SQL 执行计划，执行 SQL 语句的优化指导

培养能力	工程知识、研究、项目管理、使用现代工具		
掌握程度	★★★★	难度	高
结束条件	能够分析 SQL 语句执行计划信息，并执行优化指导。		

训练内容：
本训练任务目的是查询两个部门（'IT'和'Sales'）的部门总人数和平均工资，以下两个查询的结果是一样的。但效率不相同。
查询 1：
```
SELECT d.department_name, count(e.job_id)as "部门总人数",
avg(e.salary)as "平均工资"
from hr.departments d, hr.employees e
where d.department_id = e.department_id
and d.department_name in ('IT', 'Sales')
GROUP BY department_name;
```
查询 2：
```
SELECT d.department_name, count(e.job_id)as "部门总人数",
avg(e.salary)as "平均工资"
FROM hr.departments d, hr.employees e
WHERE d.department_id = e.department_id
GROUP BY department_name
HAVING d.department_name in ('IT', 'Sales');
```
执行上面两个比较复杂的返回相同查询结果数据集的 SQL 语句，通过分析 SQL 语句各自的执行计划，判断哪个 SQL 语句是最优的。最后将你认为最优的 SQL 语句通过 sqldeveloper 的优化指导工具进行优化指导，看看该工具有没有给出优化建议

第4章 数据库管理与配置

本章目标

知识点	理解	掌握	应用
1.掌握数据库参数查询方法		✓	✓
2.数据库配置助手dbca的使用	✓	✓	✓
3.创建插接式数据库	✓	✓	✓
4.数据库的启动与关闭		✓	✓
5.数据库的参数配置		✓	✓

训练任务

- 能够创建插接式数据库
- 能够启动和关闭数据库

知识能力点

能力点 \ 知识点	知识点1	知识点2	知识点3	知识点4	知识点5
工程知识		✓	✓		
问题分析		✓	✓		✓
设计/开发解决方案	✓	✓	✓	✓	
研究		✓	✓		
使用现代工具		✓	✓	✓	
工程与社会					✓
环境和可持续发展					
职业规范		✓	✓		
个人和团队	✓				
沟通		✓	✓		
项目管理		✓	✓	✓	
终身学习	✓				

4.1 常用的数据库配置查询方法

在配置数据库之前，需要掌握一些常用的配置查询命令，以便查询配置结果。

【示例 4-1】查看所有数据库实例

本例通过 ps 和 ls 两个 linux 命令查看有哪些 Oracle 实例。ps 命令查看正在运行的实例，ls 查看已经安装的实例，不管是否正在运行。

```
$ ps -ef | grep -i "ora_pmon"
oracle    5834    1   0 May26 ?        00: 01: 12 ora_pmon_orcl
oracle    5834    1   0 May27 ?        00: 01: 12 ora_pmon_orcl2
$ cd $ORACLE_HOME/dbs
$ ls spfile*.ora
spfileorcl2.ora  spfileorcl.ora
```

从上面的文件中可以看出，当前服务器上共安装了两个数据库：orcl 和 orcl2。

【示例 4-2】查看当前实例名称及状态以及参数

本示例通过 v$instance 查询当前实例，通过 dba_pdbs 查询所有 PDB 的状态，通过 show pdbs 快速查询所有 PDB 的状态。

```
$ sqlplus / as sysdba
SQL> SELECT instance_name, status FROM v$instance;
INSTANCE_NAME    STATUS
---------------- ------------
orcl2            OPEN
SQL> show pdbs
    CON_ID   CON_NAME                       OPEN MODE    RESTRICTED
---------- ------------------------------ ---------- ----------
         2   PDB$SEED                       READ ONLY    NO
         3   PDBORCL                        READ WRITE   NO
SQL> SELECT pdb_id, pdb_name, STATUS FROM dba_pdbs;
    PDB_ID PDB_NAME          STATUS
------- ---------------- ---------
         3   PDBORCL           NORMAL
         2   PDB$SEED          NORMAL
```

数据库的参数查询命令是 show parameter 参数名，可以查询出包含参数名的所有参数。比如 show parameter db_block 可以查询出所有参数名称中包含有 db_block 的参数值。

```
$ sqlplus / as sysdba
SQL> show parameter db_name
SQL> show parameter db_block
```

```
NAME                   TYPE      VALUE
--------------------   -------   ------------------------------
db_block_buffers       integer   0
db_block_checking      string    FALSE
db_block_checksum      string    TYPICAL
db_block_size          integer   8192
```

4.2 使用 dbca 管理数据库实例

在实际的数据库应用环境中，可能需要增加多个数据库实例，每个实例独立存储和运行，每个实例都有彼此独立的 CDB。也可能需要在一个 CDB 实例中增加多个不同的 PDB 可插接式数据库。Oracle 提供了 SQL 命令和工具 dbca（Database Configuration Assistant）来完成这个工作。

dbca 有两种工作方式，一种是界面方式，另一个种是静默方式。界面方式会弹出类似于安装 Oracle 时的界面，引导用户一步一步配置数据库，静默方式也称为命令行方式，不会弹出图形界面。界面方式的优点是直观、方便，缺点是需要一步一步确认，比较烦琐。而静默方式正好相反，由于没有图形界面，所以不直观，而优点是只要预先写好命令，就可以立即执行，工作效率非常高。

dbca 静默方式的命令行参数非常多，可以使用 dbca -help 或者 dbca -<command> -help 的方式查看命令参数各选项的信息。

```
$ dbca -help
dbca [-silent | -progressOnly] {<command> <options> } |
...
Enter "dbca -<command> -help" for more option
$ dbca -createPluggableDatabase -help
通过指定以下参数创建插接式数据库：
 -createPluggableDatabase
...
```

4.2.1 新建数据库实例

Oracle 采用 dbca 命令新建容器数据库，其他的命令是"dbca -silent -createDatabase …"。其中-silent 表示静默方式，直接生成，如果没有这个选项，将弹出对话框界面，在界面上直观选择各个参数。

【示例 4-3】dbca 静默新建数据库实例

本例以命令行静默方式为例，新建一个数据库实例 orcl2，最重要的选项是"-sid orcl2 -createAsContainerDatabase true"。

```
$ dbca -silent -createDatabase -templateName General_Purpose.dbc
```

```
    -gdbname orcl2 -sid orcl2 -createAsContainerDatabase true -responseFile
NO_VALUE -characterSet AL32UTF8 -memoryPercentage 10 -emConfiguration LOCAL
-sysPassword 123 -systemPassword 123
    复制数据库文件
    1% 已完成
    ...
    37% 已完成
    正在创建并启动 Oracle 实例
    ...
    62% 已完成
    正在进行数据库创建
    ...
    100% 已完成
    有关详细信息, 请参阅日志文件
  "/home/oracle/app/oracle/cfgtoollogs/dbca/orcl2/orcl2.log"。
```

以上操作创建了一个新的 orcl2 数据库实例,命令行中所有参数含义依次是:使用 General_Purpose.dbc 模板、全局数据库名称和数据库的系统标识符名称都叫 orcl2、createAsContainerDatabase 为 true 表示创建为容器型数据库 CDB、不使用响应文件、数据库端字符集采用 AL32UTF8、Oracle 的物理内存百分比为 10%、企业管理器配置为本地、sys 和 system 用户密码都是 123。

> 注意:在创建 CDB 实例的过程中,可能会遇到这样的错误:ORA27125 unable to create shared memory segment。遇到这个错误很可能是因为安装参数没有设置正确。见 2.1.3 配置/etc/sysctl.conf 文件。

4.2.2 删除一个容器数据库 CDB

```
$ dbca -silent -deleteDatabase -sourceDB orcl2
```

> 注意:dbca -silent 命令会无警告直接运行到最后的任务完成,因此在执行这样的命令之前一定要认真确认,否则可能会误删除数据库。

4.3 在数据库实例之间切换

在一台服务器硬件上可以安装运行多个数据库实例。当前访问的实例名称是由环境变量 $ORACLE_SID 的值决定的,要在服务器的 Shell 中切换实例,必须临时改变 $ORACLE_SID 的值为目标实例的 SID。如果要永久改变默认的$ORACLE_SID,可以在 /etc/profile 中进行修改。

【示例 4-4】切换实例

本例在 orcl 和 orcl2 两个实例之间进行登录切换。

```
$ export ORACLE_SID=orcl2
$ sqlplus / as sysdba
SQL> SELECT instance_name, status FROM v$instance;
INSTANCE_NAME    STATUS
---------------- -------------
orcl2     OPEN
SQL>exit
$ export ORACLE_SID=orcl
$ sqlplus / as sysdba
SQL> SELECT instance_name, status FROM v$instance;
INSTANCE_NAME    STATUS
---------------- -------------
Orcl      OPEN
```

注意：登录时，一定要确保实例是正确的，即确保正在管理的数据库是想要管理的数据库，否则就会张冠李戴，造成工作失误。

为了尽量避免在多实例的服务器上不犯实例名称的错误，可以修改 sqlplus 的提示符"SQL>"为"用户名@实例名>"。

方法是：只需在下面文件$ORACLE_HOME/sqlplus/admin/glogin.sql 中添加一行：

```
set sqlprompt "_user'@'_connect_identifier>"
```

即可实现这样的目的。下面是这样设置以后，登录后的效果：

```
$ sqlplus / as sysdba
...
SYS@orcl2>
```

4.4 配置插接式数据库 PDB

4.4.1 通过 SQL 语句创建插接式数据库

可以通过 SQL 语句来创建一个可插接式数据库 PDB，在创建时可以选择创建一个新的空的 PDB，不包含任何数据；也可以选择将一个现有的 PDB 克隆成另一个 PDB，所谓克隆，是指将原 PDB 的用户信息和数据信息一起复制到新的 PDB 中，这也是 Oracle 12c 的重要新特性。

【示例 4-5】创建插接式数据库 pdbtest

本示例创建一个插接式数据库 pdbtest，指定 pdbtest 的管理用户名为 user1，指定将 pdbseed 的文件作为基础，应用于 pdbtest 的文件。pdbseed 目录源于 PDB$SEED 种子数据库。

```
$ sqlplus / as sysdba
SQL> CREATE PLUGGABLE DATABASE pdbtest ADMIN USER user1
```

```
2 IDENTIFIED BY 123 FILE_NAME_CONVERT=
3 ('/home/oracle/app/oracle/oradata/orcl/pdbseed/',
4 '/home/oracle/app/oracle/oradata/orcl/pdbtest');
```

4.4.2 通过 dbca 创建插接式数据库

可以通过 dbca 工具创建插接式数据库,首先以 Oracle 用户登录,运行 dbca 命令,然后根据提示操作,详细过程如图 4-1~图 4-4。在图 4-3(a)中要注意输入正确的数据库目录、PDB 数据库名称以管理员账号信息。

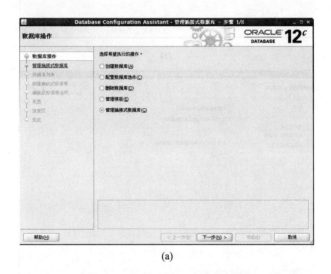

(a)

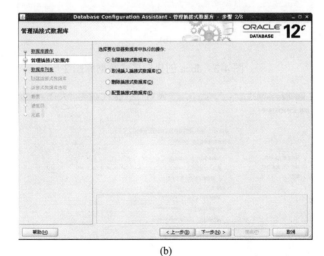

(b)

图 4-1 dbca 创建插接式数据库图 1

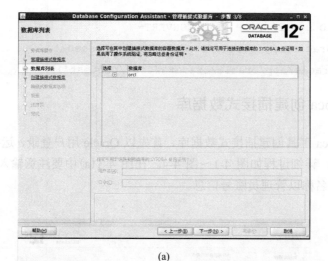

(a)

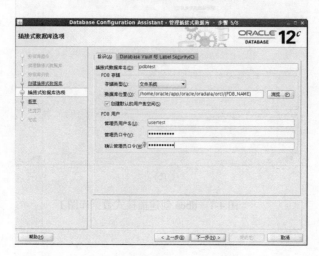

(b)

图 4-2 dbca 创建插接式数据库图 2

(a)

第 4 章 数据库管理与配置

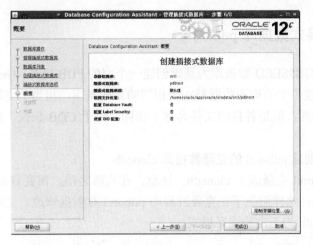

(b)

图 4-3 dbca 创建插接式数据库图 3

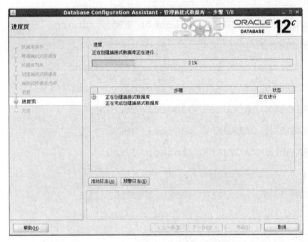

(a)

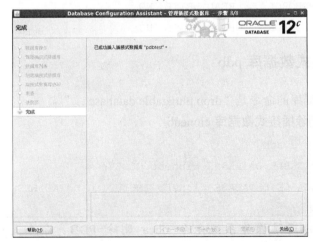

(b)

图 4-4 dbca 创建插接式数据库图 4

4.4.3 克隆插接式数据库 pdb

除了允许以 PDB$SEED 数据库为基础创建一个空的 PDB 之外，Oracle 还允许以已有的 PDB 为模板创建一个结构和数据都完全相同的新数据库，相当于克隆了数据库。新旧数据库虽然数据相同，但是各自的文件名称不相同，对于 CDB 来说，是两个完全独立的数据库。

【示例 4-6】创建 pdborcl 的克隆数据库 clonedb

本示例将 pdborcl 克隆成了 clonedb。注意，在克隆之前，需要首先关闭 pdborcl，然后打开为只读状态，这样避免了在克隆过程中 pdborcl 有数据修改。克隆完成后，需要打开 pdborcl 和 clonedb 为正常的读写模式。

```
$ sqlplus / as sysdba
SQL>ALTER PLUGGABLE DATABASE pdborcl CLOSE;
SQL>ALTER PLUGGABLE DATABASE pdborcl OPEN READ ONLY;
SQL>CREATE PLUGGABLE DATABASE clonedb FROM pdborcl
  2 file_name_convert=
  3 ('/home/oracle/app/oracle/oradata/orcl/pdborcl',
  4 '/home/oracle/app/oracle/oradata/orcl/clonedb/');
SQL>ALTER PLUGGABLE DATABASE pdborcl CLOSE;
SQL>ALTER PLUGGABLE DATABASE pdborcl OPEN;
SQL>ALTER PLUGGABLE DATABASE clonedb OPEN;
SQL>show pdbs;

   CON_ID     CON_NAME       OPEN MODE     RESTRICTED
   ----------  ----------    ----------    ----------
        2     PDB$SEED       READ ONLY     NO
        3     PDBORCL        READ WRITE    NO
        4     CLONEDB        READ WRITE    NO
```

4.4.4 删除插接式数据库 pdb

删除插接式数据库的命令是"drop pluggable database …"。

【示例 4-7】删除插接式数据库 clonedb

```
$ sqlplus / as sysdba
SQL>ALTER PLUGGABLE DATABASE clonedb CLOSE;
SQL>DROP PLUGGABLE DATABASE clonedb INCLUDING DATAFILES;
插接式数据库已删除。
```

注意：也可以通过操作系统命令 dbca 删除 PDB。命令是：$ dbca -silent -deletePluggableDatabase -sourceDB orcl -pdbName clonedb

4.4.5 插接式数据库的拔出与插入

Oracle 12c 的一个重要功能就是可以将 PDB 从一个数据库中拔出，然后插入到另一个数据库中，方便数据库的转移。

【示例 4-8】 拔出和插入 PDB

本示例首先拔出 clonedb，然后再插入 clonedb，插入后改变名称为 clonedb2。

```
$ sqlplus / as sysdba
SQL>ALTER PLUGGABLE DATABASE clonedb CLOSE;
SQL>ALTER PLUGGABLE DATABASE clonedb UNPLUG INTO
 2 '/home/oracle/app/oracle/oradata/orcl/clonedb/clonedb.xml';
SQL> show pdbs;
CON_ID    CON_NAME         OPEN MODE    RESTRICTED
-------   ------------     -----------  ------------
 2        PDB$SEED         READ ONLY    NO
 3        PDBORCL          READ WRITE   NO
 4        CLONEDB          MOUNTED
SQL> col pdb_name format a15
SQL> SELECT pdb_id,pdb_name,STATUS FROM dba_pdbs;
PDB_ID      PDB_NAME           STATUS
---------   ---------------    ---------
 3          PDBORCL            NORMAL
 2          PDB$SEED           NORMAL
 4          CLONEDB            UNPLUGGED
SQL>DROP PLUGGABLE DATABASE clonedb;
SQL>exit
```

上面的命令使用于关键字 unplug，将 clonedb 插出，并生成 clonedb.xml 文件，最后调用 drop 命令删除这个 pdb，要注意的是，由于删除的时候没有使用选项 "INCLUDING DATAFILES"，所以并未删除数据文件，clonedb 的结构和数据都完好无损。在此基础上，可以将 clonedb 重新插入到 CDB 中。

```
$ sqlplus / as sysdba
SQL>CREATE PLUGGABLE DATABASE clonedb2 USING
 2 '/home/oracle/app/oracle/oradata/orcl/clonedb/clonedb.xml' nocopy;
```

4.5 数据库的启动与关闭

Oracle 12c 的启动命令是 startup，关闭命令是 shutdown。

4.5.1 启动数据库

Oracle 12c 的数据库有容器数据库 CDB 和插接式数据库 PDB 之分。启动数据库的时候要先启动 CDB，然后启动 PDB，缺省的情况是启动 CDB 以后不会启动任何 PDB 数据库，只有靠手工逐个去启动 PDB 数据库。如果需要在启动 CDB 之后启动 PDB，可以创建一个启动触发器，在触发器中启动部分 PDB 或者全部 PDB。startup 命令有三个大的选项：

```
STARTUP db_options | cdb_options | upgrade_options
```

其中：

db_options 的语法如下：

```
[FORCE] [RESTRICT] [PFILE=filename] [QUIET] [ MOUNT [dbname]
[ OPEN [open_db_options] [dbname] ] | NOMOUNT ]
```

其中 open_db_options 的语法如下：

```
READ {ONLY | WRITE [RECOVER]} | RECOVER
```

cdb_options 的语法如下：

```
root_connection_options | pdb_connection_options
```

其中 root_connection_options 的语法如下：

```
PLUGGABLE DATABASE pdbname
[FORCE] [RESTRICT] | [OPEN [open_pdb_options]]
```

pdb_connection_options 的语法如下：

```
[FORCE] [RESTRICT] | [OPEN [open_pdb_options]]
```

其中 open_pdb_options 的选项有：READ WRITE | READ ONLY

upgrade_options 的语法如下：

```
[PFILE=filename] {UPGRADE | DOWNGRADE} [QUIET]
```

NOMOUNT：启动实例，但不装入数据库。

MOUNT：启动实例，装入数据库，但不打开数据库。

用户可以启动实例并装入数据库但不打开数据库，允许用户执行特定的维护操作。例如：重命名数据文件；添加、撤消或重命名重做日志文件；启动和禁止重做日志归档；执行数据库恢复。

RESTRICT：限制在启动时对数据库的访问。

用户可以在严格的模式下启动实例并装入数据库，这样的模式只允许 DBA 做以下的工作：执行结构维护，如重建索引；执行数据库文件的导入导出；执行数据装载；临时阻止典型用户使用数据。

FORCE：强制实例启动。

强行启动实例，即如果一个实例已经启动，则 STARTUP FORCE 重新启动。

RECOVER：启动一个实例，装入数据库，并启动全部的介质恢复。

如果用户要求介质恢复，可以启动一个实例，装入指向实例的数据库，并自动地启动恢复程序。

PFILE：从参数文件启动。

可以指定参数文件启动数据库，在数据库以参数错误无法启动的情况下特别有用。

【示例 4-9】启动一个 PDB

本例假设数据库当前处于关闭状态，并且还没有创建数据库的启动触发器，首先启动 CDB，然后手工启动 pdborcl 数据库。

```
$ sqlplus / as sysdba
SQL> STARTUP
ORACLE instance started.

Total System Global Area 1610612736 bytes
Fixed Size          2924928 bytes
Variable Size      520097408 bytes
Database Buffers  1073741824 bytes
Redo Buffers        13848576 bytes
Database mounted.
Database opened.
SQL> show pdbs
 CON_ID CON_NAME   OPEN MODE      RESTRICTED
 ------ ---------- -------------- ----------
      2 PDB$SEED   READ ONLY      NO
      3 PDBORCL    MOUNTED
```

第一次运行 startup 时，只启动了 CDB，通过 show pdbs 命令可以看出 pdborcl 数据库在 MOUNTED 状态，还没有启动。下面通过两种不种的方式启动 pdborcl。

方法 1：

```
SQL> STARTUP PLUGGABLE DATABASE pdborcl;
Pluggable Database opened.
```

方法 2：

```
SQL> ALTER SESSION SET CONTAINER=pdborcl;
Session altered.
SQL> STARTUP
Pluggable Database opened.
SQL> show pdbs
 CON_ID CON_NAME   OPEN MODE      RESTRICTED
 ------ ---------- -------------- ----------
      3 PDBORCL    READ           WRITE
SQL> exit
```

这种方式通过命令"ALTER SESSION SET CONTAINER=pdborcl;"将会话切换到

pdborcl 数据库,再执行 startup 单独启动 pdborcl 数据库,启动后,数据库 pdborcl 就处于正常的读写状态(READ WRITE)了。

【示例 4-10】一次性启动所有 PDB

如果希望一次性启动所有 PDB,只需要运行一次"ALTER PLUGGABLE DATABASE ALL OPEN;"。如果希望在启动 CDB 之后自动执行该语句,打开所有 PDB,就必须像本示例这样创建一个触发器:

```
$ sqlplus / as sysdba
SQL> CREATE OR REPLACE TRIGGER open_all_pdbs
  2  AFTER STARTUP ON DATABASE
  3  BEGIN
  4    EXECUTE IMMEDIATE 'ALTER PLUGGABLE DATABASE ALL OPEN';
  5  END open_all_pdbs;
  6  /
Trigger created.
SQL> SELECT trigger_name FROM user_triggers;
TRIGGER_NAME
------------------------------------------------------------
OPEN_ALL_PDBS
...
12 rows selected.
```

通过命令"SELECT trigger_name FROM user_triggers;"可以查看所有触发器。触发器创建好后,可以关闭数据库,再打开数据库进行测试,看是否自动打开了所有的 PDB:

```
$ sqlplus / as sysdba
SQL> SHUTDOWN IMMEDIATE
...
SQL> STARTUP
...
SQL> show pdbs
   CON_ID CON_NAME            OPEN MODE     RESTRICTED
   ------ -------------------- ---------------- ----------
        2 PDB$SEED            READ ONLY     NO
        3 PDBORCL             MOUNTED
```

可以看见,只需要在 CDB 运行一次 startup,pdborcl 也打开了(OPEN MODE=READ WRITE)。

如果要删除触发器,可以使用命令"DROP TRIGGER open_all_pdbs;"。

4.5.2 启动异常处理

如果在启动中出现异常，可以根据错误代码找到相应的解决办法。Oracle 的错误代码格式通常是"ORA-数字"，比如：

```
SQL> STARTUP
Oracle instance started.
...
ORA-03113: 通信通道的文件结尾
进程 ID: 5801
会话 ID: 1 序列号: 5
```

如果没有错误代码，或者错误代码不明确，可以进入跟踪目录：

/home/oracle/app/oracle/diag/rdbms/orcl/orcl/trace/

查看目录中的*.log 和*.trc 类型文件，这些文件中有更详细的错误说明，其中比较重要的文件是 alert_orcl.log。

4.5.3 关闭数据库

数据库的关闭使用 shutdown 命令，可以只关闭单个的 PDB，也可以通过关闭 CDB 将所有 PDB 关闭。shutdown 的命令参数是：

```
SHUTDOWN [NORMAL|TRANSACTIONAL [LOCAL]|IMMEDIATE|ABORT]
```

SHUTDOWN NORMAL：不允许新的连接、等待会话结束、等待事务结束、做一个检查点并关闭数据文件。启动时不需要实例恢复。

SHUTDOWN TRANSACTIONAL：不允许新的连接、不等待会话结束、等待事务结束、做一个检查点并关闭数据文件。启动时不需要实例恢复。local 子选项在多实例的时候有效，可以同时打开一个数据库形成多实例，local 表示关闭一个实例不会影响其他正在运行的实例。

SHUTDOWN IMMEDIATE：不允许新的连接、不等待会话结束、不等待事务结束、做一个检查点并关闭数据文件。没有结束的事务是自动 rollback。启动时不需要实例恢复。

SHUTDOWN ABORT：不允许新的连接、不等待会话结束、不等待事务结束、不做检查点且不关闭数据文件。启动时自动进行实例恢复。

【示例 4-11】关闭所有数据库

以管理员身份登录数据库，在 CDB 中运行 shutdown immediate 可以关闭所有数据库，同时关闭所有 PDB。

```
$ sqlplus / as sysdba
SQL> SHUTDOWN IMMEDIATE
Database closed.
Database dismounted.
ORACLE instance shut down.
```

```
SQL>
```

【示例 4-12】关闭一个 PDB

本示例以管理员身份登录数据库，只关闭 pdborcl，不会关闭其他 PDB，也不会关闭 CDB。

```
$ sqlplus / as sysdba
SQL> ALTER SESSION SET CONTAINER=pdborcl;
Session altered.
SQL> SHUTDOWN IMMEDIATE
Pluggable Database closed.
```

4.6 数据库参数配置

数据库管理员可以对数据库的参数进行配置，包括设置内存参数，归档路径和部分初始化参数。通过优化参数配置，可以提高数据库的效能。修改数据库参数的命令是：

ALTER SYSTEM SET 参数名=参数值 [SCOPE=BOTH|SPFILE|MEMORY];

其中 SCOPE 有三个参数，分别为：

SCOPE=BOTH 立即并永久生效，这是默认模式。

SCOPE=SPFILE 下次启动才能生效。

SCOPE=MEMORY 立即生效但下次启动时失效。

【示例 4-13】修改 sga_target 系统参数

本例是通过增加 sga_target 参数的值改善 Oracle 的 SGA 性能。通过"ALTER SYSTEM SET sga_target=1600M scope=spfile;"命令将原来的 sga_target 内存从 1536M 增加到 1600M。由于选择的选项是"scope=spfile;"，所以需要重启才能生效。通过命令"show parameter sga_;"可以查询包含 sga_ 的所有参数。

```
$ sqlplus / as sysdba
SQL> show parameter sga_;
NAME                                 TYPE            VALUE
------------------------------------ --------------- ----------------
sga_max_size                         big integer     1536M
sga_target                           big integer     1536M
unified_audit_sga_queue_size         integer         1048576
SQL> ALTER SYSTEM SET sga_target=1600M SCOPE=SPFILE;
System altered.
SQL> SHUTDOWN IMMEDIATE
SQL> STARTUP
ORACLE instance started.

Total System Global Area 1677721600 bytes
```

```
    Fixed Size                  2925120 bytes
    Variable Size             469765568 bytes
    Database Buffers 1191182336 bytes
    Redo Buffers            13848576 bytes
    Database mounted.
    Database opened.
    SQL> show parameter sga_;
    NAME                                       TYPE              VALUE
    ------------------------------------ ----------------- ----------------
    sga_max_size                               big integer       1600M
    sga_target                                 big integer       1600M
    unified_audit_sga_queue_size               integer           1048576
```

练习

一、判断题

1. 克隆 PDB 不需要新建数据文件，只需要共用原数据文件。（   ）
2. 关闭 CDB 的时候必定会关闭所有 PDB。（   ）
3. 关闭 PDB 的时候必定会关闭 CDB。（   ）
4. 在修改系统参数后，所有参数都不需要重启而立即生效。（   ）
5. SHUTDOWN IMMEDIATE 关闭数据库不会丢失用户已提交的数据。（   ）

二、单项选择题

1. 数据库启动的时候，用户启动实例、装入数据库并打开数据库，这样的启动方式是（   ）

   A. STARTUP   B. STARTUP MOUNT
   C. STARTUP FORCE   D. STARTUP NOMOUNT

2. 数据库启动的时候，用户启动实例并装入数据库但不打开数据库，这样的启动方式是（   ）

   A. STARTUP   B. STARTUP MOUNT
   C. STARTUP FORCE   D. STARTUP NOMOUNT

3. 数据库关闭的时候，不允许新的连接、不等待会话结束、不等待事务结束、做一个检查点并关闭数据文件。没有结束的事务是自动 rollback。启动时不需要实例恢复。这样的关闭方式是（   ）

   A. SHUTDOWN NORMAL   B. SHUTDOWN TRANSACTIONAL
   C. SHUTDOWN IMMEDIATE   D. SHUTDOWN ABORT

4. 数据库关闭的时候，不允许新的连接、不等待会话结束、不等待事务结束、不做检查点且没有关闭数据文件。启动时自动进行实例恢复。这样的关闭方式是（   ）

   A. SHUTDOWN NORMAL   B. SHUTDOWN TRANSACTIONAL

C. SHUTDOWN IMMEDIATE　　　　　D. SHUTDOWN ABORT

5. 修改数据库系统参数的命令格式是"ALTER SYSTEM SET 参数名=参数值 [SCOPE=BOTH|SPFILE|MEMORY];",哪个选项表示设置参数之后,立即生效但下次启动时失效?(　　)

A. SCOPE=BOTH　　　　　　　　　B. SCOPE=SPFILE
C. SCOPE=MEMORY　　　　　　　　D. 不要 SCOPE 选项

三、填空题

1. 数据库的参数查询命令是 show ＿＿＿＿＿＿ 参数名。
2. 查询所有 PDB 的命令是 show ＿＿＿＿＿＿。
3. 决定数据库登录实例的操作系统环境变量是＿＿＿＿。
4. dbca 有两种工作方式,一种是界面方式,另一个种是＿＿＿＿方式。
5. Oracle 的启动命令是＿＿＿＿,关闭命令是＿＿＿＿。

四、简答题

1. 简述创建一个 pdb 的过程。
2. 简述数据库启动命令 startup 的选项。
3. 简述数据库关闭命令 shutdown 的选项。

训练任务

1. 创建插接式数据库

培养能力	设计/开发解决方案、研究、使用现代工具、沟通		
掌握程度	★★★★	难度	高
结束条件	数据库创建成功后,能够打开数据库		
训练内容: 本训练的目的是通过 dbca 工具为某个实用订单项目 order 创建一个插接式数据库。要求: (1)静默创建,不弹出 dbca 图形界面。 (2)创建时设置管理者用户名为 sysorder,密码为 123 (3)指定数据目录为: /home/oracle/app/oracle/oradata/orcl/{PDB_NAME} (4)创建数据时创建默认表空间 users			

2. 启动和关闭数据库

培养能力	项目管理、设计/开发解决方案、使用现代工具		
掌握程度	★★★★★	难度	中
结束条件	启动和关闭数据库成功		
训练内容: 本训练的目的是训练关闭和启动 CDB 和 PDB。在启动 CDB 时,自动或者手工启动全部或者部分 PDB。在关闭数据时,关闭一个 PDB 或者所有 PDB			

第 5 章　Oracle 12c 数据库结构

本章目标

知识点	理解	掌握	应用
1.Oracle 12c 体系结构	✓	✓	
2.Oracle 12c 内存结构	✓	✓	✓
3.In-Memory 列存储	✓	✓	✓
4.服务器进程	✓		
5.定时作业任务		✓	✓

训练任务

- 能够根据实际情况调整 SGA 大小。
- 能够应用 In-Memory 列存储技术创建内存表。
- 能够维护定时作业任务。

知识能力点

能力点 \ 知识点	知识点 1	知识点 2	知识点 3	知识点 4	知识点 5
工程知识		✓	✓		✓
问题分析	✓			✓	✓
设计/开发解决方案	✓	✓	✓		
研究	✓				
使用现代工具	✓	✓	✓		✓
工程与社会	✓				✓
环境和可持续发展					✓
职业规范		✓	✓		
个人和团队	✓				
沟通		✓	✓		✓
项目管理	✓	✓	✓		
终身学习	✓			✓	

5.1 Oracle 12c 体系结构

通常,服务器可靠地管理多用户环境中的大量数据,以便用户可以同时访问相同的数据。数据库服务器还可以防止未经授权的访问,并为故障恢复提供有效的解决方案。

Oracle 数据库服务器由一个数据库和至少一个数据库实例组成。数据库是一组位于磁盘上的文件,它存储数据。数据库实例是管理数据库文件的一组内存结构。实例由共享内存区,即系统全局区 System Global Area(SGA)和一组后台进程组成。一个实例可以独立于数据库文件存在。

图 5-1 显示了一个数据库及其实例的关系。每个用户连接到实例的时候,客户端进程运行应用程序。每个客户端进程与它自己的服务器进程关联。服务器进程有自己的私有会话内存,称为程序全局区(PGA)。

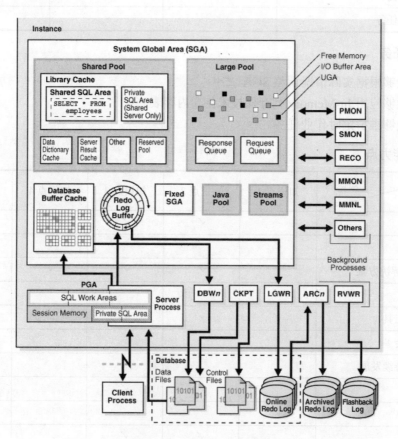

图 5-1 数据库实例和数据库

图 5-1 中,SGA 包括数据库缓冲区高速缓存(Database Buffer Cache)、重做日志缓冲区(Redo Log Buffer)、共享池(Shared Pool)、大池(Large Pool)、固定 SGA(Fixed SGA)、Java 池(Java Pool)和流池(Streams Pool)。图中 SGA 的右边是后台进程 PMON、SMON、

RECO，MMON，MMNL 等。在 SGA 下方有 DBWn，CKPT，LGWR，ARCn 和 RVWR。SGA 下方还有 PGA 和服务器进程。服务器进程连接到客户端进程。客户端进程的右侧是数据库文件（数据文件、控制文件、联机重做日志文件）、归档重做日志和闪回日志。

可以从物理和逻辑两方面看待数据库。从物理上看，物理数据可以在操作系统层面进行观察。例如，操作系统实用工具，如 Linux 的 ls 和 ps 可以列出数据库文件和 Oracle 进程。在逻辑上来看，逻辑数据仅以表（table）的形式存储于数据库中，只能通过 SQL 语句查询表的信息，而不能通过操作系统命令进行查询。

数据库具有物理结构和逻辑结构。由于物理和逻辑结构是分开的，所以可以单独管理数据的物理存储，而不影响对逻辑存储结构的访问。例如，重命名物理数据库文件不会重命名存储在该文件中的表。

5.1.1 数据库物理存储结构

数据库物理结构是指存储数据的操作系统文件。执行创建数据库语句时，创建下列文件：

1）数据文件（Data Files）

每个 Oracle 数据库都有一个或多个物理数据文件，其中包含了所有的数据库数据。逻辑数据库结构中的数据，如表和索引，在物理上都存储在数据文件中。

2）控制文件（Control Files）

每个 Oracle 数据库都有一个控制文件。控制文件包含数据库物理结构的元数据，如数据库名称和数据库文件的名称和位置。

3）联机重做日志文件（Online Redo Log Files）

每个 Oracle 数据库都有一个联机重做日志，它是由两个或多个联机重做日志文件组组成的一个联合体。联机重做日志由重做条目（也称重做日志记录）组成，它记录了数据的所有更改。

除了上述文件之外，还有一些文件也非常重要。一是参数文件和网络配置文件，二是备份文件和归档重做日志文件，这两种文件是脱机文件，用于数据库的备份和恢复。

5.1.2 逻辑存储结构

逻辑存储结构使 Oracle 数据库能够对磁盘空间进行细粒度控制。数据库的逻辑存储结构由表空间（Tablespace）、段（Segment）、区（Extents）和块（Block）4 个部分组成，图 5-2 描述了 Oracle 的逻辑结构与物理结构之间的关系。图中的鱼尾纹代表了一对多的关系，左边的逻辑列是表空间、段、区和 Oracle 数据块。每种类型与下面的类型是一对多的关系。右边是物理数据文件和操作系统的数据块。表空间与数据文件是一对多的关系。Oracle 数据块与操作系统的数据块是一对多的关系。数据文件与区（Extents）是一对多的关系。

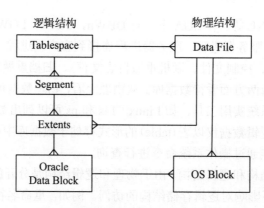

图 5-2　数据库逻辑结构与物理结构的关系

1) 表空间 (Tablespace)

数据库最大的逻辑存储单元称为表空间。表空间是段的逻辑容器。每个表空间至少由一个数据文件组成。表空间的存储空间大小等于组成该表空间的数据文件大小之和。通过 DBA_TABLESPACES、DBA_DATA_FILES 和 DBA_FREE_SPACE 这三个视图可以查询表空间、表空间的数据文件和剩余空间大小。

2) 段 (Segment)

段是由一组区 (Extents) 组成，段用于存储数据库的对象（例如表或索引）、重做数据 (Undo Data) 或临时数据 (Temporary Data)。每个需要存储的对象都仅由一个段组成。

Oracle 中的段可以分成 4 种类型：数据段、索引段、回滚段和临时段。数据段用来存储用户的数据，每个表都有一个对应的回滚段，其名称和数据表的名字相同。索引段用来存储系统和用户的索引信息。回滚段用来存储用户数据修改前的值，回退段与事务是一对多的关系，一个事务只能使用一个回退段，而一个回退段可存放一个或多个事务的回退数据。临时段用于 ORDER BY 语句的排序以及一些汇总。通过 user_extents 这个视图可以查询段的名称、段的类型以及表空间的分配等信息。

3) 区 (Extents)

区是磁盘空间分配的最小单位。磁盘按区划分，每次至少分配一个区。区存储于段中，它由连续的数据块组成。可以通过字典 dba_tablespaces 查询表空间中区的信息。可以通过字典 user_tables 查询段中区的信息。可以通过字典 user_extents 查询区的分配状况。我们可以通过以下 SQL 语句分别查询用户表、段、区的分配信息。

```
SQL>SELECT table_name, tablespace_name, min_extents, max_extents FROM user_tables;
SQL>SELECT * FROM user_extents;
```

4) 块 (Block)

数据块是数据库的最小数据组织单位与管理单位，是数据文件磁盘存储空间单位，也是数据库 I/O 的最小单位，数据块大小由 DB_BLOCK_SIZE 参数决定。一个数据块对应于磁盘上特定的字节数，Oracle 12c 默认的 DB_BLOCK_SIZE 是 8192 字节。

5.2 Oracle 12c 内存结构

当实例启动时,Oracle 数据库分配内存区域并启动后台进程。内存区存储下列信息:
(1) 程序代码。
(2) 每个连接会话(Session)的信息,不管它当前是不是活动的。
(3) 在程序执行过程中所需要的信息,例如一个提取数据行的查询语句的当前状态。
(4) 在进程间共享和传递的锁定数据信息。
(5) 缓存数据,如数据块和重做记录(这些数据也同时存在于磁盘上)。

5.2.1 基本内存结构

Oracle 数据库包含多个内存区域,其中每个都包含多个子区域组件,如图 5-1 所示,与 Oracle 数据库相关联的基本内存结构包括:

1) 系统全局区(SGA)

SGA 是一组共享内存结构,称为 SGA 组件,它包含 Oracle 数据库进程(Process)专用的数据和控制信息。所有服务器和后台进程共享 SGA。SGA 中存储的数据包括缓存数据块(Cached Data Blocks)和共享 SQL 区(Shared SQL Areas)等。

2) 程序全局区(PGA)

PGA 是一个非共享内存区,包含数据和控制专用的 Oracle 进程信息。当 Oracle 进程启动时才会创建 PGA 区域。每个服务器进程(Server Process)和后台进程(Background Process)都存在一个 PGA。

所有 PGA 的集合称为 PGA 总实例,或 PGA 实例。数据库初始化参数 pga_aggregate_limit 和 pga_aggregate_target 设置 PGA 实例的总的大小,而不是单个 PGA 的大小。

3) 用户全局区(UGA)

UGA 是与用户会话(Session)关联的内存区,存储会话的信息,比如用户的登录信息以及数据库会话所需的其他信息。

如果一个会话加载一个 PL/SQL 包(Package)到内存中,那么 UGA 就存储包的状态,即包中的所有变量的值。当包中的子程序更改变量时,包状态会也发生变化。默认情况下,在一个会话生命周期内,包中的变量值是唯一的并且是持久的。

OLAP 页面缓冲池(OLAP Page Pool)也存储在 UGA 中。它管理 OLAP 数据页,该数据页相当于数据块。页面缓冲池在 OLAP 会话开始时分配并在会话结束时释放。每当用户查询多维对象(如 cube)时,就会自动打开 OLAP 会话。

在一个会话生命周期内,UGA 都必须有效。因为这个原因,在使用共享服务连接(Shared Server)时,由于 PGA 服务于单进程,UGA 不能存储在 PGA 中,而只能存储在 SGA 中,允许任何共享的服务器进程访问它。当使用专用服务连接(Dedicated Server)时,UGA 存储在 PGA 中。

4) 软件代码区 (Software Code Areas)

软件代码区是用来存储正在运行或可以运行的 Oracle 自身代码的内存区。Oracle 自身代码的存储区不同于用户程序的存储区，它更独立并且更受保护。软件代码区的大小通常是静态的，只有当软件升级或重新安装后才会变化。这些区域所需的大小因操作系统而异。

5.2.2 PGA 概述

PGA 分为两个不同的区域：SQL 工作区（SQL Work Areas）和私有 SQL 区（Private SQL Area），每个区域的作用也不同。图 5-3 描述了 PGA 内存区的结构，以及 PGA 内存区与客户端进程的数据区之间的联系，图中的 PGA 是专用服务器会话的结构，共享服务器会话有所不同。

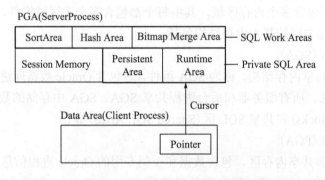

图 5-3　PGA 内存区与客户端数据区

私有 SQL 区（Private SQL Area）保存待执行的已解析 SQL 语句和其他特定会话信息的信息。当服务器进程执行 SQL 或 PL/SQL 代码时，进程使用私有 SQL 区域存储绑定变量值、查询执行状态信息和查询执行工作区。

不要将 PGA 中的私有 SQL 区与 SGA 中的共享 SQL 区（Shared SQL Area）混淆，前者存储 SQL 查询执行结果的值和数据，后者将执行计划存储在 SGA 中，相同或不同会话中的多个私有 SQL 区域可以指向 SGA 中的单个执行计划。例如，一个会话中执行 20 次"SELECT * FROM sales；"和在不同的会话中执行同一个查询 10 次可以共享相同的执行计划，但每次执行的私有 SQL 区是不共享的，并且可能包含不同的值和数据。

游标（Cursor）是指向某个私有 SQL 区的名称或句柄。在客户端，游标可以看作指针，在服务器端，游标可以看作是私有 SQL 区的状态，所以，也经常把私有 SQL 区看作游标。

5.2.3 SGA 概述

SGA 是一个读/写内存区，与 Oracle 后台进程一起形成一个数据库实例。一些服务器进程在数据库操作过程中读写 SGA。

每个数据库实例都有自己的 SGA。Oracle 数据库在实例启动时自动为 SGA 分配内存，并在实例关闭时回收内存。

【示例 5-1】 启动数据库并观察 SGA 分配

```
$ sqlplus / as sysdba
SQL> startup
ORACLE instance started.


Total System Global Area 1577058304 bytes
Fixed Size                  2924832 bytes
Variable Size             469765856 bytes
Database Buffers          922746880 bytes
Redo Buffers               13848576 bytes
In-Memory Area            167772160 bytes
Database mounted.
Database opened.
SQL> show parameter sga_
NAME                                 TYPE          VALUE
------------------------------------ ------------- ------------------
sga_max_size                         big integer   1504M
sga_target                           big integer   1504M
unified_audit_sga_queue_size         integer       1048576
```

从本示例可以看出：SGA 的大小为 1.5G。除了在启动的时候观察 SGA 之外，还可以通过系统参数 sga_target，sga_max_size，或者通过视图 v$sga 和 v$sgastat 查看 SGA 大小以及 SGA 组件的全部信息。如图 5-1 所示，SGA 的重要组件有：

1) 数据库缓冲区缓存（Database Buffer Cache）

数据库缓冲区缓存（也称为缓冲区高速缓存）是存储从数据文件中读取的数据块的副本的存储区域。缓冲区是缓冲区管理器临时缓存当前或最近使用的数据块的主内存地址。所有与数据库实例并发连接的用户共享访问缓冲区缓存。

设计数据库缓冲区缓存的目的是优化物理输入输出，提高数据更新效率。具体的过程是：当有数据更新的时候，数据库更新该缓存中的数据块，并存储更改的数据到重做日志缓冲区（Redo Log Buffer）中。提交后，数据库将重做缓冲区写入联机重做日志文件中，但不会立即将数据块写入数据文件。

2) 内存列存储（In-Memory Column Store）

内存列存储 In-Memory Column Store 是在内存中以列格式（而不是普通的行格式）存储表或者分区的一个副本，目的是提高扫描访问的速度。这种方式类似于内存数据库。

3) 重做日志缓冲区（Redo Log Buffer）

重做日志缓冲区是 SGA 中的循环缓冲区，它存储数据库的修改项，也叫重做记录。重做记录包含必要的重构和恢复信息，数据库的恢复工作可以将重做记录应用到数据文件中，从而恢复丢失的数据。后台进程日志写入进程（LGWR）将重做日志缓冲区写入磁盘上的活动联机重做日志文件组中。

4) 共享池(Shared Pool)

共享池缓存各种类型的程序数据。例如，共享池存储解析的 SQL、PL/SQL 代码、系统参数和数据字典信息。

5) 大池(Large Pool)

大池比共享池有更大的区域，它是可选的。大池的大小是通过参数 LARGE_POOL_SIZE 来决定的。它所存储的信息有 RMAN 的 I/O 缓冲区、并行查询语句执行所需的消息缓冲区、用户全局区 UGA(仅对共享服务器)和 Oracle XA 接口(用于多数据库的分布式事务)。

6) Java 池(Java Pool)

Java 池内存区存储的是所有会话的 Java 代码和 Java 虚拟机(JVM)中的数据。Java 池的大小由参数 JAVA_POOL_SIZE 设置。

5.2.4 In-Memory 列存储

在 Oracle 12c 中，SGA 区可以有内存列存储 In-Memory Column Store 的可选内存块，它以柱状格式(即列格式而不是普通的行格式)存储表或者分区的一个副本，目的是提高扫描访问的速度。这种方式类似于内存数据库。

In-Memory 列存储方式不会替换缓冲区缓存，而只是补充，这样两个内存区域可以以不同的格式存储相同的数据。默认情况下，对象只有在指定为 INMEMORY 时才会使用 In-Memory 列存储。In-Memory 列存储结构如图 5-4 所示。

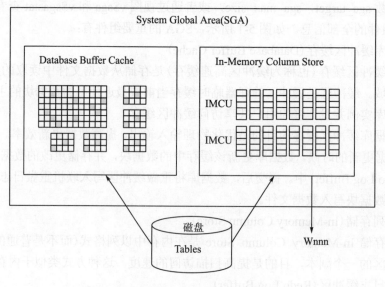

图 5-4 In-Memory 列存储

工作者进程 Worker Processes (Wnnn)将数据存储到 In-Memory 列存储区，在存储数据期间，工作者进程将行转化为列，然后将数据进行压缩，并存储到内存压缩单元

In-Memory Compression Units（IMCUs）中。Oracle 12c 数据库必须在设置两个参数 inmemory_max_populate_servers 和 inmemory_size 的值之后才能启动 In-Memory column store。其中 inmemory_max_populate_servers 参数是设置工作者进程的数量，而 inmemory_size 是设置分配的内存大小。

【示例5-2】设置 In-Memory 内存大小

本例设置 inmemory_size 为 150m，如果设置内存允许，可以设置得更大一些。参数 inmemory_max_populate_servers 控制 In-Memory 后台工作者进程 Wnnn 的个数，进程数越多，装载速度越快，但消耗的资源也更多。

```
SQL> ALTER SYSTEM SET inmemory_max_populate_servers=2 SCOPE=SPFILE;
SQL> ALTER SYSTEM SET inmemory_size=150m SCOPE=SPFILE;
SQL> SHUTDOWN IMMEDIATE
SQL> STARTUP
ORACLE instance started.

Total System Global Area 1577058304 bytes
Fixed Size                  2924832 bytes
Variable Size             469765856 bytes
Database Buffers          922746880 bytes
Redo Buffers               13848576 bytes
In-Memory Area            167772160 bytes
Database mounted.
Database opened.
SQL> show parameter inmem

NAME                                 TYPE          VALUE
------------------------------------ ------------- ----------
inmemory_clause_default              string
inmemory_force                       string        DEFAULT
inmemory_max_populate_servers        integer       0
inmemory_query                       string        ENABLE
inmemory_size                        big integer   160M
inmemory_trickle_repopulate_servers_ percent integer 1
optimizer_inmemory_aware             Boolean       RUE
```

从 startup 启动结果来看，多了一行："In-Memory Area 167772160 bytes"，表示分配了 In-Memory 列存储区。

注意：Oracle 12c 要求 inmemory_size 的值应该大于 100M，如果设置的值小于 100M，数据库无法启动（Startup），要报 ORA-64353 错误。Inmemory area 是 SGA 中的一个静态子池，占用 SGA，所以在加 inmemory_size 的时候应该相应调大 SGA，以免挤掉其他子池的空间。

设置了数据库的 In-Memory 内存之后,如果要让一个表以 In-Memory 方式访问,必须执行 "ALTER table 表名 inmemory 压缩方式",如果要取消表的 In-Memory 方式,可以执行 "ALTER table 表名 no inmemory"。设置 In-Memory 方式之后,表的访问速度会明显提高。

【示例 5-3】设置表 SALES 为 In-Memory 列存储表

本例设置表 SH.SALES 为 In-Memory 列存储表,并通过执行计划观察查询效率的提高。首先观察没有将 SALES 表设置为 In-Memory 时的执行计划,以 SH 用户登录,然后查询:

```
SQL> SET AUTOTRACE ON
SQL> SELECT * FROM sh.sales WHERE  amount_sold=23.43;
...
414 rows selected.
Execution Plan
----------------------------------------------------------
Plan hash value: 1550251865
| Id  | Operation            | Name  | Rows  | Bytes | Cost (%CPU)| Time     | Pstart| Pstop |
----------------------------------------------------------
|   0 | SELECT STATEMENT     |       |   243 |  7776 |   518   (2)| 00:00:01 |       |       |
|   1 |  PARTITION RANGE ALL |       |   243 |  7776 |   518   (2)| 00:00:01 |     1 |    28 |
|*  2 |   TABLE ACCESS FULL  | SALES |   243 |  7776 |   518   (2)| 00:00:01 |     1 |    28 |
----------------------------------------------------------
Predicate Information (IDENTIFIED BY operation id):
----------------------------------------------------------

   2 - filter("AMOUNT_SOLD"=23.43)

Statistics
----------------------------------------------------------
          0  recursive calls
          0  db block gets
       1648  consistent gets
          0  physical reads
          0  redo size
      17857  bytes sent via SQL*Net to client
```

第 5 章 Oracle 12c 数据库结构

```
       849  bytes received via SQL*Net FROM client
        29  SQL*Net roundtrips to/from client
         0  sorts (memory)
         0  sorts (disk)
       414  rows processed
```

观察未将 SALES 表设置为 In-Memory 时的执行计划，consistent gets 的值是 1648。下面将 SALES 设置为 In-Memory，并查看执行计划：

```
SQL> ALTER TABLE sales INMEMORY;
Table altered.
SQL> SELECT * FROM sh.sales WHERE  amount_sold=23.43;
...
SQL> SELECT * FROM sh.sales WHERE  amount_sold=23.43;
414 rows selected.

Execution Plan
----------------------------------------------------------
Plan hash value: 1550251865

| Id  | Operation                   | Name  | Rows  | Bytes | Cost (%CPU)| Time     | Pstart| Pstop |
-----------------------------------------------------------------------------------------
|   0 | SELECT STATEMENT            |       |   243 |  7776 |   123  (16)| 00:00:01 |       |       |
|   1 |  PARTITION RANGE ALL        |       |   243 |  7776 |   123  (16)| 00:00:01 |     1 |    28 |
|*  2 |   TABLE ACCESS INMEMORY FULL| SALES |   243 |  7776 |   123  (16)| 00:00:01 |     1 |    28 |
-----------------------------------------------------------------------------------------

Predicate Information (IDENTIFIED BY operation id):
---------------------------------------------------

   2 - inmemory("AMOUNT_SOLD"=23.43)
       filter("AMOUNT_SOLD"=23.43)


Statistics
----------------------------------------------------------
         0  recursive calls
         0  db block gets
        32  consistent gets
         0  physical reads
```

```
        0  redo size
    17857  bytes sent via SQL*Net to client
      849  bytes received via SQL*Net from client
       29  SQL*Net roundtrips to/from client
        0  sorts (memory)
        0  sorts (disk)
      414  rows processed
SQL>
```

对比观察已将 SALES 表设置为 In-Memory 时的执行计划,"TABLE ACCESS FULL"改成了"TABLE ACCESS INMEMORY FULL",而"consistent gets"的值由原来的 1648 直降为 32,查询效率显著提高了。

> 注意,语句"SELECT * FROM sh.sales WHERE amount_sold=23.43;"执行了两次,第一次查询后,SALES 表才会被装载到 In-Memory 内存中。这是因为 SALES 的 In-Memory 优先级为默认的 NONE,因此只有等第一次查询完成后,表才会被装载到 In-Memory 内存中。

在 In-Memory 列存储时,可以选择内存压缩方式和优先级。Oracle 为 In-Memory 中的对象提供了六种级别的压缩选项,可通过 MEMCOMPRESS 子句来指定,选择压缩方式的时候要注意压缩程度和访问速度成反比,比如,如果选择"不压缩",数据查询速度最快,但需要的内存空间最大。In-Memory 压缩方式见表 5-1 所示。

表 5-1 In-Memory 压缩方式

压缩方式	说明
NO MEMCOMPRESS	不压缩
MEMCOMPRESS FOR DML	最小的压缩方式,最优的 DML 性能
MEMCOMPRESS FOR QUERY LOW	查询性能最优化(默认方式)
MEMCOMPRESS FOR QUERY HIGH	平衡查询性能与存储空间,更偏向查询性能优化
MEMCOMPRESS FOR CAPACITY LOW	平衡查询性能与存储空间,更偏重存储空间优化
MEMCOMPRESS FOR CAPACITY HIGH	只从空间节省出发进行优化

除了为对象提供压缩选项外,为了控制对象的装载顺序,Oracle 定义了 5 种优先级来实现精细控制,通过 PRIORITY 子句来选择。详细信息:

表 5-2 In-Memory 优先级

装载优化级	说明
CRITICAL	数据库打开的时候立即装载对象
HIGH	等所有的 CRITICAL 对象装载完毕后，如果 IM 内存空间还有空余，才装载该对象
MEDIUM	等所有的 CRITICAL 和 HIGH 对象装载完毕后，如果 IM 内存空间还有空余，才装载该对象
LOW	等所有的 CRITICAL、HIGH 和 MEDIUM 对象装载完毕后，如果 In-Memory 内存空间还有空余，才装载该对象
NONE	如果 In-Memory 内存空间还有空余，对象仅在被第一次扫描后才被装载(默认方式)

我们可以对表 SALES 重新设置压缩方式和优先级为：以最大的压缩方式存储，以最优先的方式访问：

```
ALTER TABLE sales INMEMORY MEMCOMPRESS FOR CAPACITY HIGH PRIORITY CRITICAL;
```

Oracle 12c 观察 In-Memory 列存储的主要视图有：v$im_segments、v$im_user_segments、v$im_column_level、v$im_col_cu 和 v$inmemory_area。

5.3 服务器进程

进程是操作系统中的一种机制，它可以运行程序的一系列步骤。进程执行架构依赖于操作系统。例如，在 Windows 上 Oracle 后台进程是进程内执行的线程。在 Linux 和 UNIX 中，Oracle 进程是操作系统进程或操作系统进程中的线程。进程通常在其自己的私有内存区域中运行。大多数进程可以定期写入相关的跟踪文件，以便监控异常和进行异常处理。

如图 5-1 所示，进程分为客户端进程(Client Process)和 Oracle 进程，Oracle 进程又分为后台进程(Background Process)和服务器进程(Server Process)。

后台进程随数据库实例的启动而启动，担任数据库的维护任务，比如实例恢复、进程清理、将重做缓冲区写入磁盘等。

服务器进程是基于客户的请求而生产的。比如用于分析 SQL 查询语句，将分析结果存储到共享池(Shared Pool)中，为每个查询语句生成并执行"执行计划"，从磁盘或者数据库缓冲区缓存(Database Buffer Cache)中读数据等。

Oracle 12c 可以在多进程模式或者多线程模式下工作，由初始化参数 THREADED_EXECUTION 指定。THREADED_EXECUTION=FALSE 表示多进程模式，这是默认的模式，而 THREADED_EXECUTION=TRUE 表示多线程模式。

在多线程模式下，许多线程会被合并到相应的进程中，因而会减少 Linux 中的进程数量，这在多个实例集成到一个服务器上的环境下会很适用，如果不用多线程模型，操作系统的进程数将会很高，进程切换的效率也比线程低。

视图 V$PROCESS 可以查询 Oracle 的进程，每个进程一行。

【示例 5-4】查询进程

通过 V$PROCESS 查询操作系统的进程 ID(SPID)，线程 ID(STID)，进程名称(PNAME)。注意，PNAME 非空的是后台进程，PNAME 为空的行是与用户相关的服务器进程。还要注意 SPID 可能有重复，表示一个进程中包含了多个线程(STID)。

```
SQL> SELECT spid, stid, pname FROM v$process ORDER BY spid;
SPID                    STID                    PNAME
----------------------  ----------------------  -----
10073                   10073                   W000
10694                   10694                   W002
10705                   10705                   W001
10898                   10898                   W004
11326                   11326                   W005
12004                   12004                   W006
12687                   12687                   W007
13770                   13770
14092                   14092                   Q001
16169                   16169
42038                   42038                   PMON
42040                   42040                   PSP0
...
```

5.3.1 后台进程

Oracle 的后台进程分为强制性的后台进程、可选的后台进程和从属进程。强制性的后台进程是数据库参数文件中必须配置的。主要的强制性后台进程有：

1) 进程监控进程(PMON)

PMON 监控其他后台进程，当服务器或者调度进程异常关闭时，它能恢复进程。PMON 还负责清理数据库缓冲区缓存，释放客户进程使用的资源，比如，PMON 能重置活动事务表的状态，解除不再需要的锁，移除活动进程表中的进程 ID。

2) 系统监控进程(SMON)

SMON 负责多种系统级的清理。比如在实例启动时执行实例恢复、恢复由于文件读取或表空间脱机错误而在实例恢复期间跳过而终止的事务、清理未使用的临时段、合并相邻的数据字典所在的表空间中的空闲区(Extents)。

3) 监听器注册进程(LREG)

LREG 进程通过"Oracle Net Listener"注册数据库实例和调度进程的信息。在实例启动时，LREG 要查检监听器，确保监听器正在运行。如果监听器正在运行，LREG 传递相关参数给监听器。如果监听器没有运行，LREG 会定期监视它。

4) 数据库写入进程(DBW)

DBW 进程将有修改的数据库缓冲区的内容写入数据文件。虽然对多数系统来说，只需要一个DBW0数据库写进程就够了，但为了提高写入性能，可以增加一些进程，从DBW1到DBW9，DBWa到DBWz，以及BW36到BW99。对于单处理器系统，增加的DBW进程没有用。我们把修改过的缓冲区称为"脏"缓冲区，DBW 进程在下列条件下将脏缓冲区写入磁盘：

当扫描一定数量(由阈值决定)的描缓冲区后，服务器进程无法找到一个干净的可使用的缓冲区，启动 DBW 进程将脏缓冲区异步写到磁盘文件中。

DBW 定期写入缓冲区并推进检查点(Checkpoint)。检查点是重做线程进行实例恢复的开始位置，检查点的日志位置由时间最早的脏缓冲区位置决定。

在许多情况下，DBW 写入的磁盘块分散在磁盘中。因此，这种写操作往往比 LGWR 进程的顺序写操作慢。因此，为了提高写入性能，DBW 会尽可能进行多块的整体写操作。

5) 日志写入进程 (LGWR)

日志写入进程(LGWR)管理联机重做日志缓冲区。LGWR 将一部份缓冲区写入到联机重做日志(Online Redo Log)文件中。通过分离修改数据库缓冲区的任务，将脏缓冲区分散写入磁盘和将重做日志快速顺序写入联机重做日志文件，提高数据库的性能。在下列情况下，LGWR 将最近时间复制到重做日志缓冲区的重做记录写入联机重做日志文件：

◇ 用户提交事务。
◇ 发生重做日志切换。
◇ 距离上次 LGWR 过了 3 秒钟。
◇ 重做日志缓冲区写满了三分之一或包含了 1MB 的缓冲数据。
◇ DBW 进程将修改后的数据库缓冲区写入磁盘的时候。

在 DBW 进程将修改后的数据库缓冲区写入数据文件之前，数据库必须先将与脏缓冲区有关联的重做日志缓冲区记录写入到联机重做日志文件中，这称为先写协议(Write-Ahead Protocol)。如果 DBW 发现一些重做记录还没有写，它将通知 LGWR 先将重做记录写入磁盘，并等待 LGWR 写入磁盘完成。这就是说 DBW 与 LGWR 是两个同步的进程。

Oracle 数据库使用快速提交机制来提高提交事务的性能。当用户执行提交语句时，事务被分配一个系统改变号 SCN(System Change Number)。LGWR 将提交记录写入重做日志缓冲区，并立即将事务的重做日志记录连同提交的 SCN 写入到联机重做日志文件。尽管数据缓冲区尚未写入数据文件，Oracle 数据库还是将成功代码返回到提交事务。

> 注意：事务提交成功后，只是成功将重做日志缓冲区的内容写入到了联机重做日志文件中，数据库缓冲区中的脏缓冲区还可能没有写入到数据文件中，相关的脏缓冲区的数据块要等待 DBW 进程写入到数据文件中。

6) 检查点进程(CKPT)

检查点进程(CKPT)将检查点信息更新到控制文件和数据文件头，并通知 DBW 进程将脏数据块写入磁盘。检查点信息包括检查点位置、SCN 号、联机重做日志文件的恢复位置等。如图 5-1 所示，CKPT 不往数据文件中写入数据块，也不往联机重做日志文件中写入重做日志块。有许多情况可以产生检查点事件，比如数据库 shutdown、日志切换数据库开始备份、数据文件或者表空间的联机或者离线等。也可以手工执行检查点事件，命令是：

```
ALTER SYSTEM CHECKPOINT;
```

7) 归档进程（ARCn）

联机重做日志文件发生切换之后，归档进程 ARCn（Archiver Process）将联机重做日志文件复制到离线的归档日志文件中。这些进程收集事务重做数据并将其发送到备用数据库服务器中。只有当数据库运行在归档（ARCHIVELOG）模式下，并且设置为自动归档后，ARCn 进程才会存在。Oracle 默认在非归档模式下工作，要切换到在归档模式下工作，参见 **6.7** 重做日志文件与归档日志文件。

8) 作业队列进程（CJQ0 和 Jnnn）

队列进程运行用户的作业（Job），通常以批处理模式运行。作业是计划运行一次或多次的用户定义的任务。比如，可以使用作业队列在后台调度运行长时间更新操作。给定一个开始日期和一个时间间隔，作业队列进程能够在下一个间隔发生时运行该作业。

Oracle 数据库动态管理作业队列进程，从而使作业队列客户端在需要时使用更多的作业队列进程。当新进程空闲的时候，数据库会释放其使用的资源。

动态作业队列进程可以在给定的时间间隔同时运行多个作业。事件顺序是：

◇ 作业调度进程 CJQ0（Job Coordinator Process）由 Oracle 调度器（Oracle Scheduler）根据需要自动启动和停止。调度进程周期性地按时间顺序选择需要从系统的 JOB$ 表中运行的作业。

◇ 调度进程动态的生成工作队列从属进程（Jnnn）运行作业。

◇ 作业调度进程运行 CJQ0 进程选择的作业中的一个作业。每个作业队列进程一次完成一个作业。

◇ 当进程执行完成一个作业之后，它会执行更多的作业。如果没有待执行的作业了，那么它进入睡眠状态，直接过了一定的间隔期，它又会醒来，执行更多的作业。如果进程没有找到任何新的作业，则在预设的间隔时间结束后终止进程。

初始化参数 JOB_QUEUE_PROCESSES 代表作业队列进程可以在一个实例上并发运行的最大作业队列进程数，默认值是 1000。所以，客户端不应该假设所有的作业队列进程都可用于执行。

5.3.2 定时执行作业任务

在大中型项目中，常常会有一些复杂的或者需要重复执行的业务，比如定时备份，定时统计等。Oracle 以作业（Job）执行这类任务。Oracle 有两种程序包管理作业，分别是 DBMS_JOB 和 DBMS_SCHEDULER，在 Oracle 12c 中 DBMS_JOB 包已经过时了，建议使用 DBMS_SCHEDULER 包。普通用户需要有"create job"和"manage scheduler"的系统权限才能创建作业任务。

【示例 5-5】 给 STUDY 用户授权，让该用户可以创建作业任务

```
$ sqlplus system/***@pdborcl
SQL>GRANT create job TO study;
Grant succeeded.
SQL>GRANT manage scheduler TO study;
Grant succeeded.
```

DBMS_SCHEDULER 包使用 CREATE_JOB 过程创建作业，使用 DROP_JOB 过程删除作业。DBMS_SCHEDULER.CREATE_JOB 过程有以下参数：

(1) JOB_NAME：任务的名称，必选，任务名称必须唯一。

(2) JOB_TYPE：任务的操作类型，必选，有下列几个可选值：

◇ PLSQL_BLOCK：任务执行的是一个 PL/SQL 匿名块。

◇ STORED_PROCEDURE：任务执行的是存储过程（含 PL/SQL PROCEDURE 和 JAVA PROCEDURE）。

◇ EXECUTABLE：任务执行的是一个外部程序，比如操作系统命令。

◇ CHAIN：任务执行的是一个 CHAIN。

(3) JOB_ACTION：任务执行的操作，必选，应与 JOB_TYPE 类型中指定的参数相匹配。比如对于 PL/SQL 匿名块，此处就可以放置 PL/SQL 块的具体代码；如果是 Oracle 过程，此处应该指定过程名，注意如果过程中有 OUT 之类的输出参数，实际执行时不会有输出。

(4) START_DATE：任务初次执行的时间，本参数可为空，当为空时，表示任务立刻执行，等同于指定该参数值为 SYSDATE。

(5) REPEAT_INTERVAL：任务执行的频率，比如多长时间会被触发再次执行。本参数也可以为空，如果为空的话，就表示当前设定的任务只执行一次。REPEAT_INTERVAL 参数的语法结构比较复杂。其中最重要的是 FREQ 和 INTERVAL 两个关键字。

FREQ 关键字用来指定间隔的时间周期，可选参数有：YEARLY、MONTHLY、WEEKLY、DAILY、HOURLY、MINUTELY 和 SECONDLY，分别表示年、月、周、日、时、分、秒等单位。

INTERVAL 关键字用来指定间隔的重复次数，值的范围是 1~99。

例如：REPEAT_INTERVAL=>'FREQ=DAILY；INTERVAL=1'；表示每天执行一次，如果 INTERVAL 为 7 就表示每 7 天执行一次，相当于 'FREQ=WEEKLY；INTERVAL=1'。

一般来说，使用 DBMS_SCHEDULER.CREATE_JOB 创建一个 JOB，至少需要指定上述参数中的前 3 项。除此之外，还可以指定下列参数：

(6) NUMBER_OF_ARGUMENTS：指定该 JOB 执行时需要附带的参数的数量，默认值为 0，注意当 JOB_TYPE 列值为 PLSQL_BLOCK 或 CHAIN 时不支持附带参数，本参数必须设置为 0。

(7) END_DATE：指定任务的过期时间，默认值为 NULL。任务过期后，任务的 STATE 将自动被修改为 COMPLETED，ENABLED 被置为 FALSE。如果该参数设置为空的话，表示该任务永不过期，将一直按照 REPEAT_INTERVAL 参数设置的周期重复执行，直到

达到设置的 MAX_RUNS 或 MAX_FAILURES 值。

(8)JOB_CLASS：指定任务关联的 CLASS，默认值为 DEFAULT_JOB_CLASS。

(9)ENABLED：指定任务是否启用，默认值为 FALSE。FALSE 状态表示该任务不会被执行，除非被用户手动调用，或者用户将该任务的状态修改为 TRUE。

(10)AUTO_DROP：当该标志被置为 TRUE 时，Oracle 会在满足以下条件之一时自动删除创建的任务：

- ✧ 任务已过期。
- ✧ 任务最大运行次数已达 MAX_RUNS 的设置值。
- ✧ 任务未指定 REPEAT_INTERVAL 参数，仅运行一次。

(11)COMMENTS：设置任务的注释信息，默认值为 NULL。

【示例 5-6】创建作业

STUDY 用户创建作业 JOB_CALC，指定作业每天凌晨 1 点钟运行一次存储过程 MYPACK.CALC_ALL_TRADERECEIVABLE。每天一次的指定参数为 "repeat_interval => 'FREQ=DAILY；'"，凌晨 1 点钟由 start_date 指定。创建之后，只有当参数 enabled 为 true 的时候才能开始作业并重复循环执行。

```
$ sqlplus study/***@pdborcl
SQL>
BEGIN
  DBMS_SCHEDULER.CREATE_JOB (
    job_name => '"STUDY"."JOB_CALC"',
    job_type => 'STORED_PROCEDURE',
    job_action => 'STUDY.MYPACK.CALC_ALL_TRADERECEIVABLE',
    number_of_arguments => 0,
    start_date => TO_DATE('2016-04-01 01: 00', 'YYYY-MM-DD HH24: MI'),
    repeat_interval => 'FREQ=DAILY; ',
    end_date => NULL,
    enabled => TRUE,
    auto_drop => FALSE,
    comments => '');
END;
/
PL/SQL procedure successfully completed.
```

作业创建完成后，作业调度进程 CJQ0 就等待这个作业的下次运行时间(NEXT_RUN_DATE)到达，一旦时间到达，就立即运行，运行完成后，就等待下一次时间到达，重复运行。本例只设置了开始时间(START_DATE)，没有设置结束时间(END_DATE)，作业 JOB_CALC 会一直重复，不会停止。Oracle 12c 查询作业的相关视图有：

(1)user_scheduler_jobs：查询用户创建的所有作业，以及作业下次运行时间

(NEXT_RUN_DATE)。

(2) user_scheduler_job_log：job 日志。

(3) user_scheduler_job_run_details：job 运行日志。

(4) user_scheduler_running_jobs：正在运行的 job。

【示例 5-7】查询作业信息

本例查询用户创建的作业，注意 NEXT_RUN_DATE 表示作业下次开始执行的时间。通过这个查询，可以监控作业的运行情况。

```
SQL> COL job_name FORMAT a15
SQL> COL start_date FORMAT a30
SQL> COL next_run_date FORMAT a30
SQL> SELECT job_name, to_char(START_DATE, 'yyyy-mm-dd hh24:mi:ss')START_DATE,to_char(NEXT_RUN_DATE,'yyyy-mm-dd hh24:mi:ss')NEXT_RUN_DATE
FROM user_scheduler_jobs;
JOB_NAME         START_DATE                NEXT_RUN_DATE
---------------  ------------------------  ------------------------
JOB_CALC         2016-04-01 01:00:00       2017-04-08 01:00:00
```

如果不需要作业了，可以删除作业，删除作业使用 DBMS_SCHEDULER.DROP_JOB 过程，参数有：

job_name：任务的名称，必选。

defer：如果 defer 为 TRUE，等待正在运行的 job 结束再 drop。

force：如果 force 为 TRUE，停止正在运行的 job 后再删除。

如果同时指定 force 和 defer 为 TRUE，返回错误。如果同时指定 force 和 defer 为 FALSE，对于一个正在运行的 job，调用失败。

【示例 5-8】删除作业

```
SQL>
BEGIN
DBMS_SCHEDULER.DROP_JOB(job_name => '"STUDY"."JOB_CALC"',
                       defer => false,
                       force => false);
END;
/
```

练习

一、判断题

1. Oracle 数据库都有一个控制文件，用于存储用户数据。（   ）
2. Oracle 的最大的逻辑存储单元称为表空间。（   ）
3. Oracle 中每个需要存储的对象都仅由一个段 segment 组成。（   ）

4. 数据块(Data Block)是磁盘空间分配的最小单位。 ( )
5. 区(Extents)是数据库的最小数据组织单位与管理单位。 ( )
6. 服务器进程是基于客户的请求而生产的。 ( )

二、单项选择题

1. 系统全局区是指( )
   A. SGA   B. PGA
   C. Large Pool   D. Java Pool

2. 用来存储用户数据修改前的值的段 segment 是( )
   A. 数据段   B. 索引段
   C. 回滚段   D. 临时段

3. Oracle 12c 用于查询正在运行的 job 的视图是( )
   A. user_scheduler_jobs
   B. user_scheduler_job_log
   C. user_scheduler_job_run_details
   D. user_scheduler_running_jobs

4. In-Memory 压缩方式中哪个是只从空间节省出发进行优化：( )
   A. MEMCOMPRESS FOR DML
   B. MEMCOMPRESS FOR QUERY LOW
   C. MEMCOMPRESS FOR CAPACITY LOW
   D. MEMCOMPRESS FOR CAPACITY HIGH

5. 下面哪个 In-Memory 优先级表示数据库打开时立即装载对象( )
   A. CRITICAL   B. HIGH
   C. MEDIUM   D. LOW

三、填空题

1. 数据库实例是管理数据库文件的一组_____。
2. Oracle 数据库具有物理结构和_____。
3. 联机重做日志是由两个或多个联机重做日志_____组成的一个联合体。
4. 数据库的逻辑存储结构由表空间、段、区和_____4 个部分组成的。
5. Oracle 的段可以分成 4 种类型：_____、索引段、回滚段和临时段。
6. Oracle 12c 可以工作在多进程模式或者_____模式下。
7. Oracle 后台进程分为_____的后台进程、可选的后台进程和从属进程。
8. _____进程将联机重做日志文件复制到离线的归档日志文件中。

四、简答题

1. Oracle 数据库的物理存储结构中包括哪些文件，各有何作用？
2. In-Memory 列存储的特点是什么？有哪些作用？
3. 主要的强制性后台进程有哪些？
4. 简述 Oracle 写入进程(DBW)的工作方式。
5. 简述 Oracle 日志写入进程(LGWR)的工作方式。

训练任务

1. 增加 SGA 空间大小

培养能力	设计/开发解决方案、研究		
掌握程度	★★★	难度	中
结束条件	增加 SGA 大小后,重启数据库		

训练内容:
本训练的背景是当前 SGA 空间不够大,需要增加 SGA 空间的大小。首先需要查看当前的 SGA 空间的大小,然后适当增加一点。修改 SGA 空间的大小需要用到两个系统参数: sga_target 和 sga_max_size

2. 应用 In-Memory 列存储技术创建内存表

培养能力	工程能力、问题分析、使用现代工具		
掌握程度	★★★★	难度	高
结束条件	创建 In-Memory 表成功后,测试数据访问速度		

训练内容:
本训练的背景是有一张需要频繁地被访问的表,需要将该表设计为 In-Memory 类型表。然后插入数据,查询数据,测试数据访问速度

3. 维护定时作业任务

培养能力	工程能力、问题分析、使用现代工具、沟通		
掌握程度	★★★★★	难度	高
结束条件	创建作业成功后,观察作业运行情况		

训练内容:
本训练的背景是项目需要每天定时执行一个存储过程,需要创建一个作业来定时执行这个存储过程,创建作业成功后,要定时监控这个作业是否正常运行,运行结果是否达到预期的效果

第6章 数据库存储管理

本章目标

知识点	理解	掌握	应用
1.表空间和数据文件的管理	✓	✓	✓
2.控制文件的管理	✓	✓	
3.重做日志文件的作用	✓	✓	
4.归档模式及归档文件	✓	✓	
5.系统参数设置及参数文件的管理		✓	✓

训练任务

- 能够完成表空间和数据文件的日常维护工作。
- 能够完成重做日志文件组的管理工作。
- 能够设置各种系统参数。

知识能力点

能力点 \ 知识点	知识点1	知识点2	知识点3	知识点4	知识点5
工程知识	✓				✓
问题分析	✓		✓	✓	✓
设计/开发解决方案	✓	✓		✓	
研究	✓		✓		
使用现代工具	✓	✓		✓	✓
工程与社会	✓				✓
环境和可持续发展			✓		✓
职业规范		✓			✓
个人和团队	✓				
沟通		✓			✓
项目管理	✓			✓	
终身学习	✓			✓	

6.1 表空间和数据文件的管理

数据库可以划分为多个逻辑存储单元，存储单元也称为表空间(Tablespace)，这是因为表空间的主要存储对象就是表。表空间是由一个或者多个数据文件组成的，而一个数据文件只能属于一个空间，表空间本身没有物理大小，表空间的逻辑大小等于其包含的数据文件大小之和，如图 6-1 所示。

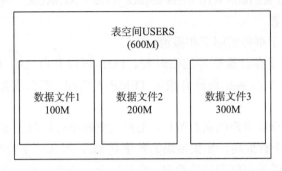

图 6-1 表空间与数据文件

表空间 USERS 由 3 个数据文件组成，其大小为 600M，是 3 个数据文件大小之和。表空间由多文件组成带来两大好处：一是数据存储数量可以突破操作系统单个文件大小的限制；二是将多文件放到不同的磁盘上，有助于提高数据的读写效率。

Oracle 12c 中有根容器数据库 CDB，以及多个 PDB 数据库，CDB 及每个 PDB 都有自己的一套表空间和文件系统。如图 6-2 所示，每个数据库中都有 SYSTEM(系统)表空间、SYSAUX(系统辅助)表空间和 TEMP(临时)表空间，CDB 中还独有 UNDO(还原)表空间，在其他 PDB 中还允许创建用户自己的表空间。

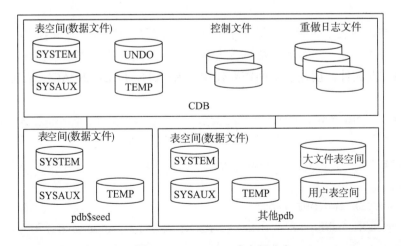

图 6-2 Oracle 12c 表空间分布

SYSTEM 表空间是由系统安装的时候自动创建的,不能被删除,不能离线。SYSTEM 表空间包含数据库的数据字典表,一些程序单元对象(如存储过程、函数、包和触发器等)也保存在 SYSTEM 表空间中。通过命令:"SELECT * FROM dba_segments WHERE tablespace_name='SYSTEM'"可以查询 SYSTEM 表空间内存储的对象及其拥有者。

SYSAUX 表空间也是在系统安装的时候自动创建的,不能被删除,不能离线。SYSAUX 表空间存储一些数据库组件产生的对象,不存储在 SYSTEM 表空间中,目的是减轻 SYSTEM 表空间的负荷,避免了因反复创建一些相关对象及组件引起的 SYSTEM 表空间碎片的问题。通过命令:"SELECT * FROM dba_segments WHERE tablespace_name='SYSAUX'"可以查询 SYSAUX 表空间存储的对象及其拥有者。

UNDO 表空间用于事务的回滚和撤消。

TEMP 表空间用于存储数据库的临时表。PDB 可以有自己的 TEMP 表空间,也可以不用,在创建 PDB 的时候如果没有指定 TEMP 表空间,那么就会公用 CDB 的 TEMP 表空间。

用户表空间用于存储用户的私有数据,用户表空间中可以包含多个数据文件。每个数据文件最多包含 4M 个数据块,如果每块的容量是 8K,那么一个数据文件最大是 32GB。

大文件表空间也用于存储用户的数据,它只包含一个单独的大文件。大文件是指文件最多包含 4G 个数据块,如果每块的容量是 8K,那么一个数据文件最大是 32TB,所以相对而言,用户表空间可以叫小文件表空间。

> 注意:在整个 Oracle 12c 数据库中,有多个与 SYSTEM,TEMP 等同名的表空间,但由于它们属于 CDB 或者分属不同的 PDB,所以尽管表空间名称相同,但由于它们对应的数据文件是不同的,因此根本不是相同的表空间,这点在数据库管理时特别容易混淆。

6.2 创建表空间

Oracle 在创建表空间时将完成两个工作,一方面是在数据字典和控制文件中记录新建的表空间信息。另一方面是在操作系统中创建指定大小的操作系统文件,并作为与表空间对应的数据文件。创建表空间需要使用 CREATE TABLESPACE 语句,其基本语法如下:

```
CREATE [ TEMPORARY | UNDO | BIGFILE ] TABLESPACE tablespace_name
[
   DATAFILE | TEMPFILE 'file_name' SIZE size K | M [ REUSE ]
   [
      AUTOEXTEND OFF | ON
      [ NEXT number K | M MAXSIZE UNLIMITED | number K | M ]
   ]
   [ , ...]
```

```
    ]
    [ MININUM EXTENT number K | M ]
    [ BLOCKSIZE number K]
    [ ONLINE | OFFLINE ]
    [ LOGGING | NOLOGGING ]
    [ FORCE LOGGING ]
    [ DEFAULT STORAGE storage ]
    [ COMPRESS | NOCOMPRESS ]
    [ PERMANENT | TEMPORARY ]
    [
        EXTENT MANAGEMENT DICTIONARY | LOCAL
        [ AUTOALLOCATE | UNIFORM SIZE number K | M ]
    ]
    [ SEGMENT SPACE MANAGEMENT AUTO | MANUAL ];
```

语法中各参数的说明如下：

1) TEMPORARY | UNDO| BIGFILE

表空间的类型。TEMPORARY 表示创建临时表空间；UNDO 表示创建还原表空间；BIGFILE 表示创建为大文件表空间。如果不指定类型，则表示创建的表空间为永久性表空间。

2) tablespace_name 表空间的名称

3) DATAFILE | TEMPFILE 'file_name'

指定与表空间相关联的数据文件。一般使用 DATAFILE，如果是创建临时表空间，则需要使用 TEMPFILE；file_name 指定文件名与路径。如果不是大文件表空间，可以为一个表空间指定多个数据文件。

4) SIZE size

数据文件的大小，要指定单位，比如 100K，100M。

5) REUSE

如果指定的数据文件已经存在，则使用 REUSE 可以清除并重新创建该数据文件。如果文件已存在，又没有指定 REUSE，创建表空间时会报错。

6) AUTOEXTEND OFF | ON

指定数据文件是否自动扩展。OFF 表示不自动扩展；ON 表示自动扩展。默认情况下为 OFF。

7) NEXT number

如果指定数据文件为自动扩展，NEXT 用于指定数据文件每次扩展的大小。

8) MAXSIZE UNLIMITED | number

如果指定数据文件为自动扩展，MAXSIZE 用于指定数据文件的最大尺寸。如果指定 UNLIMITED，则表示大小无限制，默认为此选项。

9) MININUM EXTENT number

表空间中的区可以分配的最小的尺寸。

10）BLOCKSIZE number

如果创建的表空间需要另外设置数据块大小，而不是采用初始化参数 db_block_size 指定的数据块大小，则可以使用此子句进行设置。此子句仅适用于永久性表空间。

11）ONLINE | OFFLINE

指定表空间的状态为在线（ONLINE）或离线（OFFLINE）。如果为 ONLINE，则表空间可以使用；如果为 OFFLINE，则表空间不可使用。默认为 ONLINE。

12）LOGGING | NOLOGGING

指定存储在表空间中的数据库对象的任何操作是否产生日志。LOGGING 表示产生；NOLOGGING 表示不产生。默认为 LOGGING。

13）FORCE LOGGING

此选项用于强制表空间中的数据库对象的任何操作都产生日志，将忽略 LOGGING 或 NOLOGGING 子句。

14）DEFAULT STORAGE storage

指定保存在表空间中的数据库对象的默认存储参数。当然，数据库对象也可以指定自己的存储参数。

15）COMPRESS | NOCOMPRESS

指定是否压缩数据段中的数据。COMPRESS 表示压缩；NOCOMPRESS 表示不压缩。数据压缩发生在数据块层次中，以便压缩数据块内的行，消除列中的重复值。默认为 COMPRESS。

16）PERMANENT | TEMPORARY

指定表空间中数据对象的保存形式。PERMANENT 表示持久保存；TEMPORARY 表示临时保存。

17）EXTENT MANAGEMENT DICTIONARY | LOCAL

指定表空间的管理方式。DICTIONARY 表示采用数据字典的形式管理；LOCAL 表示采用本地化管理形式管理。默认为 LOCAL。

18）AUTOALLOCATE | UNIFORM SIZE number

指定表空间中的区（EXTENT）大小。AUTOALLOCATE 表示区大小由 Oracle 自动分配，不能指定大小；UNIFORM SIZE number 表示表空间中的所有区大小相同，都为指定值。默认为 AUTOALLOCATE。

19）SEGMENT SPACE MANAGEMENT AUTO | MANUAL

指定表空间中段的管理方式。AUTO 表示自动管理方式；MANUAL 表示手动管理方式。默认为 AUTO。

【示例 6-1】在 PDBORCL 数据库中创建一个表空间

本示例通过 SYSTEM 用户在 PDBORCL 数据库中创建一个表空间 USERS02，由两个数据文件组成，每个文件的初始大小都是 100M，自动增长，每次扩展 50M，文件大小无限制。

```
$ sqlplus system/***@pdborcl
```

```
SQL>CREATE TABLESPACE users02 DATAFILE
'/home/oracle/app/oracle/oradata/orcl/pdborcl/pdbtest_users02_1.dbf'
  SIZE 100M AUTOEXTEND ON NEXT 50M MAXSIZE UNLIMITED,
'/home/oracle/app/oracle/oradata/orcl/pdborcl/pdbtest_users02_2.dbf'
  SIZE 100M AUTOEXTEND ON NEXT 50M MAXSIZE UNLIMITED
EXTENT MANAGEMENT LOCAL SEGMENT SPACE MANAGEMENT AUTO;
tablespace created.
SQL>
```

> 注意：从创建表空间的命令看，表空间只属于某个数据库，并不属于哪个特定用户。事实上，表空间只是存储空间，用户需要分配空间配额后才具有使用空间的权利。

6.3 查看表空间信息

通过 dba_tablespaces，dba_free_space 可以查询表空间的信息；通过 dba_data_files 和 dba_temp_files 查询数据文件和临时文件的名称、文件 ID、文件大小以及文件所属的表空间名称等属性。

【示例 6-2】查询 PDBORCL 数据库的所有表空间

本示例查询 PDBORCL 数据库的所有表空间。从结果可以看出，所有表空间都在线，TEMP 表空间是临时(TEMPORARY)表空间。

```
$ sqlplus system/***@pdborcl
SQL> SELECT TABLESPACE_NAME,STATUS,CONTENTS,LOGGING
2 FROM dba_tablespaces;
TABLESPACE_NAME                STATUS    CONTENTS    LOGGING
------------------------------ --------- ----------- ---------
SYSTEM                         ONLINE    PERMANENT   LOGGING
SYSAUX                         ONLINE    PERMANENT   LOGGING
TEMP                           ONLINE    TEMPORARY   NOLOGGING
USERS                          ONLINE    PERMANENT   LOGGING
EXAMPLE                        ONLINE    PERMANENT   NOLOGGING
USERS02                        ONLINE    PERMANENT   LOGGING
7 rows selected.
SQL>
```

【示例 6-3】统计 PDBORCL 数据库中所有表空间的使用情况

```
$ sqlplus system/***@pdborcl
SQL> SELECT a.tablespace_name "表空间名", Total "大小",
2 free "剩余", ( total - free )"使用",
3 Round(( total - free )/ total, 4)* 100 "使用率%"
```

```
4  from (SELECT tablespace_name, Sum(bytes)free
5         FROM  dba_free_space group BY tablespace_name)a,
6        (SELECT tablespace_name, Sum(bytes)total FROM dba_data_files
7          group BY tablespace_name)b
8  where  a.tablespace_name = b.tablespace_name;

表空间名                  大小              剩余             使用            使用率%
--------------------  ----------      ----------     ----------    ----------
SYSAUX                692060160        58589184       633470976     91.53
USERS                 79953920          6750208        73203712     91.56
SYSTE                 356515840        71827456       284688384     79.85
EXAMPLE               1359216640       66715648       1292500992    95.09
USERS02               209715200        139329536      70385664      33.56
6 rows selected.
```

6.4 设置表空间

6.4.1 修改表空间名称

有时候我们可能会需要修改表空间的名称，比如做数据迁移的时候。修改表空间的命令是："ALTER TABLESPACE 原名称 RENAME TO 新名称；"。

【示例 6-4】修改表空间 users02 的名称为 users03

```
$ sqlplus system/***@pdborcl
SQL> ALTER TABLESPACE users02 RENAME TO users03;
Tablespace altered.
```

6.4.2 修改表空间大小

表空间的大小是在创建时已经定义的，随着数据的增加，表空间可能无法承担更多的数据，此时需要对表空间的大小进行修改。增加表空间的大小有两种办法，一是增加表空间中的文件的大小，二是增加新文件进入表空间。

【示例 6-5】增加表空间大小

本示例修改表空间 USERS02 中数据文件 pdbtest_users02_1.dbf 的大小，从 100M 修改为 120M，再给表空间增加一个数据文件 pdbtest_users02_3.dbf，大小是 50M，并且自动增长。通过查询 dba_data_files 对比查询修改前后的大小。

```
$ sqlplus system/***@pdborcl
SQL> COL file_name FORMAT a70
SQL> SELECT FILE_NAME,BYTES FROM dba_data_files
```

```
  2 WHERE  tablespace_name='USERS02';
  FILE_NAME                                                         BYTES
  ----------------------------------------------------------------- ----------
/home/oracle/app/oracle/oradata/orcl/pdborcl/pdbtest_users02_1.dbf 104857600
  /home/oracle/app/oracle/oradata/orcl/pdborcl/pdbtest_users02_2.dbf 104857600
  SQL> ALTER database datafile
'/home/oracle/app/oracle/oradata/orcl/pdborcl/pdbtest_users02_1.dbf' RESIZE 120M;
  Database altered.
  SQL> ALTER tablespace users02 add datafile
'/home/oracle/app/oracle/oradata/orcl/pdborcl/pdbtest_users02_3.dbf'  size 50M autoextend on;
  Tablespace altered.
  SQL> SELECT FILE_NAME, BYTES FROM dba_data_files WHERE tablespace_name='USERS02';
  FILE_NAME                                                         BYTES
  ----------------------------------------------------------------- ----------
  /home/oracle/app/oracle/oradata/orcl/pdborcl/pdbtest_users02_3.dbf 52428800
  /home/oracle/app/oracle/oradata/orcl/pdborcl/pdbtest_users02_1.dbf 125829120
  /home/oracle/app/oracle/oradata/orcl/pdborcl/pdbtest_users02_2.dbf 104857600
```

> 注意：如果希望减小表空间的大小，要注意 RESIZE 不能减小到不足以存储现有数据的程度，这个操作是安全的，最多提示错误"ORA-03297：文件包含在请求的 RESIZE 值以外使用的数据"，而不会损失数据。Oracle 也不能直接删除表空间中的文件，只有通过将文件离线（Offline），然后在操作系统中删除文件，但是这样做并不安全，因为文件中可能包括有表数据，删除后会令数据库崩溃。总之，表空间的大小最好只增不减。

6.4.3 切换表空间状态

表空间有脱机状态（Offline）、联机状态（Online），联机状态又分为读写状态（Read Write）和只读状态（Read Only）。在脱机状态时，表空间不能读写。在只读状态时只能读不能写，在读写状态时可以读和写。

【示例 6-6】切换表空间的状态

本示例不断切换表空间 USERS02 的状态，路径是：Offline→Online→Read Only→Read

Write,并通过 dba_tablespaces 查询表空间 USERS02 的状态。

```
$ sqlplus system/***@pdborcl
SQL> ALTER tablespace users02 OFFLINE;
Tablespace altered.
SQL> SELECT tablespace_name, status FROM dba_tablespaces
  2 WHERE  tablespace_name='USERS02';
TABLESPACE_NAME                STATUS
------------------------------ ---------
USERS02                        OFFLINE
SQL> ALTER tablespace users02 ONLINE;
SQL> SELECT tablespace_name, status FROM dba_tablespaces
  2 WHERE  tablespace_name='USERS02';
TABLESPACE_NAME                STATUS
------------------------------ ---------
USERS02                        ONLINE
SQL> ALTER tablespace users02 READ ONLY;
Tablespace altered.
SQL> SELECT tablespace_name, status FROM dba_tablespaces
  2 WHERE  tablespace_name='USERS02';
TABLESPACE_NAME                STATUS
------------------------------ ---------
USERS02                        READ ONLY
SQL> ALTER tablespace users02 READ WRITE;
Tablespace altered.
SQL> SELECT tablespace_name, status FROM dba_tablespaces
  2 WHERE  tablespace_name='USERS02';
TABLESPACE_NAME                STATUS
------------------------------ ---------
USERS02                        ONLINE
```

> 注意:表空间的"读写"状态,并不显示为"Read Write",而是显示为"ONLINE",这点容易引起混淆。

6.5 删除表空间

对于不需要使用的表空间,可以删除它。删除表空间的命令格式是:

DROP TABLESPACE 表空间名称 [INCLUDING CONTENTS [AND DATAFILES]]

其中：

INCLUDING CONTENTS 表示删除表空间时删除其中的段，但保留数据文件。

AND DATAFILES 表示删除表空间后，再删除数据文件。

【示例 6-7】删除 PDBORCL 中的表空间 USERS02 以及对应文件

```
$ sqlplus system/***@pdborcl
SQL> DROP TABLESPACE users02 INCLUDING CONTENTS AND DATAFILES;
Tablespace dropped.
SQL>
```

> 注意：删除表空间就会删除表的数据，删除的时候 Oracle 也没有确认提示，所以要特别小心。

6.6 控制文件的管理

每个 Oracle 数据库都有一个控制文件，只属于 CDB，PDB 中不需要控制文件。控制文件是一个非常小的二进制文件，保存了如下内容。

（1）数据库名称和标识。

（2）数据库创建时的时间戳。

（3）表空间名称。

（4）数据文件和日志文件的名称和位置。

（5）当前日志文件序列号。

（6）最近检查点信息。

（7）恢复管理器信息。

控制文件在数据库启动的 MOUNT 阶段被读取，由此可以看出控制文件的重要性。所以 Oracle 采用克隆（也叫多路复用）的形式，将控制文件生成为几个动态备份，每个文件内容完全一致，但放在不同的地方存储，以防止控制文件的失效造成数据库无法启动，也就是说，控制文件在物理上是几个不同的文件，但在逻辑上是一个文件。当成功启动数据库后，在数据库的运行过程中，数据库服务器要不断地修改控制文件中的内容。所以在数据库被打开阶段，控制文件必须是可读写的。

多路复用控制文件的设置是由系统参数 control_files 设置的。可以通过修改参数 control_files 的值维护控制文件的动态备份。

【示例 6-8】增加控制文件

Oracle 缺省是两个控制文件互为备份，本示例增加一个控制文件，形成 3 个控制文件互为备份的局面。首先以 sys 登录到 CDB，查询参数 control_files 的原始值，然后修改 control_files 的值，即在值的字符串中增加一个文件名称 control03.ctl。

但要注意增加了名称之后，文件本身并不存在，还需要将已有的控制文件复制为新文件 control03.ctl，操作时要注意先关闭数据库，再复制。

```
$ sqlplus / as sysdba
```

```
SQL> show parameter control_files;

NAME                                 TYPE        VALUE
------------------------------------ ----------- --------------------
control_files                        string      /home/oracle/app/oracle/oradata/orcl/control01.ctl,
/home/oracle/app/oracle/fast_recovery_area/orcl/control02.ctl
SQL> ALTER SYSTEM SET control_files='/home/oracle/app/oracle/oradata/orcl/control01.ctl',
'/home/oracle/app/oracle/fast_recovery_area/orcl/control02.ctl',
'/home/oracle/app/oracle/fast_recovery_area/orcl/control03.ctl'
SCOPE=SPFILE;
System altered.
SQL> SHUTDOWN IMMEDIATE
SQL> host cp /home/oracle/app/oracle/fast_recovery_area/orcl/control02.ctl
/home/oracle/app/oracle/fast_recovery_area/orcl/control03.ctl
SQL> STARTUP
SQL> show parameter control_files;
NAME                                 TYPE        VALUE
------------------------------------ ----------- --------------------
control_files                        string      /home/oracle/app/oracle/oradata/orcl/control01.ctl,
/home/oracle/app/oracle/fast_recovery_area/orcl/control02.ctl,
/home/oracle/app/oracle/fast_recovery_area/orcl/control03.ctl
```

> 注意，本示例中的命令"host cp"表示调用 Linux 的 cp 命令复制文件，文件复制成功后，startup 打开数据库成功，再查询参数 control_files 的值，就有 3 个控制文件了。

除了多文件保证机制之外，还可以通过 rman 备份控制文件。见第 13 章备份与恢复。

Oracle 还允许当其他方法无法恢复控件文件的时候，通过重建控制文件的方式重建新的控制文件。重建控制文件的命令是"CREATE CONTROLFILE …"，该命令非常复杂，不容易编写，但是我们可以在数据库正常运行期间跟踪并导出控制文本的重建命令脚本。

【示例 6-9】导出控制文件的重建命令脚本

本示例将生成一个控制文件的跟踪文件 control_trace.trc，该文件是文本文件，里面有重建控制文件的命令。生成后使用 Linux 的 cat 命令查看该文件的内容。

```
$ sqlplus / as sysdba
SQL> ALTER DATABASE BACKUP CONTROLFILE TO TRACE
2 AS '/home/oracle/control_trace.trc';
Database altered.
```

```
SQL> !cat /home/oracle/control_trace.trc
...
```

只要保存好这个文本文件,也能恢复控制文件。只是要注意两点,一是如果数据库发生了结构性的改变,需要重新生成这个文件,以反映数据库最新情况;二是通过这种方式恢复的控制文件可能会丢失最新的数据。

6.7 重做日志文件与归档日志文件

重做日志(Redo Log)用于保存数据库的所有变化信息,归档日志(Archive Log)是重做日志的持久性存储,是可选的,数据库只有工作在归档日志的模式下,才会使用归档日志文件。Oracle 12c 仅由 CDB 管理重做日志,PDB 不管理重做日志。

6.7.1 重做日志与归档日志的基本概念

重做日志文件记录了对数据库修改的信息,包括用户对数据修改和数据库管理员对数据库结构的修改。在发生数据故障的时候,利用重做日志文件和数据库备份文件共同恢复数据库,数据故障一般有两种情况:介质损坏和用户误操作。

每个数据库至少有两个日志文件组,每组由 1 个或者多个日志文件组成,同一组中的多个日志文件互为备份(即文件大小和内容完全相同),这样的设计是为了在日志文件组内某个日志文件损坏后及时提供备份,所以同一组的日志文件一般应当存放在不同磁盘上,如图 6-3 所示,在线重做日志文件组#1 由两个文件 Redo_A1 和 Redo_B1 组成,这两个文件互为备份,内容相同,但存储在不同的磁盘中。

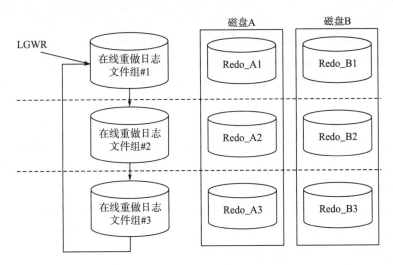

图 6-3 重做日志写入过程与重做日志文件组的结构

联机重做日志文件组是循环使用的。当第一个重做日志文件组达到一定数量时,就会停止写入而转向第二个重做日志文件组;第二个满转向第三个;第三个满就向第一个写入,循环往复。所以重做日志文件组中的信息是会被覆盖的,是暂时存放的。由于数据库恢复需要从重做日志文件组中读取重做记录,因此数据库的恢复能力也是有限的,被覆盖的重做数据是无法恢复的。

Oracle 可以设置为归档模式,在归档模式下,重做日志中的信息在被覆盖之前会被自动转存到归档日志文件中,这样做的代价是增加了数据库的写操作,更多地消耗存储空间,但是换来的好处就是多存储了重做数据,提高了数据恢复能力。

为了提高数据库的恢复能力,可以增加重做日志组,或者增加重做日志文件的大小。归档日志文件使用的最大磁盘空间也是受限的,由系统参数 db_recovery_file_dest_size 的值决定,增加这个参数的值,也可以提高数据恢复能力。

【示例 6-10】查看重做日志文件

本例通过 v$log 查看日志文件组,通过 v$logfile 查看日志文件,从查询结果来看,共有三个重做日志文件组,每个组仅有一个文件。

```
$ sqlplus / as sysdba
SQL> SELECT group#, archived, status FROM v$log;
    GROUP# ARC STATUS
---------- --- ----------------
         1 YES INACTIVE
         2 YES INACTIVE
         3 NO  CURRENT
SQL> col member format a50
SQL> SELECT group#, member FROM v$logfile;
    GROUP# MEMBER
---------- --------------------------------------------------
         3 /home/oracle/app/oracle/oradata/orcl/redo03.log
         2 /home/oracle/app/oracle/oradata/orcl/redo02.log
         1 /home/oracle/app/oracle/oradata/orcl/redo01.log
```

联机重做日志文件组有 6 种状态(STATUS):INACTIVE、ACTIVE、CURRENT、UNUSED、CLEARING 和 CLEARING_CURRNT。

UNUSED:表示该联机重做日志文件组对应的文件还从未被写入过数据,通常刚刚创建的联机重做日志文件组会显示成这一状态。当日志切换到这一组时,就会改变状态。

CURRENT:表示当前正在使用的日志文件组,并且是活动的。

ACTIVE:表示该组是活动的但不是当前组,实例恢复时需要这组日志。如果处于这一状态,表示虽然当前并未使用,该文件中内容尚未归档,或者没有全部写入数据文件,一旦需要实例恢复,必须借助该文件中保存的内容。

INACTIVE:表示实例恢复已不再需要这组联机重做日志组了。表示对应的联机重做日志文件中的内容已被妥善处理,该组联机重做日志当前处于空闲状态。

CLEARING:表示该组重做日志文件正被重建(重建后状态会变成 UNUSED)。

CLEARING_CURRENT:表示该组重做日志重建时出现错误。

【示例 6-11】查看归档模式及归档日志文件所在目录

本示例通过 ARCHIVE LOG LIST 查看数据库是否工作在归档模式，通过参数 db_recover 查看归档文件最大占用空间大小以及归档目录，通过 v$recovery_file_dest 查询归档空间的使用情况。

```
SQL> ARCHIVE LOG LIST
Database log mode               Archive Mode
Automatic archival              Enabled
Archive destination             USE_DB_RECOVERY_FILE_DEST
Oldest online log sequence      274
Next log sequence to archive    276
Current log sequence            276
SQL> show parameter db_recover
NAME                                 TYPE        VALUE
------------------------------------ ----------- --------------------
db_recovery_file_dest                string
/home/oracle/app/oracle/fast_recovery_area
db_recovery_file_dest_size           big integer 4560M
SQL> col name format a50
SQL> SELECT NAME, SPACE_LIMIT, SPACE_USED FROM v$recovery_file_dest;
NAME                                           SPACE_LIMIT SPACE_USED
---------------------------------------------- ----------- ----------
/home/oracle/app/oracle/fast_recovery_area     4781506560  212244992
SQL>
```

Oracle 默认是在非归档模式下运行的，要将数据库切换到归档模式下工作，可以使用命令"ALTER DATABASE ARCHIVELOG；"，如果想切换回到非归档模式，可以使用命令"ALTER DATABASE NOARCHIVELOG；"。

【示例 6-12】将数据库切换到归档模式。

必须以 sysdba 身份登录到 CDB，不能在 PDB 数据库中操作。登录后，首先查询一下当前的归档日志方式和参数：

```
$ sqlplus / as sysdba
...
SQL> ARCHIVE LOG LIST;
Database log mode               No Archive Mode
Automatic archival              Disabled
Archive destination             USE_DB_RECOVERY_FILE_DEST
Oldest online log sequence      192
Current log sequence            194
```

当前的归档日志方式是非归档模式"No Archive Mode"，自动归档为"Disabled"。

在设置为归档模式之前必须关闭数据库，再以"STARTUP MOUNT"的方式开启数据库，设置完归档模式后，再打开数据库"ALTER DATABASE OPEN；"。

```
SQL> SHUTDOWN IMMEDIATE
...
SQL> STARTUP MOUNT
...
SQL> ALTER DATABASE ARCHIVELOG;
Database altered.
SQL> archive log list;
Database log mode              Archive Mode
Automatic archival             Enabled
Archive destination            USE_DB_RECOVERY_FILE_DEST
Oldest online log sequence     192
Next log sequence to archive   194
Current log sequence           194
SQL> ALTER DATABASE OPEN;
Database altered.
```

设置完成后，当前的归档日志方式是归档模式"Archive Mode"，自动归档为"Enabled"。归档日志文件存储的位置是由 Archive Destination 参数决定的，这里的值是 USE_DB_RECOVERY_FILE_DEST，这是另一个参数，它的值是目录：/home/oracle/app/oracle/fast_recovery_area，可以定期观察这个目录中的归档日志文件。

> 注意：数据库处于归档模式后，可以存储更多的重做数据，数据库可以恢复到更早的时间点的数据。但也会增大服务器的资源开销。

6.7.2 重做日志组管理

重做日志组的日常管理包括在重做日志文件组中增加日志文件，删除日志文件组，删除日志文件等。

【示例 6-13】在重做日志文件组中增加文件

本示例通过命令"ALTER DATABASE ADD LOGFILE MEMBER …"在重做日志文件组 1 中增加一个日志文件 redo01_2.log，增强该文件组的可靠性。增加完成后，文件组 1 就有两个互为备份的文件了。

```
$ sqlplus / as sysdba
SQL> COL member FORMAT a50
SQL> SELECT group#, member FROM v$logfile;
    GROUP# MEMBER
---------- --------------------------------------------------
         3 /home/oracle/app/oracle/oradata/orcl/redo03.log
```

```
            2 /home/oracle/app/oracle/oradata/orcl/redo02.log
            1 /home/oracle/app/oracle/oradata/orcl/redo01.log
    SQL> ALTER DATABASE ADD LOGFILE MEMBER
'/home/oracle/app/oracle/oradata/orcl/redo01_2.log' TO GROUP 1;
    Database altered.
    SQL> SELECT group#, member FROM v$logfile;
        GROUP# MEMBER
    ---------- --------------------------------------------------
            3 /home/oracle/app/oracle/oradata/orcl/redo03.log
            2 /home/oracle/app/oracle/oradata/orcl/redo02.log
            1 /home/oracle/app/oracle/oradata/orcl/redo01.log
            1 /home/oracle/app/oracle/oradata/orcl/redo01_2.log
    SQL>
```

【示例6-14】增加重做日志文件组

本示例增加一个日志文件组4，在文件组中分配两个日志文件，大小为50M。增加重做日志组成功的必要条件是这两个日志文件都不存在。

```
    $ sqlplus / as sysdba
    SQL> COL member FORMAT a50
    SQL> SELECT group#, member FROM v$logfile;
        GROUP# MEMBER
    ---------- --------------------------------------------------
            3 /home/oracle/app/oracle/oradata/orcl/redo03.log
            2 /home/oracle/app/oracle/oradata/orcl/redo02.log
            1 /home/oracle/app/oracle/oradata/orcl/redo01.log
            1 /home/oracle/app/oracle/oradata/orcl/redo01_2.log
    SQL> ALTER DATABASE ADD LOGFILE GROUP 4
('/home/oracle/app/oracle/oradata/orcl/redo04.log',
'/home/oracle/app/oracle/oradata/orcl/redo04_2.log')SIZE 50M;
    SQL> SELECT group#, member FROM v$logfile;
        GROUP# MEMBER
    ---------- --------------------------------------------------
            3 /home/oracle/app/oracle/oradata/orcl/redo03.log
            2 /home/oracle/app/oracle/oradata/orcl/redo02.log
            1 /home/oracle/app/oracle/oradata/orcl/redo01.log
            1 /home/oracle/app/oracle/oradata/orcl/redo01_2.log
            4 /home/oracle/app/oracle/oradata/orcl/redo04.log
            4 /home/oracle/app/oracle/oradata/orcl/redo04_2.log
    6 rows selected.
```

如果要删除文件组，可以使用命令"ALTER DATABASE DROP LOGFILE GROUP 组号；"，前提是被删除组的状态不是 current。删除重做日志文件组后，还需要手工删除日志组的文件。

6.8 参数文件

Oracle 参数文件存储 Oracle 的系统参数，包括名称定义参数、静态参数和动态参数等。数据库参数文件分为初始参数文件 pfile（Initialization Parameter File）和服务器参数文件：spfile（Server-Side Parameter File）。参数文件存储在目录$ORACLE_HOME/dbs 中。

pfile 是文本文件，对 Oracle 来说是只读的，需要在 Oracle 外部的操作系统中修改；pfile 中的参数值修改后，只能重启数据库才有效，因而全部都是静态的。pfile 文件的文件名是 init$ORACLE_SID.ora（每个数据库实例一个）和 init.ora，init$ORACLE_SID.ora 优先，如果没有 init$ORACLE_SID.ora 才会使用 init.ora。

spfile 是二进制文件，对 Oracle 来说是可读写的，通过命令"ALTER SYSTEM SET 参数名=参数值；"方式修改，因为是二进制格式，所以不能在 Oracle 外部修改。spfile 中的参数值可以通过 Oracle 内部命令修改，有些参数修改之后立即生效，称为动态参数，有一些必须重启数据库才生效，称为静态参数。spfile 的文件名是 spfile$ORACLE_SID.ora（每个数据库实例一个）和 spfile.ora，spfile$ORACLE_SID.ora 优先，如果没有 spfile$ORACLE_SID.ora 才会使用 spfile.ora。

当数据库实例启动时，系统只需要从 4 个参数文件中的一个参数文件中读取参数值，读取参数文件的顺序是：spfile$ORACLE_SID.ora，spfile.ora，init$ORACLE_SID.ora 和 init.ora。

Oracle 的部分参数可以在 PDB 中单独设置。如果在 PDB 中设置了，就以 PDB 中设置的参数为准，如果没有在 PDB 中设置，就采用 CDB 中的参数值。可以通过 v$pdbs 和 pdb_spfile$查看 PDB 以及 PDB 的服务器参数设置。

【示例 6-15】查看所有参数值

本示例通过命令"SHOW PARAMETER SPFILE"查看参数 spfile 的值，该参数的值表示参数文件的全路径名称，通过"SHOW PARAMETER"查看所有参数的值。

```
SQL> SHOW PARAMETER SPFILE
NAME                                 TYPE        VALUE
------------------------------------ ----------- --------------------
Spfile                               string
/home/oracle/app/oracle/product/12.1.0/dbhome_1/dbs/spfileorcl.ora
SQL> SHOW PARAMETER
...
```

6.8.1 修改 spfile 参数值

修改 spfile 参数值的命令是：ALTER SYSTEM SET parameter=value [CONTAINER=ALL | CURRENT] [SCOPE=MEMORY|SPFILE|BOTH]

其中：

CONTAINER=ALL：对所有数据库起作用。

CONTAINER=CURRENT：仅修改当前数据库的参数，对其他数据库不起作用。

SCOPE=MEMORY：修改后当前实例起作用，重启数据库不起作用。

SCOPE=SPFILE：修改后当前实例不起作用，下次重启才起作用，只允许动态变量。

SCOPE=BOTH：修改后当前实例起作用，下次重启也起作用，只允许动态变量。

当 Oracle 在以 spfile 参数启动时，默认值是 BOTH，而以 pfile 启动时，默认值是 MEMORY，也就是只修改当前值。所以如果不能确定启动的参数文件，可以用 SHOW PARAMETER SPFILE 查看。

如果 startup 命令不带参数启动，默认调用 spfile 参数文件，指定以初始参数文件 pfile 启动数据库的命令是：

startup pfile='/$ORACLE_HOME/dbs/init$ORACLE_SID.ora'

> 注意：不能以指定服务器参数文件 spfile 来启动数据库，可以先将 spfile 转换成 pfile，再用这个命令来达到目的。

【示例 6-16】修改 CDB 的系统参数

本例修改参数 db_recovery_file_dest_size 的值，从 4560M 增加到 4600M。在 CDB 中修改该参数，会影响整个 PDB。

```
$ sqlplus / as sysdba
SQL> SHOW PARAMETER db_recovery_file_dest_size ;
NAME                                 TYPE        VALUE
------------------------------------ ----------- ------------------
db_recovery_file_dest_size           big integer 4560M
SQL> ALTER SYSTEM SET db_recovery_file_dest_size=4600M scope=both;
System altered.
SQL> show parameter db_recovery_file_dest_size ;
NAME                                 TYPE        VALUE
------------------------------------ ----------- ------------------
db_recovery_file_dest_size           big integer 4600M
```

6.8.2 从 spfile 创建 pfile

有时候我们希望直接查看 spfile 二进制参数文件中的参数，可以通过"CREATE PFILE"将 spfile 转存为 pfile。pfile 文本文件可作为 startup 命令的启动参数文件。注意 pfile 中只包含 CDB 的部分参数，不包含 PDB 参数。

【示例 6-17】从 spfile 创建 pfile

本例从 spifle 中创建 pfile。新创建的文件是文本文件，可以在操作系统中直接查看，这里用的是 Linux 的 cat 命令。

```
$ sqlplus / as sysdba
SQL> CREATE PFILE='pfilesid.ora' FROM spfile;
File created.
SQL> !cat $ORACLE_HOME/dbs/pfilesid.ora
orcl.__data_transfer_cache_size=0
orcl.__db_cache_size=922746880
...
```

练习

一、判断题

1. 一个表空间只能由一个数据文件组成。　　　　　　　　　　　　　　　（　）
2. 一个数据文件只能属于一个表空间。　　　　　　　　　　　　　　　　（　）
3. 一个重做日志文件组中可以有多个文件，这些文件内容互不相同。　　　（　）
4. 归档日志文件只有在归档模式下才会生成。　　　　　　　　　　　　　（　）
5. 修改系统参数不允许从 PDB 内部操作，只能在 CDB 上操作。　　　　　（　）
6. 服务器参数文件 spfile 是文本类型的文件。　　　　　　　　　　　　　（　）

二、单项选择题

1. 查询表空间数据文件的视图是（　　）
   A. dba_tablespaces　　　　　　　　　B. dba_free_space
   C. dba_data_files　　　　　　　　　　D. dba_temp_files
2. 删除表空间 users02 及其数据文件的正确的命令是（　　）
   A. DROP TABLESPACE users02；
   B. DROP TABLESPACE users02 INCLUDING CONTENTS；
   C. DROP TABLESPACE users02 INCLUDING DATAFILES；
   D. DROP TABLESPACE users02 INCLUDING CONTENTS AND DATAFILES；
3. 控制文件在数据库启动的（　　）阶段被读取的
   A. STARTUP MOUNT；　　　　　　　B. STARTUP NOMOUNT；
   C. STARTUP FORCE；　　　　　　　D. STARTUP；
4. 将数据库切换到归档日志模式的命令是（　　）
   A. ALTER DATABASE ARCHIVELOG；
   B. ALTER DATABASE NOARCHIVELOG；
   C. ALTER DATABASE OPEN；
   D. ARCHIVE LOG LIST；

第 6 章 数据库存储管理

5. 修改 spfile 参数值的命令是 ALTER SYSTEM SET parameter=value[SCOPE=MEMORY|SPFILE|BOTH]，其中，修改后只是当前示例起作用，重启数据库就不起作用的选项是（　　）

A. SCOPE=MEMORY　　　　　　　B. SCOPE=SPFILE
C. SCOPE=BOTH　　　　　　　　 D. 不需要 SCOPE 选项

三、填空题

1. 表空间的状态有两种：在线和离线，其中在线又分为只读和_____两种。
2. 表空间大小的计算方法是它包含的数据文件大小_____。
3. 不管是 CDB 还是 PDB 都有_____表空间、_____表空间和临时表空间。
4. 多路复用控制文件的设置是由系统参数_____设置的。可以通过修改该参数的值维护控制文件的动态备份。
5. 数据库参数文件分为初始化参数文件 pfile 和_____参数文件 spfile。

四、简答题

1. 表空间的作用是什么？它和数据文件的关系是什么？
2. 表空间有哪些类型？每个类型的作用是什么？
3. 重做日志文件组和归档日志文件的作用是什么？

训练任务

1. 新建一个用户类型表空间

培养能力	工程能力、设计/开发解决方案、项目管理		
掌握程度	★★★★	难度	中
结束条件	表空间创建成功，在表空间上创建一个测试表		
训练内容： 根据现有服务器的磁盘具体情况以及项目的数据存储空间要求，创建一个用户类型的表空间，供项目存储数据。表空间创建成功之后，在表空间上创建一些测试表，并往表中插入大量数据，看表空间能否容纳这些数据。在创建表空间的时候要注意： (1) 表空间名称要直观、合理。 (2) 合理设置参数：数据文件的数量、文件名称和大小、存储磁盘及目录、数据文件的空间是否自动扩展			

2. 在 CDB/PDB 中分别设置系统参数

培养能力	工程能力、问题分析、环境和可持续发展		
掌握程度	★★★	难度	中
结束条件	参数设置成功，并查看 PDB 中所有参数值		
训练内容： 在 CDB 根数据库和 pdborcl 插接式数据库中分别设置各自的 open_cursors 参数值，并查看 pdborcl 中单独设置的所有参数值			

第 7 章 用户及权限管理

本章目标

知识点	理解	掌握	应用
1.Oracle 权限的分类		✓	
2.Oracle 角色的分类	✓	✓	
3.管理用户	✓	✓	✓
4.概要文件的功能	✓	✓	

训练任务

● 掌握角色、权限、用户的管理能力，并在用户之间共享对象。

知识能力点

能力点＼知识点	知识点 1	知识点 2	知识点 3	知识点 4
工程知识	✓	✓		
问题分析	✓	✓	✓	✓
设计/开发解决方案			✓	✓
研究	✓	✓		
使用现代工具			✓	✓
工程与社会	✓	✓		
环境和可持续发展			✓	✓
职业规范				✓
个人和团队	✓	✓	✓	
沟通				
项目管理	✓	✓		✓
终身学习	✓	✓		✓

7.1 权　　限

权限是指用户访问数据库系统和资源的能力。Oracle 权限分为系统权限(System Privileges)和对象权限(Object Privileges)。

7.1.1 系统权限

系统权限是指在系统级控制数据库的存取和使用的机制，即执行某种 SQL 语句的能力。如是否能启动、停止数据库，是否能修改数据库参数，是否能连接到数据库，是否能创建、删除、更改方案对象(如表、索引、视图、过程)等。它是针对某一类方案对象或非方案对象的某种操作的全局性能力。视图 system_privilege_map 中包括了 Oracle 数据库中的所有系统权限，常用的系统权限如表 7-1 所示。

表 7-1　Oracle 系统权限一览表

类型/系统权限	说明
群集权限	
CREATE CLUSTER	在自己的方案中创建、更改和删除群集
CREATE ANY CLUSTER	在任何方案中创建群集
ALTER ANY CLUSTER	在任何方案中更改群集
DROP ANY CLUSTER	在任何方案中删除群集
数据库权限	
ALTER DATABASE	运行 ALTER DATABASE 语句，更改数据库的配置
ALTER SYSTEM	运行 ALTER SYSTEM 语句，更改系统的初始化参数
AUDIT SYSTEM	运行 AUDIT SYSTEM 和 NOAUDIT SYSTEM 语句，审计 SQL
AUDIT ANY	运行 AUDIT 和 NOAUDIT 语句，对任何方案的对象进行审计
索引权限	
CREATE ANY INDEX	在任何方案中创建索引 注意：没有 CREATE INDEX 权限，CREATE TABLE 权限包含了 CREATE INDEX 权限
ALTER ANY INDEX	在任何方案中更改索引
DROP ANY INDEX	任何方案中删除索引
过程权限	
CREATE PROCEDURE	在自己的方案中创建过程、函数和包
CREATE ANY ROCEDURE	在任何方案中创建过程、函数和包
ALTER ANY PROCEDURE	在任何方案中更改过程、函数和包

续表

类型/系统权限	说明
DROP ANY PROCEDURE	在任何方案中删除过程、函数或包
EXECUTE ANY PROCEDURE	在任何方案中执行或者引用过程
概要文件权限	
CREATE PROFILE	创建概要文件
ALTER PROFILE	更改概要文件
DROP PROFILE	删除概要文件
角色权限	
CREATE ROLE	创建角色
ALTER ANY ROLE	更改任何角色
DROP ANY ROLE	删除任何角色
GRANT ANY ROLE	向其他角色或用户授予任何角色 注意：没有对应的 REVOKE ANY ROLE 权限
回退段权限	
CREATE ROLLBACK SEGMENT	创建回退段 注意：没有对撤销段的权限
ALTER ROLLBACK SEGMENT	更改回退段
DROP ROLLBACK SEGMENT	删除回退段
序列权限	
CREATE SEQLENCE	在自己的方案中创建、更改、删除和选择序列
CREATE ANY SEQUENCE	在任何方案中创建序列
ALTER ANY SEQUENCE	在任何方案中更改序列
DROP ANY SEQUENCE	在任何方案中删除序列
SELECT ANY SEQUENCE	在任何方案中从任何序列中进行选择
会话权限	
CREATE SESSION	创建会话，登录进入(连接到)数据库
ALTER SESSION	运行 ALTER SESSION 语句，更改会话的属性
ALTER RESOURCE COST	更改概要文件中的计算资源消耗的方式
RESTRICTED SESSION	在数据库处于受限会话模式下连接到数据
同义词权限	
CREATE SYNONYM	在自己的方案中创建、删除同义词

续表

类型/系统权限	说明
CREATE ANY SYNONYM	在任何方案中创建专用同义词
CREATE PUBLIC SYNONYM	创建公共同义词
DROP ANY SYNONYM	在任何方案中删除同义词
DROP PUBLIC SYNONYM	删除公共同义词
表权限	
CREATE TABLE	在自己的方案中创建、更改和删除表
CREATE ANY TABLE	在任何方案中创建表
ALTER ANY TABLE	在任何方案中更改表
DROP ANY TABLE	在任何方案中删除表
COMMENT ANY TABLE	在任何方案中为任何表、视图或者列添加注释
SELECT ANY TABLE	在任何方案中选择任何表中的记录
INSERT ANY TABLE	在任何方案中向任何表插入新记录
UPDATE ANY TABLE	在任何方案中更改任何表中的记录
DELETE ANY TABLE	在任何方案中删除任何表中的记录
LOCK ANY TABLE	在任何方案中锁定任何表
FLASHBACK ANY TABLE	允许使用 AS OF 子句对任何方案中的表、视图执行一个 SQL 语句的闪回查询
表空间权限	
CREATE TABLESPACE	创建表空间
ALTER TABLESPACE	更改表空间
DROP TABLESPACE	删除表空间，包括表、索引和表空间的群集
MANAGE TABLESPACE	管理表空间，使表空间处于 ONLINE（联机）、OFFLINE（脱机）、BEGIN BACKUP（开始备份）、END BACKUP（结束备份）状态
UNLIMITED ABLESPACE	不受配额限制地使用表空间，只能将 UNLIMITED TABLESPACE 授予账户而不能授予角色
用户权限	
CREATE USER	创建用户
ALTER USER	更改用户
BECOME USER	当执行完全装入时，成为另一个用户
DROP USER	删除用户
视图权限	
CREATE VIEW	在自己的方案中创建、更改和删除视图

续表

类型/系统权限	说明
CREATE ANY VIEW	在任何方案中创建视图
DROP ANY VIEW	在任何方案中删除视图
COMMENT ANY TABLE	在任何方案中为任何表、视图或者列添加注释
FLASHBACK ANY TABLE	允许使用 AS OF 子句对任何方案中的表、视图执行一个 SQL 语句的闪回查询
触发器权限	
CREATE TRIGGER	在自己的方案中创建、更改和删除触发器
CREATE ANY TRIGGER	在任何方案中创建触发器
ALTER ANY TRIGGER	在任何方案中更改触发器
DROP ANY TRIGGER	在任何方案中删除触发器
ADMINISTER DATABASE TRIGGER	允许创建 ON DATABASE 触发器。在能够创建 ON DATABASE 触发器之前，还必须先拥有 CREATE TRIGGER 或 CREATE ANY TRIGGER 权限
专用权限	
SYSOPER（系统操作员权限）	STARTUP SHUTDOWN ALTER DATABASE MOUNT/OPEN ALTER DATABASE BACKUP CONTROLFILE ALTER DATABASE BEGIN/END BACKUP ALTER DATABASE ARCHIVELOG RECOVER DATABASE RESTRICTED SESSION CREATE SPFILE/PFILE
SYSDBA（系统管理员权限）	SYSOPER 的所有权限，并带有 WITH ADMIN OPTION 子句 CREATE DATABASE RECOVER DATABASE UNTIL
其他权限	
ANALYZE ANY	对任何方案中的任何表、群集或者索引执行 ANALYZE 语句
GRANT ANY OBJECT PRIVILEGE	授予任何方案上的任何对象上的对象权限，注意：没有对应的 REVOKE ANY OBJECT PRIVILEGE
GRANT ANY PRIVILEGE	授予任何系统权限，注意：没有对应的 REVOKE ANY PRIVILEGE
SELECT ANY ICTIONARY	允许从 sys 用户所拥有的数据字典表中进行选择

7.1.2 对象权限

Oracle 数据库的对象主要是指：表、索引、视图、序列、同义词、过程、函数、包以及触发器。创建对象的用户拥有该对象的所有对象权限，不需要授予。所以，对象权限的设置实际上是对象的所有者授予其他用户操作该对象的某种权利。不同类型的对象有不同的对象权限，见表 7-2。

表 7-2　对象权限分类表

对象权限	说明
ALTER	允许被授权者修改对象
DELETE	允许被授权者删除对象
EXECUTE	允许被授权者执行过程(Procedure)、操作符(Operator)或者类型(Type)
INDEX	允许被授权者在表上创建索引或者锁定该表
INSERT	允许被授权者将数据插入表或者视图，授权可以精确到表的列
READ	只能在目录上授予。允许被授权者读取指定目录中的 BFILE
REFERENCES	允许被授权者创建引用该表的参照完整性约束，只能授予用户，而不能授予角色
SELECT	允许被授权者在表、视图或者序列上执行 SELECT 语句。不能授予列的访问权限
UPDATE	允许被授权者更改表或者视图中的数据值
ON COMMIT REFRESH	允许被授权者在实体化视图的提交时立即刷新的权限
QUERY REWRITE	允许被授权者在实体化视图的查询时重写的权限
DEBUG	允许被授权者调试过程的权限
FLASHBACK	允许被授权者有闪回的权限

一个典型的授权命令是："GRANT SELECT，UPDATE ON table1 TO user1；"，表示将表 table1 的 SELECT 和 UPDATE 权限授予用户 user1。

7.2　角　　色

如果有一组用户，他们的所需的权限是一样的，当对他们的权限进行管理的时候要执行很多重复的授权命令，会很不方便。有一个很好的解决办法就是：角色。角色(Role)是一组权限的集合，将角色赋给一个用户，这个用户就拥有了这个角色中的所有权限，对角色进行维护，会影响到角色的所有用户。

Oracle 12c 的角色分为公共角色(Common Role)和本地角色(Local Role)。在 CDB 中只能创建公共角色，在 PDB 中只能创建本地角色。Oracle 12c 规定，公共角色的名称必须以 C##三个字母开头。公共角色存在于所有容器(根 CDB 和所有的 PDB)中，本地角色只在特定的 PDB 中存在，同样的本地角色名称可以在多个 PDB 中存在，但它们之间没有关系。

7.2.1 系统预定义角色

在安装数据库时，安装过程会调用$ORACLE_HOME/rdbms/admin 目录中的 sql.bsq、sql 文件用于创建 Oracle 的内置角色。Oracle 的内置角色都是公共角色。可以通过 role_sys_privs 查询内置角色名称以及角色包含的系统权限。

【示例 7-1】查询所有内置角色以及角色包含的系统权限

```
$ sqlplus / as sysdba
SQL> COL role FORMAT a20
SQL> COL privilege FORMAT a20
SQL> COL common FORMAT a7
SQL> SELECT * FROM role_sys_privs;
…
524 rows selected.
SQL> SELECT * FROM role_sys_privs WHERE  role='CONNECT';
ROLE                 PRIVILEGE            ADM COMMON
-------------------- -------------------- --- -------
CONNECT              SET CONTAINER        NO  YES
CONNECT              CREATE SESSION       NO  YES
SQL>
```

通过这个示例可以看出，CONNECT 角色包含两个权限"SET CONTAINER"和"CREATE SESSION"。并且都是公共的角色，这是因为 COMMON=YES。

> 注意：可以添加其他系统权限到一个内置角色中，甚至也可以添加其他角色到内置角色中。但不能删除内置角色。

比较常用的预定义角色有：dba、connect、resource 等。其中 dba 角色拥有全部特权，是系统最高权限，只有 dba 才可以创建数据库结构。resource 角色具有应用开发人员所需要的权限，比如建立存储过程，触发器等，拥有 resource 角色的用户只可以创建实体，不可以创建数据库结构。拥有 connect 权限的用户只可以登录 Oracle，不可以创建实体和数据库结构。一般来说对于普通用户授予 connect，resource 权限就够了。典型的授权一个普通用户的命令是："GRANT connect，resource TO user1；"。

7.2.2 创建公共角色

除了预定义角色之外，还可以创建用户角色。用户角色分为适合于 CDB 和所有 PDB 的公共角色和只适用于一个 PDB 的本地角色。通过自定义角色，可以自由组合访问权限。对于角色的管理，通常有以下 4 个命令。

1) 创建角色：命令格式是

```
CREATE ROLE role_name [IDENTIFIED BY password]
```

其中 password 是角色的密码。新创建的角色没有包含任何权限和其他角色，是一个空的角色，必须通过授权命令(GRANT)将其他权限或者角色赋予新角色。如果要收回赋予的权限，必须使用 REVOKE 命令。

2) 授权：命令格式是

```
GRANT 权限或者角色 TO 被授权的角色
[WITH ADMIN OPTION][CONTAINER=ALL | CURRENT]
```

其中权限或者角色可以是多个，用逗号分隔。

WITH ADMIN OPTION：可把此角色授予其他用户或角色，当收回这个权限时，对已经授予其他用户或角色的权限不会因传播无效。

CONTAINER=CURRENT：只授权给当前的容器，这是默认选项。

CONTAINER=ALL：授权给所有 CDB 和 PDB。

3) 收回权限：命令格式是

```
REVOKE 用逗号分隔的权限或者角色 FROM 角色 [CONTAINER=ALL | CURRENT]
```

4) 删除角色：命令格式是

```
DROP ROLE 角色名
```

Oracle 提供了以下视图查询角色及其包含的权限：

dba_roles：查询所有角色。

dba_sys_privs：查询授予用户和其他角色的系统权限。

dba_role_privs：查询授予用户和其他角色的角色。

dba_tab_privs：查询数据库中对象的所有授权。

【示例 7-2】创建自定义公共角色 c##con_res

本示例创建新角色 c##con_res，并给它分配一些系统权限和角色。这个新角色实际具有了系统角色 connect 和 resource 的权限之和。其中 set container，create session 权限和 em_express_all 角色属于 connect 角色，其他属于 resource 角色。创建完成后，通过 dba_sys_privs 和 dba_role_privs 查询 c##con_res 包含的所有权限和角色。注意，where 后面的'C##CON_RES'要大写。另外，由于是创建公共角色，所以本示例必须在 CDB 中执行。

```
$ sqlplus / as sysdba
SQL> CREATE ROLE c##con_res ;
Role created.
SQL> GRANT SET CONTAINER, CREATE SESSION, CREATE TABLE, CREATE CLUSTER,
  2  CREATE OPERATOR, CREATE INDEXTYPE, CREATE TYPE, CREATE SEQUENCE,
  3  CREATE TRIGGER, CREATE PROCEDURE
  4  TO c##con_res container = all;
Grant succeeded.
SQL> GRANT EM_EXPRESS_ALL TO c##con_res CONTAINER=ALL;
Grant succeeded.
SQL> COL common FORMAT a7
SQL> SELECT * FROM dba_sys_privs WHERE  grantee='C##CON_RES';
GRANTEE    PRIVILEGE                                ADM COMMON
```

```
---------- ---------------------------------------- --- ---
C##CON_RES CREATE TABLE                              NO  YES
C##CON_RES CREATE OPERATOR                           NO  YES
C##CON_RES CREATE TYPE                               NO  YES
C##CON_RES SET CONTAINER                             NO  YES
C##CON_RES CREATE CLUSTER                            NO  YES
C##CON_RES CREATE TRIGGER                            NO  YES
C##CON_RES CREATE SESSION                            NO  YES
C##CON_RES CREATE INDEXTYPE                          NO  YES
C##CON_RES CREATE PROCEDURE                          NO  YES
C##CON_RES CREATE SEQUENCE                           NO  YES

10 rows selected.
SQL> col grantee format a20
SQL> SELECT * FROM dba_role_privs WHERE grantee='C##CON_RES';
GRANTEE     GRANTED_ROLE     ADM DEL DEF   COMMON
----------- ---------------- ---- --- ----- ----
C##CON_RES  EM_EXPRESS_ALL   NO  NO  YES   YES
```

上述查询之后，可以看出 c##con_res 内部原来是公共的权限仍然是公共权限（COMMON=YES），这是因为在 GRANT 授权的时候使用了选项"CONTAINER=ALL"，否则角色中的权限就只是本地权限了，在其他 PDB 中不能起作用。紧接上面的操作，下面在 CDB 和 PDB 中分别测试 COMMON(公共)特性：

```
SQL> COL oracle_maintained FORMAT a20
SQL> COL role FORMAT a20
SQL> SELECT * FROM dba_roles WHERE role='C##CON_RES';
ROLE              PASSWORD AUTHENTICAT COMMON ORACLE_MAINTAINED
----------------- -------- ----------- ------ ---------------------
C##CON_RES        NO       NONE        YES    N
SQL>ALTER SESSION SET CONTAINER=pdborcl;
Session altered.
SQL> SELECT * FROM dba_roles WHERE role='C##CON_RES';
ROLE              PASSWORD AUTHENTICAT COMMON ORACLE_MAINTAINED
----------------- -------- ----------- ------ ---------------------
C##CON_RES        NO       NONE        YES    N
SQL> SELECT * FROM dba_role_privs WHERE grantee='C##CON_RES';
GRANTEE     GRANTED_ROLE     ADM DEL DEF   COMMON
----------- ---------------- ---- --- ----- ----
C##CON_RES  EM_EXPRESS_ALL   NO  NO  YES   YES
```

从查询 dba_roles 可以看出，c##con_res 是公共角色（COMMON=YES），所以在所有数据库中都可以查询得到，在 PDB 中也能够查询到公共角色中包含的公共权限和角色。

7.2.3 创建本地角色

与创建自定义公共角色相似，可以创建自定义的本地角色。不能在 CDB 中创建本地角色，本地角色的适用范围仅仅是创建它的 PDB。所以在创建本地角色的时候，不能有选项 CONTAINER=ALL。本地角色命名的时候不能在前面加 c##。

【示例 7-3】 创建自定义本地角色 con_res

本示例在 PDBORCL 中创建一个本地角色 con_res，将 connect 和 resource 两个角色赋予它。

```
$ sqlplus system/***@pdborcl
SQL> CREATE ROLE con_res;
Role created.
SQL> GRANT connect,resource TO con_res;
Grant succeeded.
SQL> COL grantee FORMAT a20
SQL> COL granted_role FORMAT a20
SQL> COL common FORMAT a7
SQL> COL role FORMAT a20
SQL> COL oracle_maintained FORMAT a15
SQL> SELECT * FROM dba_roles WHERE role='CON_RES';
ROLE                 PASSWORD   AUTHENTICAT COMMON  ORACLE_MAINTAIN
-------------------- --------   ----------- ------- ---------------
CON_RES              NO         NONE        NO      N
SQL> SELECT * FROM dba_role_privs WHERE grantee='CON_RES';
GRANTEE              GRANTED_ROLE         ADM DEL DEF COMMON
-------------------- -------------------- --- --- --- -------
CON_RES              RESOURCE             NO  NO  YES NO
CON_RES              CONNECT              NO  NO  YES NO
SQL> SELECT * FROM dba_sys_privs WHERE grantee='CON_RES';
no rows selected
```

通过 dba_roles 和 dba_role_privs 的查询可以看出，con_res 角色确实包含了两个角色，并且角色 con_res 和其包含的角色都是本地的（COMMON=NO）。由于只将两个角色授予了 con_res，而没有将权限授予它，所以 dba_sys_privs 的查询结果为空。

7.2.4 删除自定义角色

删除角色的命令格式是"DROP ROLE 角色名"。

【示例 7-4】 删除自定义本地角色 con_res

```
$ sqlplus system/***@pdborcl
SQL> DROP ROLE con_res;
Role dropped.
```

> 注意：角色删除后，原来拥有该角色的用户就不再拥有该角色了，相应的权限也就没有了。

7.3 用 户 管 理

除了 Oracle 自身的用户之外，对于具体的应用项目，还应该创建相应的用户账户。跟角色一样，Oracle 12c 的用户也分为公共用户（Common User）和本地用户（Local User）。在 CDB 中只能创建公共用户，在 PDB 中只能创建本地用户。公共用户的名称也必须以 c##三个字母开头。公共用户存在于所有容器（根 CDB 和所有的 PDB）中，本地用户只有在特定的 PDB 中存在，相同的本地用户名可以在多个 PDB 中存在，但它们之间没有关系。

Oracle 预设了两个公共管理用户，分别是 sys 和 system，都是 Oracle 的管理员，都可以访问和管理所有 PDB，但是 system 没有 sysdba 权限，只有 sys 拥有包括 sysdba 权限在内的所有权限，sys 才是 Oracle 最高级别管理员。sys 用户可以以 sysdba 和 sysoper 方式登录，system 只能以 normal 和 sysoper 方式登录。以 sysoper 方式登录之后，用户会变成 public。要查看用户的所有权限，在用户登录之后，运行以下命令：

```
SQL> SELECT * FROM session_privs;
SQL> SELECT * FROM session_roles;
```

7.3.1 创建公共用户

Oracle 创建用户的命令的基本语法是：

```
CREATE USER 用户名 IDENTIFIED BY 密码
[DEFAULT TABLESPACE 表空间名称]
[TEMPORARY TABLESPACE 表空间名称]
[ QUOTA 数字[K|M] | UNLIMITED ON 表空间名称
QUOTA 数字[K|M] | UNLIMITED ON 表空间名称...]
[PROFILE 概要文件名称 | DEFAULT]
[PASSWORD EXPIRE]
[ACCOUNT LOCK | UNLOCK];
```

上述语法的参数说明如下。

1）IDENTIFIED BY 密码

创建用户同时为其指定密码。

2）DEFAULT TABLESPACE 表空间名称

用户存储默认使用的表空间，当用户创建对象没有设置表空间时，就将保存在默认表空间下。

3) TEMPORARY TABLESPACE 表空间名称

用户使用的临时表空间。

4) QUOTA 数字 [K|M] | UNLIMITED ON 表空间名称

用户在表空间上的使用限额，可以指定多个表空间的限额，如果设置为 UNLIMITED，则表示不设置限额。

5) PROFILE 概要文件名称 | DEFAULT

用户操作的资源文件，如果不指定则使用默认配置资源文件。

6) PASSWORD EXPIRE

用户密码失效，如果设置，用户在第一次使用时必须修改密码。

7) ACCOUNT LOCK | UNLOCK

用户是否为锁定状态，LOCK 表示锁定，UNLOCK 为默认值，表示未锁定。处于锁定状态的用户不能登录。

【示例 7-5】创建公共用户

本示例创建一个公共用户 C##USER1，并将自定义角色 C##CON_RES 授权给该用户，使得 C##USER1 可以登录 CDB 以及所有 PDB。

```
$sqlplus / as sysdba
SQL> CREATE USER c##user1 IDENTIFIED BY 123 CONTAINER=ALL;
User created.
SQL> COL username FORMAT a20
SQL> SELECT username, con_id FROM cdb_users
  2  WHERE  username = 'C##USER1';
USERNAME                 CON_ID
-------------------- ----------
C##USER1                      4
C##USER1                      3
C##USER1                      1
SQL> GRANT c##con_res TO c##user1 CONTAINER=ALL;
SQL> ALTER USER c##user1 DEFAULT ROLE c##con_res CONTAINER=ALL;
User altered.
SQL> col granted_role format a15
SQL> col granted format a15
SQL> SELECT * FROM cdb_role_privs WHERE  grantee='C##USER1';
GRANTEE              GRANTED_ROLE    ADM DEL DEF COMMON  CON_ID
-------------------- --------------- --- --- --- ------- ----------
C##USER1             C##CON_RES      NO  NO  YES YES          1
C##USER1             C##CON_RES      NO  NO  YES YES          4
C##USER1             C##CON_RES      NO  NO  YES YES          3
```

创建用户 c##user1 之后，通过 cdb_users 查询可知，该用户存在于 3 个数据库中。CON_ID 为 1 的数据库是 CDB，CON_ID 为 3，4 表示 PDB。如果在创建命令中将 CONTAINER=ALL 改成 CONTAINER=CURRENT，那么创建的公共用户就只属于 CDB，对所有的 PDB 都不起作用。

通过 GRANT 将 c##con_res 公共角色授予 c##user1，并设置为默认角色。Oracle 中默认角色的意思是当该用户登录时，自动具有 DEFAULT ROLE 中所包含的权限，其他角色的权限要通过手工运行命令"SET ROLE 角色名"来获得。所以如果将一个角色授予了某个用户的时候，没有设置为默认角色，那么角色就不会自动起作用。接下来，设置用户的默认表空间为 system，再对 system 表空间的空间配额为 100M。

```
SQL> ALTER USER c##user1 DEFAULT TABLESPACE system CONTAINER=ALL;
User altered.
SQL> ALTER USER c##user1 QUOTA 100M ON system;
User altered.
```

注意：也可以设置 system 以外的其他表空间为公共用户的表空间，但必须保证每个数据库都有这个相同名称的表空间。

用户和权限都分配好之后，就要测试新用户登录了。下面测试 c##user1 登录 PDBORCL，登录之后，通过 session_privs 和 session_roles 查询当前 session 包含的权限和角色，然后创建表并插入数据。

```
$ sqlplus c##user1/123@pdborcl
SQL> COL role FORMAT a40
SQL> SELECT * FROM session_roles;
…
SQL> SELECT * FROM session_privs;
…
SQL> CREATE TABLE t1 (id number);
Table created.
SQL> INSERT INTO t1(id)VALUES(1);
1 row created.
SQL> col table_name format a20;
SQL> col tablespace_name format a20;
SQL> SELECT table_name, tablespace_name FROM user_tables;
TABLE_NAME           TABLESPACE_NAME
-------------------- --------------------
T1                   SYSTEM
```

如果还有其他 PDB，也可以登录到其他 PDB 做相似的操作，看看是不是也可以这样的操作，真正起到"公共"用户的作用。

7.3.2 授予用户对象权限

Oracle 的对象都有拥有者（Owner），通常拥有者就是创建该对象的用户。拥有者对其拥有的对象具有所有权限。除此之外，Oracle 允许将对象的某些权限授予其他用户。这就是在用户之间共享对象的权限。和系统权限一样，还是通过 GRANT 命令授权，通过 REVOKE 命令取消授权。详细的对象权限见表 7-2 对象权限分类表。

```
GRANT object_priv|ALL [(columns)] ON object
TO {user|role|PUBLIC}[WITH GRANT OPTION];
```

其中：
ALL：所有对象权限。
PUBLIC：授给所有的用户。
WITH GRANT OPTION：允许用户再次给其他用户授权。

【示例 7-6】授予对象权限

本示例演示 HR 用户将表 JOBS 的 SELECT 权限和 UPDATE 权限授予 SCOTT 用户。SELECT 权限只能是整个表的所有列，而 UPDATE 权限可以是表的是部分列。本例 HR 仅将 jobs.min_salary 和 jobs.max_salary 的 UPDATE（修改）权限授予 SCOTT，这样 SCOTT 用户就只能修改 HR.JOBS 的这两列。授权之后通过查询 USER_TAB_PRIVS_MADE 和 USER_COL_PRIVS_MADE 分别查看 hr 授予出去的对象整体权限和对象的列权限：

```
$ sqlplus hr/123@pdborcl
SQL> GRANT SELECT ON jobs TO scott;
Grant succeeded.
SQL> GRANT UPDATE(min_salary, max_salary)ON jobs TO scott;
Grant succeeded.
SQL> COL grantee FORMAT a10
SQL> COL table_name FORMAT a10
SQL> COL grantor FORMAT a10
SQL> COL column_name FORMAT a15
SQL> SELECT grantee, table_name, grantor, privilege FROM
2 USER_TAB_PRIVS_MADE WHERE  grantee='SCOTT';
GRANTEE    TABLE_NAME GRANTOR    PRIVILEGE
---------- ---------- ---------- ----------------
SCOTT      JOBS       HR         SELECT
SQL> SELECT grantee, table_name, column_name, grantor, privilege
2 FROM USER_COL_PRIVS_MADE WHERE  grantee='SCOTT';
GRANTEE    TABLE_NAME   COLUMN_NAME       GRANTOR     PRIVILEGE
---------- ------------ ----------------- ----------- ----------------
SCOTT      JOBS         MIN_SALARY        HR          UPDATE
```

| SCOTT | JOBS | MAX_SALARY | HR | UPDATE |

HR 用户授权完成之后，可以用 SCOTT 用户登录，测试一下授权是否成功，能不能进行授权之外的操作：

```
$ sqlplus scott/123@pdborcl
SQL> SELECT count(*) FROM hr.jobs;
  COUNT(*)
----------
        19
SQL> UPDATE hr.jobs SET min_salary=min_salary;
19 rows updated.
SQL> UPDATE hr.jobs SET job_title=job_title;
ERROR at line 1:
ORA-01031: 权限不足
SQL> DELETE FROM hr.jobs;
ERROR at line 1:
ORA-01031: 权限不足
```

从 SCOTT 的操作来看，授权之内的操作是可以的，授权之外的操作(如修改 job_title 列，删除 JOBS 表的数据)是不允许的。被授权用户可以通过 USER_TAB_PRIVS 和 USER_COL_PRIVS_RECD 分别查询被授予的对象权限和列权限。

> 注意：用户只能将表的所有列的 SELECT 权限授予其他用户，如果用户希望将表的部分列授予其他用户，可以通过创建只有部分列的视图，然后将视图授予其他用户。

7.3.3 用户的其他常用操作

对用户的常用操作有修改密码、锁定用户/解除锁定用户、删除用户等。除了修改密码可以是用户自己之外，锁定、解锁和删除操作都不能是用户自己，而必须由 sys 或者 system 用户来操作。

【示例 7-7】用户常用操作

本示例演示了修改用户密码、锁定用户、解除锁定用户和删除用户的操作。

```
$ sqlplus / as sysdba
SQL> ALTER USER c##user1 IDENTIFIED BY 123;
User altered.
SQL> ALTER USER c##user1 ACCOUNT LOCK;
User altered.
SQL> ALTER USER c##user1 ACCOUNT UNLOCK;
User altered.
SQL> DROP USER c##user1 CASCADE;
User dropped.
```

> 注意：删除用户的命令后有选项"CASCADE"，这表示级联删除用户的所有对象。如果用户是公共用户，那么 CDB 以及每个 PDB 中的该用户对象都将被删除！因此，在使用 CASCADE 选项时要非常小心。

7.3.4 监视用户

作为系统管理员，有时可能会想监视其他用户在线情况；查看某用户最近使用的查询语句；可能会希望强制中止一些用户会话。一般过程是：通过 v$open_cursor 查询最近使用的语句，通过 v$session 找出某个用户会话的 sid 和 serial#，然后 KILL（终止）这个会话。

【示例 7-8】监视用户

监视用户 HR 的查询语句，强行终止 HR 的会话

```
$ sqlplus / as sysdba
SQL> SELECT user_name , address , sql_text FROM v$open_cursor WHERE user_name='HR' and cursor_type='OPEN';
USER_NAME  ADDRESS           SQL_TEXT
---------- ----------------- ---------------------------------------
HR         00000000AC997728  SELECT * FROM EMPLOYEES
HR         00000000761A35B8  SELECT * FROM jobs
SQL> col machine format a10
SQL> col terminal format a10
SQL> col username format a10
SQL> SELECT username,sid,serial#,status,machine,terminal,logon_time FROM v$session WHERE username='HR';
USERNAME   SID    SERIAL# STATUS   MACHINE    TERMINAL  LOGON_TIME
---------- ------ ------- -------- ---------- --------- -----------
HR         256    36013   INACTIVE oracle-pc  pts/4     2017-05-13
SQL> ALTER SYSTEM KILL SESSION '256, 36013';
System altered.
```

用户进程被 KILL 之后，如果被 KILL 的用户继续操作的话，Oracle 会提示"ORA-00028：您的会话已被终止"。如果是进程死锁，这个命令可能无法终止会话，这时就应该使用操作系统命令来终止进程了。

7.4 概要文件

概要文件是限制用户口令以及用户使用系统资源的文件。每个用户都必须有限制它的概要文件，Oracle 缺省的概要文件是 DEFAULT。用户口令的限制参数见表 7-3，系统资源参数见表 7-4。

表 7-3 口令限制参数

参数名称	说明
FAILED_LOGIN_ATTEMPTS	当连续登录失败次数达到该参数指定值时,用户被加锁
PASSWORD_LIFE_TIME	口令的有效期(天),默认值为 UNLIMITED
PASSWORD_REUSE_TIME	口令被修改后原有口令隔多少天后可以被重新使用,默认值为 UNLIMITED
PASSWORD_REUSE_MAX	口令被修改后原有口令被修改多少次才允许被重新使用
PASSWORD_VERIFY_FUNCTION	口令校验函数
PASSWORD_LOCK_TIME	账户被锁定时,加锁天数
PASSWORD_GRACE_TIME	口令过期后,继续使用原口令的宽限期(天)

表 7-4 系统资源限制参数

参数名称	说明
SESSION_PER_USER	允许一个用户同时创建 SESSION 的最大数量
CPU_PER_SESSION	每个 SESSION 允许使用 CPU 的时间数,单位为 ms
SESSION_PER_USER	用户的最大并发会话数
CPU_PER_CALL	限制每次调用 SQL 语句期间,CPU 的时间总量
CONNECT_TIME	每个 SESSION 的连接时间数,单位为分钟
IDLE_TIME	每个 SESSION 的超时时间,单位为分钟
LOGICAL_READS_PER_SESSION	为了防止笛卡儿积的产生,可以限定每一个用户最多允许读取的数据块数
LOGICAL_READS_PER_CALL	每次调用 SQL 语句期间最多允许读取的数据块数
PRIVATE_SGA	SGA 私有区域的最大容量

Oracle 12c 的概要文件也分为公共概要文件和本地概要文件。在 CDB 中只能创建公共概要文件,在 PDB 中只能创建本地概要文件。公共概要文件的名称必须以 C##三个字母开头。公共概要文件作用于公共用户,本地概要文件作用于本地用户。

7.4.1 创建概要文件

Oracle 安装的时候就有缺省的概要文件,名称为 DEFAULT。在新建用户的时候如果不指定概要文件就会使用 DEFAULT 概要文件,并受其中参数的约束。也可以创建自定义的概要文件,指定自定义的参数值。创建概要文件的命令是:

```
CREATE PROFILE 概要文件名称 LIMIT 参数名 参数值 …
```

概要文件创建之后,可以指定给用户,让用户资源受到这个概要文件的限制。概要文件指定给用户的命令是：

```
ALTER USER 用户名 PROFILE 概要文件名称
```

Oracle 的系统参数 resource_limit 决定概要文件是否起作用。值为 true 的时候，用户自定义的概要文件才能发挥作用。可以通过以下命令设置该参数为 true：

```
ALTER SYSTEM SET RESOURCE_LIMIT=TRUE;
```

> 注意：resource_limit 参数也是各个 PDB 各自拥有的，设置了一个 PDB，并不意味着其他 PDB 也设置了，需要每个 PDB 分别设置。

【示例 7-9】创建概要文件并指定给用户 HR

本示例由 SYSTEM 用户在 PDBORCL 中创建一个本地概要文件 profile1，并指定给用户 HR。概要文件中设置了两个参数的值。其他未指定参数的值自动设置为"DEFAULT"。通过 dba_profiles 查询概要文件的所有参数。

```
$ sqlplus system/***@pdborcl;
SQL> CREATE PROFILE profile1 LIMIT
  2  FAILED_LOGIN_ATTEMPTS 2
  3  SESSIONS_PER_USER 3;
Profile created.
SQL> COL profile FORMAT a10
SQL> COL resource_name FORMAT a30
SQL> COL resource FORMAT a10
SQL> COL limit FORMAT a10
SQL> SELECT * FROM dba_profiles WHERE profile='PROFILE1';
PROFILE    RESOURCE_NAME                  RESOURCE    LIMIT      COM
---------- ------------------------------ ----------  ---------- ---
PROFILE1   COMPOSITE_LIMIT                KERNEL      DEFAULT    NO
PROFILE1   SESSIONS_PER_USER              KERNEL      3          NO
PROFILE1   CPU_PER_SESSION                KERNEL      DEFAULT    NO
PROFILE1   CPU_PER_CALL                   KERNEL      DEFAULT    NO
PROFILE1   OGICAL_READS_PER_SESSION       KERNEL      DEFAULT    NO
PROFILE1   LOGICAL_READS_PER_CALL         KERNEL      DEFAULT    NO
PROFILE1   IDLE_TIME                      KERNEL      DEFAULT    NO
PROFILE1   CONNECT_TIME                   KERNEL      DEFAULT    NO
PROFILE1   PRIVATE_SGA                    KERNEL      DEFAULT    NO
PROFILE1   FAILED_LOGIN_ATTEMPTS          PASSWORD    2          NO
PROFILE1   PASSWORD_LIFE_TIME             PASSWORD    DEFAULT    NO
PROFILE1   PASSWORD_REUSE_TIME            PASSWORD    DEFAULT    NO
PROFILE1   PASSWORD_REUSE_MAX             PASSWORD    DEFAULT    NO
PROFILE1   PASSWORD_VERIFY_FUNCTION       PASSWORD    DEFAULT    NO
PROFILE1   PASSWORD_LOCK_TIME             PASSWORD    DEFAULT    NO
```

```
PROFILE1    PASSWORD_GRACE_TIME           PASSWORD    DEFAULT    NO
16 rows selected.
SQL> ALTER USER hr PROFILE profile1;
User altered.
```

现在，HR 用户的概要文件变成了 profile1，这个概要文件中有一项限制是"failed_login_attempts 2"表示如果用户连续两次登录失败，用户将被锁定。假定用户 HR 的正确密码是 123，现在两次登录，故意输入错误的密码，第 3 次登录即使密码正确也会失败，必须用 SYSTEM 用户解除锁定之后，才能登录。

```
$ sqlplus hr/wrongpass@pdborcl
ERROR:
ORA-01017: 用户名/口令无效； 登录被拒绝
$ sqlplus hr/wrongpass@pdborcl
ERROR:
ORA-01017: 用户名/口令无效； 登录被拒绝
$ sqlplus hr/123@pdborcl
ERROR:
ORA-28000: the account is locked
```

7.4.2 修改概要文件

如果需要修改概要文件中的参数，可以使用命令：

```
ALTER PROFILE 概要文件名|DEFAULT LIMIT 参数名 参数值 ...
```

其中 DEFAULT 为缺省的概要文件名，比如：

```
SQL> ALTER PROFILE profile1 LIMIT FAILED_LOGIN_ATTEMPTS 3;
Profile altered.
```

7.4.3 删除概要文件

删除概要文件，可以使用命令：

```
DROP PROFILE 概要文件名 [CASCADE]
```

只能删除未指定用户的概要文件。如果概要文件已经指定了用户，就不能删除，除非在命令中增加选项 CASCADE，这样会让这个概要文件作用范围内的所有用户的概要文件恢复为默认的概要文件 DEFAULT。

【示例 7-10】 删除概要文件 profile1

```
$ sqlplus system/123@pdborcl;
SQL> DROP PROFILE profile1 CASCADE;
Profile droped.
SQL> SELECT username, profile FROM dba_users WHERE username='HR';
USERNAME    PROFILE
```

```
--------   ----------
HR         DEFAULT
```

练习

一、判断题

1. 连接到数据库的权限是对象权限。 （ ）
2. 创建、删除、更改表的权限是系统权限。 （ ）
3. 一个用户将表授予其他用户访问，这是对象权限的授权。 （ ）
4. 可以删除自定义角色，但不能删除内置角色。 （ ）
5. 角色中只能包含权限，不能包含其他角色。 （ ）
6. 本地用户只能在 pdb 中创建。 （ ）
7. 可以在授予 SELECT 对象权限时指定表的部分列。 （ ）
8. 可以在授予 UPDATE 对象权限时指定表的部分列。 （ ）

二、单项选择题

1. 下列权限中属于对象权限的是（   ）
   A. DROP USER   B. CREATE TABLE
   C. INSERT   D. SYSDBA
2. 下列权限中属于系统权限的是（   ）
   A. UPDATE   B. CREATE TABLE
   C. INSERT   D. DELETE
3. 创建用户时指定用户的登录密码的选项是（   ）
   A. IDENTIFIED BY   B. PASSWORD
   C. PASSWORD EXPIRE   D. ACCOUNT LOCK
4. 如果没有为创建的用户赋予任何概要文件，那么：（   ）
   A. 该用户不能创建   B. 该用户不能与数据库连接
   C. 该用户没有概要文件   D. 该用户被赋予 DEFAULT 概要文件
5. 在 Oracle 预定义角色中，（   ）角色表示授予开发人员的权限
   A. EXP_FULL_DATABASE   B. SELECT_CATALOG_ROLE
   C. CONNECT   D. RESOURCE

三、填空题

1. Oracle 权限分为 _____ 和 _____。
2. 公共角色名称和公共用户名称都必须以_____三个字母开头。
3. 创建用户时设置表空间的关键词有 QUOTA，作用是_____。
4. 为用户或角色授予权限时需要使用_____关键词。
5. 取消用户或角色授予的权限时需要使用_____关键词。
6. Oracle 的两个预定义系统管理用户是 sys 和_____。

四、简答题

1. Oracle 数据库的系统权限和对象权限有什么区别？
2. 概要文件有什么作用，对用户有哪些方面的约束？
3. Oracle 12c 的公共用户和本地用户有什么区别？

训练任务

1. 掌握管理角色、权限、用户的能力，并在用户之间共享对象。

培养能力	工程能力、设计/开发解决方案、沟通、项目管理		
掌握程度	★★★	难度	中
结束条件	创建角色、用户成功，并测试成功		
训练内容： Oracle 有一个开发者角色 resource，可以创建表、过程、触发器等对象，但是不能创建视图。本训练要求在 pdborcl 插接式数据中创建一个新的本地角色 con_res_view，该角色包含 connect 和 resource 角色，同时也包含 CREATE VIEW 权限，这样任何拥有 con_res_view 的用户就同时拥有这三种权限。创建角色之后，再创建用户 new_user，给用户分配表空间，设置限额为 50M，授予 con_res_view 角色。最后测试：用新用户 new_user 连接数据库、创建表，插入数据，创建视图，查询表和视图的数据			

第 8 章 数据库的对象管理

本章目标

知识点	理解	掌握	应用
1.表管理	✓	✓	✓
2.分区表管理	✓	✓	✓
3.索引、簇表、视图	✓	✓	✓
4.序列、同义词	✓	✓	
5.XML 表	✓	✓	

训练任务

- 掌握创建表和管理表
- 熟悉创建分区表
- 熟悉创建索引

知识能力点

能力点 \ 知识点	知识点 1	知识点 2	知识点 3	知识点 4	知识点 5
工程知识	✓	✓			✓
问题分析	✓	✓	✓		✓
设计/开发解决方案	✓	✓			
研究	✓	✓	✓	✓	✓
使用现代工具	✓	✓			✓
工程与社会		✓			✓
环境和可持续发展	✓		✓		
职业规范	✓	✓			
个人和团队	✓				✓
沟通		✓			✓
项目管理	✓	✓	✓		✓
终身学习	✓				

8.1 表

表是最基本的数据库对象,在关系型数据库中,由行和列的二维结构组成。其中行表示数据记录信息,列表示记录的属性,也称字段,每个字段都有类型和长度。

8.1.1 数据类型

在设计表结构时,需要选择适当的数据类型以节省存储空间,Oracle 中的数据类型有:字符型、数值型、日期/时间型、大对象(LOB)型和 ROWID 型。

1) 字符型

Oracle 中常用的字符型数据类型如表 8-1 所示。

表 8-1 字符型数据类型

数据类型	说明
CHAR(n)	固定长度字符串,n 表示存储字符数量,长度<=2000 字节
VARCHAR2(n)	可变长字符串,n 表示存储字符数量,长度<=4000 字节
NCHAR(n)	根据字符集而定的固定长度字符串,n 表示存储的字符数量,长度<=2000 字节
NVARCHAR2(n)	根据字符集而定的可变长度字符串,n 表示存储的字符数量,长度<=4000 字节
LONG	可变长度字符串,长度<=2GB
RAW	可变长度二进制字符串,长度<=2000 字节
LONG RAW	可变长度二进制字符串,长度<2GB

2) 数值型

Oracle 中常用的数值型数据类型如表 8-2 所示。

表 8-2 数值型数据类型

数据类型	说明
NUMBER(p, s)	数值类型,p 表示精度,s 表示小数位数
FLOAT	浮点数类型,相当于 NUMBER(38),双精度
DECIMAL(p, s)	数值类型,p 表示精度,s 表示小数位数
INTEGER	整数类型
SMALLINT	短整数类型
REAL	实数类型,相当于 NUMBER(63),精度更高

3) 日期/时间型

Oracle 中常用的日期/时间型数据类型如表 8-3 所示。

表 8-3 日期/时间型数据类型

数据类型	说明
DATE	日期类型，精确到秒
TIMESTAMP(n)	日期类型，精确到微秒，n(0~9)表示微秒的精确范围，默认为 6
INTERVAL YEAR(n) TO MONTH	按年、月存储一个时间段，n 表示年份位数，默认值为 2
INTERVAL DAY(n) TO SECOND(m)	按日、时、分、秒存储一个时间段，n 表示天的位数，默认值为 2；m(0~9)表示微秒精确范围，默认值为 6

4) 大对象(LOB)型

Oracle 中常用的大对象型数据类型如表 8-4 所示。

表 8-4 大对象型数据类型

数据类型	说明
BFILE	存放在数据库外的二进制数据，指向服务器文件系统上的文件定位器，长度<=4G
BLOB	二进制数据，长度<=4G
CLOB	字符数据，长度<=4G
NCLOB	根据字符集而定的字符数据，长度<=4G

5) ROWID 型

Oracle 中常用的 ROWID 型数据类型如表 8-5 所示。

表 8-5 ROWID 型数据类型

数据类型	说明
ROWID	数据表中记录的唯一行号
UROWID	通用的 ROWID，既可以保存物理的 ROWID，也可以保存逻辑的 ROWID

8.1.2 创建表

本节介绍创建表的基本方法。

1) 使用设计器创建表

打开 Oracle SQL Developer，连接到指定数据库，右击表节点，选择[新建表]，打开创建表的对话框，如图 8-1 所示。

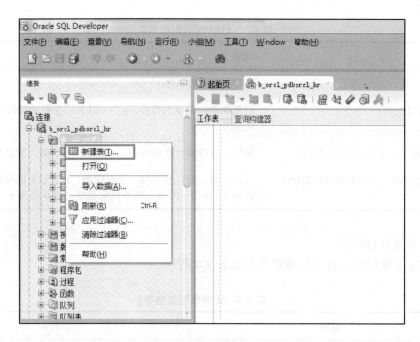

(a)

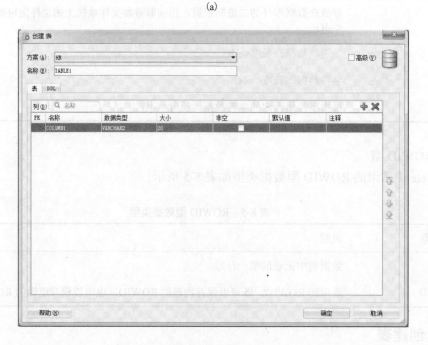

(b)

图 8-1　创建表

在创建表的对话框中，可修改表的名称。单击按钮 ➕ 可添加新的列，在列的定义行中，可直接修改列的名称、类型、大小，并可设置是否为空和是否为主键。

若想详细定义列的属性，可选择右上角的【高级】复选框，打开表的高级设置对话框，如图 8-2 所示。

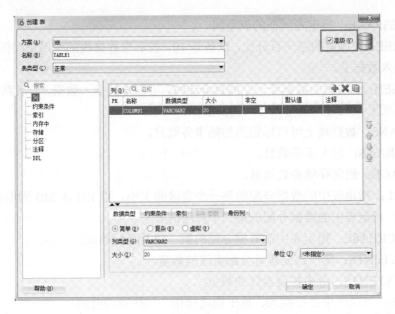

图 8-2 表的高级设置

在表的高级设置对话框中，还可以进行约束条件、索引、存储和分区等设置。

2）使用 SQL 语句创建表

虽然使用设计器创建表直观简单，但在程序中动态创建表，就只能使用 SQL 语句来完成。可使用 CREATE TABLE 语句来创建表，其语法格式如下：

```
CREATE TABLE [schema.]<table_name>(
<column_name> <data_type> [DEFAULT <expression>] [CONSTRAINT]
[, <column_name> <data_type> [DEFAULT <expression>] [CONSTRAINT]]
[, …]
)[TABLESPACE <tablespace_name>] |
[PCTFREE n] |
[PCTUSED n] |
[INITRANS n] |
[MAXTRANS n] |
[STORAGE(INITAIL n|NEXT n|PCTINCREASE n|MINEXTENTS n| MAXEXTENTS n)] |
[CACHE];
```

参数说明：

schema：指定表所属的用户名或所属的用户模式名称。

table_name：要创建表的名称。

column_name：表中包含列的名称。

data_type：列的数据类型。

DEFAULT：设置列的默认值。

CONSTRAINT：设置列约束，即列值必须满足的规则。

TABLESPACE：指定表所在的表空间。

PCTFREE：块保留的空间百分比，默认是 10，表示当数据块的可用空间低于 10%后，就不可以插入数据了。

PCTUSED：当块里的数据低于空间百分比时，可以重新插入数据，一般默认值是 40，即：当数据低于 40%时，又可以写入新的数据。

INITRANS：数据块上可以标记的初始事务数目。

MAXTRANS：最大事务数目。

STORAGE：指定存储参数信息。

INITIAL：指定表中的数据分配的第一个盘区的大小，以 KB 或 MB 为单位。

NEXT：指定表中的数据分配的第二个盘区的大小。

PCTINCREASE：指定表中的数据分配的第三个及其后的盘区大小。

MINEXTENTS：第一次创建时分配盘区的数量。

MAXEXTENTS：最多扩展盘区的数量。

CACHE：对缓存块进行换入、换出调度操作，加快查询时间。

【示例 8-1】创建表

在 STUDY 用户中创建表 products。

```
SQL> CREATE TABLE study.products
  2  (
  3    product_id VARCHAR2(40 BYTE)NOT NULL
  4  , product_name VARCHAR2(40 BYTE)NOT NULL
  5  , product_type VARCHAR2(40 BYTE)NOT NULL
  6  , CONSTRAINT products_pk PRIMARY KEY(product_id)ENABLE
  7  )
  8  LOGGING
  9  TABLESPACE "USERS"
 10  PCTFREE 10
 11  INITRANS 1
 12  STORAGE
 13  (
 14    INITIAL 65536
 15    NEXT 1048576
 16    MINEXTENTS 1
 17    MAXEXTENTS 2147483645
 18    BUFFER_POOL DEFAULT
 19  );
表已创建。
```

在 products 表中有三个字段：product_id、product_name 和 product_type，每个字段都非空，主键是 product_id。定义 products 表中的参数说明如下：

LOGGING：对表的所有操作都记录到重做日志中。
TABLESPACE "USERS"：把表创建到 USERS 表空间中。
PCTFREE 10：表示当数据块的可用空间低于 10%后，就不可以插入数据了。
INITRANS 1：数据块上初始事务数目为 1。
INITIAL 65536：数据分配的第一个盘区的大小为 65536KB。
NEXT 1048576：数据分配的第二个盘区的大小为 1048576KB。
MINEXTENTS 1：第一次创建表时只分配 1 个盘区。
MAXEXTENTS 2147483645：最多可扩展到 2147483645 个盘区。
BUFFER_POOL DEFAULT：表数据存储在 DEFAULT 缓冲池中。

8.1.3 修改、删除表

表创建后，可以修改、删除表，主要有以下几种操作：

1）修改表名

可以使用 ALTER TABLE … RENAME TO 语句修改表名，其语法格式如下：

```
ALTER TABLE table_name RENAME TO new_table_name;
```

其中参数说明：

table_name：原表名。

new_table_name：新表名。

2）添加列

可以使用 ALTER TABLE…ADD 语句在表中添加列，其语法格式如下：

```
ALTER TABLE table_name ADD col_name data_type;
```

其中参数说明：

table_name：表名。

col_name：要添加列名。

data_type：列的数据类型。

【示例 8-2】添加列

在 study 用户的 products 表中添加列 product_desc。

```
SQL> DESC products;
名称               空值        类型
-------------    --------   -------------
PRODUCT_ID       NOT NULL   VARCHAR2(40)
PRODUCT_NAME     NOT NULL   VARCHAR2(40)
PRODUCT_TYPE     NOT NULL   VARCHAR2(40)
SQL> ALTER TABLE products ADD product_desc VARCHAR2(100);
Table PRODUCTS 已变更。
SQL> COMMIT;
SQL> DESC products;
```

```
名称                     空值          类型
-------------       --------    -------------
PRODUCT_ID              NOT NULL    VARCHAR2(40)
PRODUCT_NAME    NOT     NULL        VARCHAR2(40)
PRODUCT_TYPE    NOT     NULL        VARCHAR2(40)
PRODUCT_DESC                        VARCHAR2(100)
```

3）修改列

可使用 ALTER TABLE…RENAME COLUMN TO 语句修改列名，其语法格式如下：

```
ALTER TABLE table_name RENAME COLUMN oldcol_name TO newcol_name;
```

其中参数说明：

table_name：表名。

oldcol_name：旧列名。

newcol_name：新列名。

可使用 ALTER TABLE…MODIFY 语句修改列类型、默认值等，其语法格式如下：

```
ALTER TABLE table_name MODIFY col_name new_datatype;
ALTER TABLE table_name MODIFY col_name DEFAULT default_value;
```

其中参数说明：

table_name：表名。

col_name：列名。

new_datatype：新类型。

default_value：默认值。

【示例 8-3】修改列

在 study 用户的 products 表中修改列 product_desc。

```
SQL> DESC products;
名称                 空值          类型
-------------   --------    -------------
PRODUCT_ID      NOT NULL    VARCHAR2(40)
PRODUCT_NAME    NOT NULL    VARCHAR2(40)
PRODUCT_TYPE    NOT NULL    VARCHAR2(40)
PRODUCT_DESC                VARCHAR2(100)
SQL> ALTER TABLE products RENAME COLUMN product_desc TO product_des;
SQL> ALTER TABLE products MODIFY product_des VARCHAR2(200);
SQL> ALTER TABLE products MODIFY product_des
  2  DEFAULT 'THIS IS A TEST';
SQL> COMMIT;
SQL> DESC products;
名称                 空值          类型
-------------   --------    -------------
```

```
PRODUCT_ID       NOT NULL    VARCHAR2(40)
PRODUCT_NAME     NOT NULL    VARCHAR2(40)
PRODUCT_TYPE     NOT NULL    VARCHAR2(40)
PRODUCT_DES                  VARCHAR2(200)
```

4) 删除列

可使用 ALTER TABLE…DROP 语句删除列，其语法格式如下：

```
ALTER TABLE table_name DROP COLUMN col_name;
```

其中参数说明：

table_name：表名。

col_name：列名。

【示例 8-4】删除列

在 study 用户的 products 表中删除列 product_des。

```
SQL> DESC products;
名称                空值         类型
-------------      --------     -------------
PRODUCT_ID         NOT NULL    VARCHAR2(40)
PRODUCT_NAME       NOT NULL    VARCHAR2(40)
PRODUCT_TYPE       NOT NULL    VARCHAR2(40)
PRODUCT_DESC                   VARCHAR2(100)
SQL> ALTER TABLE products DROP COLUMN product_des;
Table PRODUCTS 已变更。
SQL> COMMIT;
SQL> DESC products;
名称                空值         类型
-------------      --------     -------------
PRODUCT_ID         NOT NULL    VARCHAR2(40)
PRODUCT_NAME       NOT NULL    VARCHAR2(40)
PRODUCT_TYPE       NOT NULL    VARCHAR2(40)
```

5) 删除数据表

可使用 DROP TABLE 语句删除表，其语法格式如下：

```
DROP TABLE table_name [CASCADE CONSTRAINTS] [PURGE];
```

其中参数说明：

table_name：表名。

CASCADE CONSTRAINTS：表示删除表的同时也删除该表的视图、索引、约束和触发器等。

PURGE：表示表删除成功后释放占用的资源。

8.1.4 表的约束

约束是对数据进行限制的一种机制，根据约束的用途，可以将约束分为：

1) 主键约束(PRIMARY KEY)

主键约束用于指定表的主键，主键可由一列或多列组成。

2) 外键约束(FOREIGN KEY)

外键约束用于指定表的外键，外键引用另外一个表(或同一个表)中的主键。

3) 唯一性约束(UNIQUE)

唯一性约束用于指定一列只能存储唯一值。

4) 非空约束(NOT NULL)

非空约束用于指定一列不允许为空。

5) 检查约束(CHECK)：

检查约束用于指定一列的值必须满足某种条件。

【示例 8-5】CHECK 约束

在 study 用户的 products 表中增加一个 CHECK 约束。

```
SQL> ALTER TABLE products
  2  ADD CONSTRAINT products_chk1 CHECK
  3  (product_type IN ('耗材','手机','电脑'))
  4  ENABLE;
表已更改。
SQL> COMMIT;
```

说明：在 products 表增加一个 CHECK 约束，限制列 product_type 只能取 "耗材"、"手机" "电脑" 中的值。

【示例 8-6】唯一约束

在 study 用户的 products 表中增加一个唯一约束。

```
SQL> ALTER TABLE products
  2  ADD CONSTRAINT products_uni1 UNIQUE(product_name) ENABLE;
表已更改。
SQL> COMMIT;
```

说明：在 products 表增加唯一约束，限制列 product_name 取值不能重复。

8.2 分 区 表

当表中的数据量不断增大，查询数据的速度就会变慢，应用程序的性能就会下降，这时就应该考虑对表进行分区。表进行分区后，逻辑上表仍然是一张完整的表，只是将表中的数据在物理上存放到多个表空间(物理文件上)，这样查询数据时，不至于每次都扫描整张表。

8.2.1 分区类型

在 Oracle 中提供了多种分区的类型，常用的分区类型有：
1）范围分区

范围分区是将数据根据范围映射到每一个分区，范围是在创建分区时由指定的分区键决定的，分区键经常采用日期。

【示例 8-7】按订单日期范围分区

在 study 用户中创建分区表 orders，按订单日期分区。

```
SQL> CREATE TABLE study.orders
2   (
3   order_id NUMBER(10, 0)NOT NULL
4   , customer_name VARCHAR2(40 BYTE)NOT NULL
5   , customer_tel VARCHAR2(40 BYTE)NOT NULL
6   , order_date DATE NOT NULL
7   , employee_id NUMBER(6, 0)NOT NULL
8   , discount NUMBER(8, 2)DEFAULT 0
9   , trade_receivable NUMBER(8, 2)DEFAULT 0
10  )
11  TABLESPACE USERS
12  PCTFREE 10
13  INITRANS 1
14  STORAGE
15  (
16   BUFFER_POOL DEFAULT
17  )
18  PARTITION BY RANGE (order_date)
19  (
20   PARTITION partition_before_2016 VALUES LESS THAN (
21    TO_DATE(' 2016-01-01 00: 00: 00', 'SYYYY-MM-DD HH24: MI: SS',
22    'NLS_CALENDAR=GREGORIAN'))TABLESPACE USERS,
23   PARTITION partition_before_2017 VALUES LESS THAN (
24    TO_DATE(' 2017-01-01 00: 00: 00', 'SYYYY-MM-DD HH24: MI: SS',
25    'NLS_CALENDAR=GREGORIAN'))TABLESPACE USERS02
26  );
SQL> COMMIT;
```

说明：在建立 orders 表时按订单日期（ORDER_DATE）进行了范围分区，将 2016 年以前的数据存储在 USERS 表空间中，将 2106 年的数据存储在 USERS02 表空间中。

【示例 8-8】 按订单号范围分区

在 study 用户中创建分区表 orders，按订单号分区。

```
SQL> CREATE TABLE study.orders
  2  (
  3   order_id NUMBER(10,0)NOT NULL
  4   , customer_name VARCHAR2(40 BYTE)NOT NULL
  5   , customer_tel VARCHAR2(40 BYTE)NOT NULL
  6   , order_date DATE NOT NULL
  7   , employee_id NUMBER(6,0)NOT NULL
  8   , discount NUMBER(8,2)DEFAULT 0
  9   , trade_receivable NUMBER(8,2)DEFAULT 0
 10  )
 11  PARTITION BY RANGE (order_id)
 12  (
 13    PARTITION ord_part1 VALUES LESS THAN (100000)
 14     TABLESPACE USERS,
 15    PARTITION ord_part2 VALUES LESS THAN (200000)
 16     TABLESPACE USERS02
 17  );
SQL> COMMIT;
```

说明：在建立 orders 表时按订单号进行了范围分区，将 100000 号以前的数据存储在 users 表空间中，将 100000~200000 号之间的数据存储在 USERS02 表空间中。

> 注意：当使用范围分区时，须考虑以下几个规则：
> （1）每一个分区都必须有一个 VALUES LESS THEN 子句，它指定了一个不包括在该分区中的上限值。分区键的任何值等于或者大于这个上限值的记录都会被加入到下一个高一些的分区中。
> （2）所有分区，除了第一个，都会有一个隐式的下限值，这个值就是此分区的前一个分区的上限值。
> （3）在最高的分区中，MAXVALUE 被定义。MAXVALUE 代表了一个不确定的值。这个值高于其他分区中的任何分区键的值，也可以理解为高于任何分区中指定的 VALUE LESS THEN 的值，同时包括空值。

2）列表分区

根据某列的值来进行分区，指定列的值不宜过多。

【示例 8-9】 按产品类型列表分区

在 study 用户中创建分区表 products，按产品类型分区。

```
SQL> CREATE TABLE study.products
  2  (
  3    product_id VARCHAR2(40 BYTE)NOT NULL
```

```
  4  , product_name VARCHAR2(40 BYTE)NOT NULL
  5  , product_type VARCHAR2(40 BYTE)NOT NULL
  6  , CONSTRAINT products_pk PRIMARY KEY(product_id)ENABLE
  7  )
  8  PARTITION BY LIST(product_type)
  9  (
 10   PARTITION  pd_part1 VALUES ('门套','门楣')TABLESPACE  USERS,
 11   PARTITION  pd_part2 VALUES ('双开门','推拉门') TABLESPACE USERS02
 12  );
SQL> COMMIT;
```

说明：在建立 products 表时按产品类型进行了列表分区，将"门套""门楣"类型的数据存储在 users 表空间中，将"双开门""推拉门"类型的数据存储在 USERS02 表空间中。

3）散列分区

在列值上使用散列算法，以确定将数据放入哪个分区中。当列的值没有合适的条件时，建议使用散列分区。

【示例 8-10】按 ID 散列分区

在 study 用户中创建分区表 test，按 id 散列分区。

```
SQL> CREATE TABLE study.test
  2  (
  3   id NUMBER(8),
  4   information VARCHAR2(100)
  5  )
  6  PARTITION BY HASH(id)
  7  (
  8   PARTITION test_part1 TABLESPACE USERS,
  9   PARTITION test_part2 TABLESPACE USERS02,
 10  );
SQL> COMMIT;
```

说明：在建立 TEST 表时按 ID 进行了散列分区，数据将自动根据散列算法结果存储到不同的表空间中。

4）引用分区

引用分区（Reference Partition）是针对主外键关联建立的分区。主表分区之后，借助引用分区可以实现按照主表分区的方式对从表进行分区，这样从表就会继承主表的分区机制（即使从表中没有主表分区对应的列）。

【示例 8-11】在主表 orders 和从表 order_details 之间建立引用分区

在 study 用户中创建两个表：orders（订单表）和 order_details（订单详表），两个表通过列 order_id 建立主外键关联。orders 表按范围分区进行存储，order_details 使用引用分区进行存储。

创建 orders 表的部分语句是：

```
SQL> CREATE TABLE orders
2   (
3     order_id NUMBER(10, 0) NOT NULL
4   , customer_name VARCHAR2(40 BYTE) NOT NULL
5   , customer_tel VARCHAR2(40 BYTE) NOT NULL
6   , order_date DATE NOT NULL
7   , employee_id NUMBER(6, 0) NOT NULL
8   , discount NUMBER(8, 2) DEFAULT 0
9   , trade_receivable NUMBER(8, 2) DEFAULT 0
10  )
11  TABLESPACE USERS
12  PCTFREE 10 INITRANS 1
13  STORAGE (  BUFFER_POOL DEFAULT  )
14  NOCOMPRESS NOPARALLEL
15  PARTITION BY RANGE (order_date)
16  (
17    PARTITION PARTITION_BEFORE_2016 VALUES LESS THAN (
18      TO_DATE(' 2016-01-01 00:00:00', 'SYYYY-MM-DD HH24:MI:SS',
19      'NLS_CALENDAR=GREGORIAN'))
20    NOLOGGING
21    TABLESPACE USERS
22    PCTFREE 10
23    INITRANS 1
24    STORAGE
25    (
26      INITIAL 8388608
27      NEXT 1048576
28      MINEXTENTS 1
29      MAXEXTENTS UNLIMITED
30      BUFFER_POOL DEFAULT
31    )
32    NOCOMPRESS NO INMEMORY
33    , PARTITION PARTITION_BEFORE_2017 VALUES LESS THAN (
34      TO_DATE(' 2017-01-01 00:00:00', 'SYYYY-MM-DD HH24:MI:SS',
35      'NLS_CALENDAR=GREGORIAN'))
36    NOLOGGING
```

```
37    TABLESPACE USERS02
38    ...
39 );
```

说明：在建立 orders 表时按订单日期（ORDER_DATE）进行了范围分区，将 2016 年以前的数据存储在 USERS 表空间中，将 2106 年的数据存储在 USERS02 表空间中。

创建 order_details 表的部分语句如下：

```
SQL> CREATE TABLE order_details
 2  (
 3    id NUMBER(10, 0) NOT NULL
 4  , order_id NUMBER(10, 0) NOT NULL
 5  , product_id VARCHAR2(40 BYTE) NOT NULL
 6  , product_num NUMBER(8, 2) NOT NULL
 7  , product_price NUMBER(8, 2) NOT NULL
 8  , CONSTRAINT order_details_fk1 FOREIGN KEY (order_id)
 9    REFERENCES orders ( order_id )
10    ENABLE
11 )
12 TABLESPACE USERS
13 PCTFREE 10 INITRANS 1
14 STORAGE (   BUFFER_POOL DEFAULT )
15 NOCOMPRESS NOPARALLEL
16 PARTITION BY REFERENCE (order_details_fk1)
17 (
18    PARTITION PARTITION_BEFORE_2016
19    NOLOGGING
20    TABLESPACE USERS  --必须指定表空间，否则会将分区存储在用户的默认表空间中
21    ...
22    )
23  NOCOMPRESS NO INMEMORY,
24  PARTITION PARTITION_BEFORE_2017
25    NOLOGGING
26    TABLESPACE USERS02
27    ...
28    )
29  NOCOMPRESS NO INMEMORY
30 );
```

说明：由于 order_details 是 orders 的从表，通过引用分区语句"PARTITION BY

REFERENCE（order_details_fk1）"，利用外键 order_details_fk1 关联到主表 orders，使从表按主表的分区方案与主表存储在同一分区中。

> 注：在从表 order_details 中，虽然没有订单日期列（ORDER_DATE），但由于建立了引用分区，其数据也是按主表 orders 的日期范围进行分区存储。

8.2.2 分区表的维护

在 Oracle 中提供了强大的分区表维护操作，常用的维护操作包括：添加分区、删除分区、截断分区、合并分区、拆分分区、重命名表分区、交换分区、移动分区等。

【示例 8-12】 分区表维护操作示例。

```
--添加分区
SQL> ALTER TABLE products ADD PARTITION partition_before_2018
  2  VALUES LESS THAN(TO_DATE('2018-01-01', 'YYYY-MM-DD'));
--删除分区
SQL> ALTER TABLE products DROP PARTITION partition_before_2018;
--截断分区：删除分区中的数据
SQL> ALTER TABLE products TRUNCATE PARTITION partition_before_2018;
--合并分区：结果分区将采用较高分区的界限
SQL> ALTER TABLE products MERGE
  2  PARTITIONS partition_before_2017, partition_before_2018
  3  INTO PARTITION partition_before_2018;
```

8.3 索　引

索引是关系型数据库的一个基本对象，通过建立索引，用户可以快速访问数据库表的中特定信息。

Oracle 支持以下几种主要的索引结构：

1) B-树索引

B-树索引是 Oracle 中最常用的、默认的索引结构。B-树索引由根节点块、分支节点块和叶节点块组成。在树结构中，位于最底层的块称为叶节点块，每个叶子节点包含被索引列值和行所对应的 rowid，指向表里的数据行；在叶节点块上面是分支节点块，包含了索引项列(关键字)范围值和另一索引块(分支节点块或叶节点块)地址；B-树的根节点块只有一个，它是位于树的最顶端的分支节点。B-树索引结构如图 8-3 所示，B 表示分支节点，L 表示叶子节点。

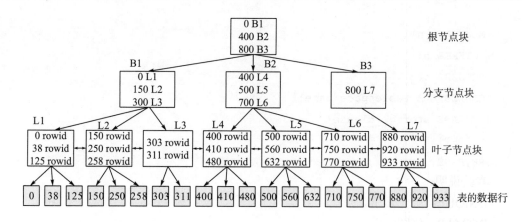

图 8-3 B-树索引结构图

2) 表簇索引

表簇索引即 B-tree 簇索引,将簇键键值与相应数据所在数据块地址(DBA)关联进行数据定位,而普通 B-树索引是将索引键值与数据的 rowid 进行关联。

3) 全局和本地索引

全局和本地索引是分区索引的两种类型。本地索引使用与表相同的分区键和范围界限进行分区,可以是 B-树或位图索引,如果是 B-树索引,它可以是唯一或不唯一的索引;全局索引是在一个索引分区中包含来自多个表分区的键,在创建全局分区索引时,必须定义分区键的范围和值,全局索引只能是 B-树索引。

4) 反向索引

反向索引是 B-树索引的一种分支,反转 B-树索引中索引列的键值字节,当索引列值递增且批量插入数据时,可使索引分布均匀。如:数据 1234、1235 和 1236 就被存储为 4321、5321 和 6321,使得索引会为每次新插入的行去更新不同索引块。

5) 位图索引

位图索引是通过使用位图,标识被索引的列值,管理与数据行的对应关系。位图索引非常适合于决策支持系统(Decision Support System,DSS)和数据仓库,不宜用于通过事务处理应用程序访问的表。

6) 函数索引

函数索引是根据函数对数据列计算的结果作为索引键值建立的索引结构。

8.3.1 创建索引

创建索引可使用 CREATE INDEX 语句,其基本的语法格式如下:

```
CREATE [UNIQUE|BITMAP] INDEX [schema.]<index_name>
ON [schema.]<table_name>(<column_name>[ASC|DESC],
[, <column_name>[ASC|DESC], …, [column_expression]])
|CLUSTER [schema.]<cluster_name>
[INITRANS n]
```

```
[MAXTRANS n]
[PCTFREE n]
[PCTUESD n]
[TABLESPACE tablespace_name]
[STORAGE storage_clause]
[NOSORT]
[REVERSE]
```

参数说明：

UNIQUE：唯一索引。

BITMAP：位图索引。

schema：指定索引（或表）所属的用户名。

index_name：要创建索引的名称。

table_name：索引所在表的名称。

column_name：要创建索引的列。

ASC|DESC：升序或降序。

column_expression：基于函数的索引查询时，对应的 SQL 函数或自定义函数创建的表达式。

CLUSTER：创建 cluster_name 簇索引，不能为散列簇创建簇索引。

TABLESPACE：指定索引所在的表空间。

NOSORT：数据库中的行以升序保存，在创建索引时不必对行排序。若索引列或多列的行不以升序保存，Oracle 会返回错误。

REVERSE：反向索引。NOSORT 不能与 REVERSE 一起指定。

INITRANS、MAXTRANS、PCTFREE、PCTUESD、STORAGE：与创建表类似。

注意：通常将索引与表存储到不同的表空间，以提高检索的性能。

【示例 8-13】创建 B-树索引

在 study 用户的 employees 表中按 employee_id 和 name 创建 B-树索引。

```
SQL> CREATE UNIQUE INDEX employees_pk ON employees (employee_id ASC)
  2  TABLESPACE USERS;
SQL> CREATE INDEX employees_index1_name ON employees (name ASC)
  2  TABLESPACE USERS;
SQL> COMMIT;
SQL> COL index_name FORMAT a25;
SQL> COL table_name FORMAT a15;
SQL> COL index_type FORMAT a15;
SQL> SELECT index_name,table_name,index_type,uniqueness FROM
  2  user_indexes WHERE table_name = 'EMPLOYEES';
INDEX_NAME                TABLE_NAME      INDEX_TYPE  UNIQUENESS
------------------------  --------------- ----------- ----------
```

| EMPLOYEES_PK | EMPLOYEES | NORMAL | UNIQUE |
| EMPLOYEES_INDEX1_NAME | EMPLOYEES | NORMAL | NONUNIQUE |

说明：在 employees 表按 employee_id 创建了唯一索引，按 name 创建了非唯一索引。通过 user_indexes 表可以查看 employees 表中的索引信息。

【示例 8-14】 建本地分区索引

在 study 用户的 orders 表中创建本地分区索引 orders_index_date。

```
SQL> CREATE INDEX orders_index_date ON orders (order_date ASC)
  2  LOCAL
  3  (
  4  PARTITION partition_before_2016
  5  TABLESPACE USERS
  6  PCTFREE 10
  7  INITRANS 2
  8  STORAGE
  9  (
 10     INITIAL 8388608
 11     NEXT 1048576
 12     MINEXTENTS 1
 13     MAXEXTENTS UNLIMITED
 14     BUFFER_POOL DEFAULT
 15  )
 16  NOCOMPRESS
 17  , PARTITION partition_before_2017
 18  TABLESPACE USERS02
 19  PCTFREE 10
 20  INITRANS 2
 21  STORAGE
 22  (
 23     INITIAL 8388608
 24     NEXT 1048576
 25     MINEXTENTS 1
 26     MAXEXTENTS UNLIMITED
 27     BUFFER_POOL DEFAULT
 28  )
 29  NOCOMPRESS
 30  )
 31  STORAGE
 32  (
```

```
33    BUFFER_POOL DEFAULT
34    )
35  NOPARALLEL;
SQL> COMMIT;
```

【示例 8-15】创建全局分区索引

在 study 用户的 orders 表中创建全局分区索引 orders_pk。

```
SQL> CREATE UNIQUE INDEX orders_pk ON orders (order_id ASC)
  2  GLOBAL PARTITION BY HASH (order_id)
  3  (
  4    PARTITION index_partition1 TABLESPACE USERS
  5  NOCOMPRESS
  6  , PARTITION index_partition2 TABLESPACE USERS02
  7  NOCOMPRESS
  8  )
  9  NOLOGGING
 10  TABLESPACE USERS
 11  PCTFREE 10
 12  INITRANS 2
 13  STORAGE
 14  (
 15    INITIAL 65536
 16    NEXT 1048576
 17    MINEXTENTS 1
 18    MAXEXTENTS UNLIMITED
 19    BUFFER_POOL DEFAULT
 20  )
 21  NOPARALLEL;
SQL> COMMIT;
```

说明：如果按范围分区，可以使用"GLOBAL PARTITION BY RANGE（日期字段）"短语。

【示例 8-16】创建基于函数的索引

在 study 用户的 employees 表中创建基于函数的索引 emp_upper_idx。

```
SQL> CREATE INDEX emp_upper_idx ON employees (UPPER(last_name));
SQL> SELECT * FROM employees WHERE UPPER(last_name)='KING';
```

说明：在上面的查询中，将按函数索引查找，以提升查询效率。

8.3.2 修改、删除索引

使用 ALTER INDEX 语句可以修改索引，使用 DROP INDEX 语句可以删除索引，主要操作的语法格式如下：

```
--设置索引不可用
ALTER INDEX index_name UNUSABLE;
--重建索引
ALTER INDEX index_name REBUILD;
--修改索引名
ALTER INDEX old_index_name RENAME TO new_index_name;
--删除索引
DROP INDEX index_name;
```

8.4 簇 表

簇（Cluster）是保存表数据的一种可选方式，它由共享相同数据块的一组表组成。创建簇后，用户可以在簇中创建表，这些表称为簇表。

8.4.1 簇的概念

我们使用 Oracle 数据库中 HR 用户自带的两个表 departments 和 employees 来阐述簇的概念。departments 表的结构如表 8-6 所示，employees 表的结构如表 8-7 所示。

表 8-6　departments 表结构（主要字段）

列名	数据类型	说明
DEPARTMENT_ID	NUMBER	部门编号
DEPARTMENT_NAME	VARCHAR2(30)	部门名称
LOCATION_ID	NUMBER	位置编号

表 8-7　employees 表结构（主要字段）

列名	数据类型	说明
EMPLOYEE_ID	NUMBER	员工编号
LAST_NAME	VARCHAR2(25)	员工姓名
DEPARTMENT_ID	NUMBER	所在部门编号

如果把表 departments 和表 employees 组合成簇，Oracle 数据库会把这两个表中每个部门的所有记录在物理上保存到相同的数据块中，如图 8-4 所示。从图中可以看到，簇表比非簇表更节省存储空间，并方便执行几个表的联合查询。

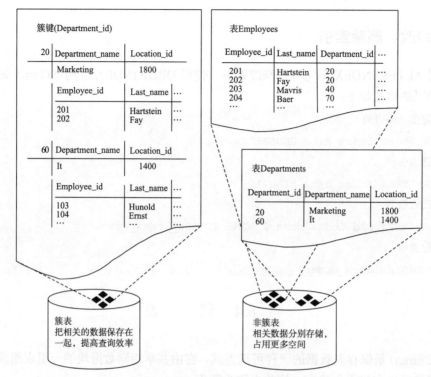

图 8-4 簇表与非簇表数据存储对比

Oracle 中对簇表的使用需要注意以下几点：

（1）簇里的表主要用于连接查询，如果表经常进行独立查询，或经常进行插入和更新操作，不适宜于创建簇表。

（2）簇键是表的一列或多列的组合，为簇表所共有。在创建簇表时需要指定簇键列，每个簇键值在簇和簇索引中只存储 1 次。

簇索引与簇表不同，它并不存在于簇中，而是与普通索引一样需要具有独立的存储空间。

8.4.2 创建簇表

创建簇表的顺序一般是：簇→簇表→簇索引→数据。

1）创建簇

创建簇的基本语法格式如下：

```
CREATE CLUSTER <cluster_name>
(<column date_type> [, column datatype]...)
[PCTUSED n]
[PCTFREE n]
[SIZE n]
[INITRANS n]
[MAXTRANS n]
```

```
[TABLESPACE tablespace]
[STORAGE storage]
```

参数说明:

cluster_name:簇名。

column:簇键列名。

date_type:簇键数据类型。

SIZE:指定估计的平均簇键,以及与其相关的行所需的字节数。

PCTFREE、PCTUSED、INITRANS、MAXTRANS、STORAGE、TABLESPACE:与创建表的参数类似。

【示例 8-17】创建簇

在 study 用户中创建簇 emp_dept_cluster,用于保存部门和员工数据。

```
SQL> CREATE CLUSTER study.emp_dept_cluster (dep_id NUMBER )
  2    PCTUSED 60
  3    PCTFREE 10
  4    SIZE 1024
  5    TABLESPACE USERS
  6    STORAGE (
  7      INITIAL 128 k
  8      NEXT 128 k
  9      MINEXTENTS 2
 10      MAXEXTENTS 20
 11    );
```

2)创建簇表

在 CREATE TABLE 语句中使用 CLUSTER 子句,可以创建簇表。

【示例 8-18】创建簇表

在 study 用户的 emp_dept_cluster 簇中创建簇表 dep_clu 和 emp_clu。

```
SQL>CREATE TABLE dep_clu (
  2   dep_id number,
  3   dep_name varchar2 (20)
  4  )
  5  CLUSTER emp_dept_cluster(dep_id);
SQL>CREATE TABLE emp_clu (
  2   emp_id NUMBER,
  3   emp_name VARCHAR2(20),
  4   job_id VARCHAR2(10),
  5   dep_id NUMBER
  4  )
  5  CLUSTER emp_dept_cluster(dep_id);
```

3)创建簇索引

【示例 8-19】 创建簇索引

在 study 用户的 emp_dept_cluster 簇中创建簇索引 index_clu。

```
SQL>CREATE INDEX index_clu ON CLUSTER emp_dept_cluster;
```

注意：若不创建索引，则在插入数据时会报错：ORA-02032：clustered tables cannot be used before the cluster index is built

8.4.3　查看簇信息

可以使用 dba_clusters 视图查看簇信息，dba_clusters 中的主要字段如表 8-8 所示：

表 8-8　DBA_CLUSTERS 主要字段属性

字段名	数据类型	说明
OWNER	VARCHAR2(30)	簇的所有者
CLUSTER_NAME	VARCHAR2(30)	簇名
TABLESPACE_NAME	VARCHAR2(513)	簇所属的表空间名
CLUSTER_TYPE	VARCHAR2(5)	簇的类型，B-树索引或者散列簇

【示例 8-20】 查询簇信息

使用 system 用户，以 sysdba 的身份登录后查询簇信息。

```
SQL> COL cluster_name FORMAT a20;
SQL> COL owner FORMAT a20;
SQL> COL tablespace_name FORMAT a20;
SQL> SELECT cluster_name, owner, tablespace_name, cluster_type
  2  FROM dba_clusters;
CLUSTER_NAME         OWNER   TABLESPACE_NAME      CLUSTER_TYPE
-------------------- ------  -------------------  ---------------
SMON_SCN_TO_TIME_AUX SYS     SYSAUX               INDEX
C_COBJ#              SYS     SYSTEM               INDEX
C_FILE#_BLOCK#       SYS     SYSTEM               INDEX
C_MLOG#              SYS     SYSTEM               INDEX
C_OBJ#               SYS     SYSTEM               INDEX
C_OBJ#_INTCOL#       SYS     SYSTEM               INDEX
C_RG#                SYS     SYSTEM               INDEX
C_TOID_VERSION#      SYS     SYSTEM               INDEX
C_TS#                SYS     SYSTEM               INDEX
C_USER#              SYS     SYSTEM               INDEX
```

```
EMP_DEPT_CLUSTER              STUDY     USERS                 INDEX
```
已选择 11 行。

说明：从上面的信息可以看出，study 用户在 users 表空间中创建了一个簇 emp_dept_cluster。

8.4.4 管理簇

簇创建好后，可以修改和删除簇。

1) 修改簇

可以使用 ALTER CLUSTER 语句修改簇，其基本语法格式如下：

```
ALTER CLUSTER <cluster_name>
[PCTUSED n]
[PCTFREE n]
[SIZE n]
[INITRANS n]
[MAXTRANS n]
[TABLESPACE tablespace]
[STORAGE storage]
```

注意：只有拥有 ALTER ANY CLUSTER 系统权限的用户才可以修改簇。

【示例 8-21】修改簇

修改簇 emp_dept_cluster 的 PCTUSED 属性。

```
SQL> ALTER CLUSTER emp_dept_cluster PCTUSED 100;
```

2) 删除簇

可以使用 DROP CLUSTER 语句删除簇，其基本语法格式如下：

```
DROP CLUSTER <cluster_name>
[INCLUDING TABLES[CASCADE CONSTRAINTS]];
```

参数说明：

cluster_name：簇名。

INCLUDING TABLES：删除簇时将其中的簇表一起删除。

CASCADE CONSTRAINTS：删除外键约束。

8.5 视 图

Oracle 的视图有普通视图和物化视图，都是从一个或几个实体表（或视图）导出的表，产生视图的表叫做该视图的基表。

普通视图本身是不包含任何真实数据的虚拟表，不占用物理存储空间。视图只有逻辑定义，每次使用时只是重新执行一条查询的 SQL 语句，基表中才存储视图的真实数据。

物化视图是一个含有真实数据的物理表，要占用相应的物理存储空间，它是远程数据的本地副本,或者用来生成基于数据表求和的汇总表。物化视图存储的是基于远程表的数据,也称为快照。

视图的作用有:

(1) 提供各种数据表现形式。可以使用不同的方式将基表的数据展现在用户面前。

(2) 隐藏数据的逻辑复杂性并简化查询语句。多表查询语句一般是比较复杂的,容易写错,如果创建一个视图,用户就可以直接对这个视图进行"简单查询"而获得结果。

(3) 提供某些安全性保证。视图提供了一种可以控制的方式,即可以让不同的用户看见不同的列,而不允许访问那些敏感的列,这样就可以保证敏感数据不被用户看见。

(4) 简化用户权限的管理。可以将视图的权限授予用户,而不必将基表中某些列的权限授予用户,这样就简化了用户权限的定义。

8.5.1 创建普通视图

可使用 CREATE 语句创建普通视图,其基本的语法格式如下:

```
CREATE [OR REPLACE] [FORCE|NOFORCE] VIEW [schema.]<view_name>
[(column1,column2,...)]
AS
SELECT ...
[WITH CHECK OPTION]
[CONSTRAINT <constraint_name>]
[WITH READ ONLY];
```

参数说明:

OR REPLACE:如果存在同名的视图,则使用新视图"替代"已有的视图。

FORCE:强制创建视图,不考虑基表是否存在,也不考虑是否具有使用基表的权限。

NOFORCE:表示要创建视图的基表必须存在,否则无法创建,是默认参数。

column1,column2,...:视图的列名,列名的个数必须与 SELECT 子查询中列的个数相同;如果 SELECT 子查询中包含函数或表达式,则必须为其定义列名。

WITH CHECK OPTION:指定对视图执行的 DML 操作必须满足 SELECT 子查询的条件,即通过视图进行的增删改操作时,必须是 SELECT 子查询所能查询到的数据。默认情况下,在增删改之前是不会检查这些行是否能被 SELECT 子查询检索到。

CONSTRAINT:设置列约束。

WITH READ ONLY:创建的视图只能用于查询数据,而不能用于更改数据。

> 注意:要在当前用户中创建视图,用户必须具有 CREATE VIEW 系统权限;要在其他用户中创建视图,用户必须具有 CREATE ANY VIEW 系统权限。

Oracle 普通视图定义可以分为:

1) 简单视图定义

简单视图定义是指基于单个表建立的,不包含任何函数、表达式和分组数据的视图。

2）连接视图定义

连接视图定义是指基于多个表所创建的视图，即定义视图的查询是一个连接查询。主要目的是为了简化连接查询。

3）复杂视图定义

复杂视图定义是指包含函数、表达式、或分组数据的视图。主要目的是为了简化查询，并不用于执行 DML 操作。

【示例 8-22】创建连接视图

在 HR 用户中创建 employees 表、departments 表和 jobs 表联合查询的连接视图 v_emp_dept。

```
SQL> CREATE OR REPLACE VIEW v_emp_dept
  2  AS
  3  SELECT e.employee_id,e.last_name,j.job_title,d.department_name
  4  FROM departments d JOIN employees e
  5  ON d.department_id=e.department_id JOIN jobs j
  6  ON e.job_id=j.job_id;
视图已创建。
SQL> COMMIT;
SQL> SELECT * FROM v_emp_dept WHERE job_title='Programmer';
EMPLOYEE_ID LAST_NAME   JOB_TITLE      DEPARTMENT_NAME
----------- ----------- -------------- ------------------
       107  Lorentz     Programmer     IT
       106  Pataballa   Programmer     IT
       105  Austin      Programmer     IT
       104  Ernst       Programmer     IT
       103  Hunold      Programmer     IT
```

8.5.2 操作普通视图

普通视图本身属于一个 Oracle 对象，可以像实体表那样被查询和删除，还可以在视图上定义一个新的视图。但由于普通视图中的数据是虚拟的，对数据更新时会存在一定限制。操作视图说明：

1）简单视图

简单视图可以通过视图对基表中的数据进行增、删、改操作。

2）连接视图

一般不对连接视图做 DML 操作。如果创建连接视图的 SELECT 查询不包含下面的结构，这样的连接视图是可更新的。

①集合运算符(UNION，INTERSECT，MINUS)；

②DISTINCT 关键字；

③GROUP BY，ORDER BY，CONNECT BY 或 START WITH 子句；

④子查询；

⑤分组函数；

⑥需要更新的列不是由"列表达式"定义的；

⑦基表中所有 NOT NULL 列均属于该视图。

3）复杂视图

复杂视图主要用于执行查询操作，不用于执行 DML 操作。

【示例 8-23】在 HR 用户中通过连接视图 V_EMP_DEPT 插入数据。

```
SQL> INSERT INTO v_emp_dept VALUES(999,'Eric','Programmer','IT');
SQL 错误：ORA-01776: 无法通过联接视图修改多个基表
```

8.5.3 普通视图的更改与删除

普通视图定义好之后，是可以更改的，但视图更改（或重定义）需注意以下两点：

(1) 由于视图只是一个虚表，其中没有数据，所以更改视图只是改变数据字典中对该视图的定义信息，不会影响视图基表。

(2) 视图之后，依赖于该视图的所有视图和 PL/SQL 程序都将变为 INVALID（失效）状态。

可以通过执行 CREATE OR REPLACE VIEW 语句来更改视图。

【示例 8-24】更改视图

在 HR 用户中将视图 v_emp_dept 改为只读。

```
SQL> CREATE OR REPLACE VIEW v_emp_dept
2   AS
3   SELECT e.employee_id,e.last_name,j.job_title,d.department_name
4   FROM departments d JOIN employees e
5   ON d.department_id=e.department_id JOIN jobs j
6   ON e.job_id=j.job_id
7   WITH READ ONLY;
视图已创建。
```

> 注意：视图修改后，会保留视图上的权限，但与该视图相关的存储过程和视图会失效。

可以通过执行 DROP VIEW 语句来删除视图。其语法格式如下：

```
DROP VIEW <view_name>
```

> 注意：视图被删除后，其他引用该视图的视图及存储过程等都会失效。

8.5.4 创建物化视图

可使用 CREATE MATERIALIZED VIEW 语句创建物化视图，其基本的语法格式如下：

```
CREATE MATERIALIZED VIEW [schema.]<view_name>
[BUILD <IMMEDIATE|DEFERRED>]
[REFRESH <FAST|COMPLETE|FORCE>
[ON <DEMAND|COMMIT>]
[START WITH date1]
[NEXT date2]
[WITH <PRIMARY KEY|ROWID>]]
AS
SELECT ...
```

参数说明：

BUILD <IMMEDIATE|DEFERRED>：BUILD IMMEDIATE 是在创建物化视图的时候就生成数据；BUILD DEFERRED 则在创建时不生成数据，以后根据需要再生成数据。默认为 BUILD IMMEDIATE。

REFRESH <FAST|COMPLETE|FORCE>：REFRESH FAST（快速刷新）是采用增量刷新的机制，只将自上次刷新以后对基表进行的所有操作刷新到物化视图中去，该刷新方式必须要创建基于基表的视图日志；REFRESH COMPLETE（完全刷新）是删除视图中所有的记录，然后根据物化视图查询语句的定义重新生成物化视图；REFRESH FORCE 是默认的数据刷新方式，Oracle 会自动判断是否满足快速刷新的条件，如果满足则进行快速刷新，否则进行完全刷新。

ON <DEMAND|COMMIT>：ON DEMAND 指物化视图在用户需要的时候进行刷新；ON COMMIT 指一旦基表进行了事务提交（COMMIT），则立刻刷新，更新物化视图，使得数据和基表一致。

START WITH date1：第一次刷新时间。

NEXT date2：刷新时间间隔。

WITH <PRIMARY KEY|ROWID>：创建基于 PRIMARY KEY，或基于 ROWID 的物化视图。

【示例 8-25】创建物化视图

在 HR 用户中创建物化视图 mv_materialized_emp。

```
SQL> CREATE MATERIALIZED VIEW mv_materialized_emp
  2  REFRESH FORCE ON DEMAND
  3  START WITH SYSDATE Next SYSDATE+1
  4  AS
  5  SELECT * FROM employees;
```

说明：上面创建的物化视图相当于 employees 表的一个副本，采用 FORCE 方式每天刷新一次数据。

> 注意：可以使用 ALTER MATERIALIZED VIEW 修改物化视图；使用 DROP MATERIALIZED VIEW 删除物化视图；使用 CREATE INDEX 在物化视图上建立索引等。

8.6 序　　列

序列(SEQUENCE)是 Oracle 数据库中很重要的一个对象，利用它可以生成唯一的整数，主要用于自动地生成主键值。

8.6.1 创建序列

可以使用 CREATE SEQUENCE 语句创建序列，其基本语法格式如下：

```
CREATE SEQUENCE <sequence_name>
[INCREMENT BY n]
[START WITH n]
[{MAXVALUE n|NOMAXVALUE}]
[{MINVALUE n|NOMINVALUE}]
[{CYCLE|NOCYCLE}]
[{CACHE n|NOCACHE}];
```

参数说明：

sequence_name：创建序列名称。

INCREMENT BY：用于定义序列的步长，如果省略，则默认为 1。如果 n 为负数，则代表序列的值是按照此步长递减。

START WITH：定义序列的初始值(即产生的第一个值)，默认为 1。

MAXVALUE：定义序列生成器能产生的最大值。选项 NOMAXVALUE 是默认选项，代表没有最大值定义，这时对于递增序列，系统能够产生的最大值是 10 的 27 次方；对于递减序列，最大值是-1。

MINVALUE：定义序列生成器能产生的最小值。选项 NOMINVALUE 是默认选项，代表没有最小值定义，这时对于递减序列，系统能够产生的最小值是-10 的 26 次方；对于递增序列，最小值是 1。

CYCLE|NOCYCLE：表示当序列生成器的值达到限制值后是否循环。CYCLE 代表循环，NOCYCLE 代表不循环。

CACHE：定义存放序列的内存块的大小，默认为 20。NOCACHE 表示不对序列进行内存缓冲。

【示例 8-26】创建序列

在 study 用户中使用默认值创建序列 seq_test。

```
SQL> CREATE SEQUENCE seq_test
```

序列已创建。

8.6.2 使用序列

序列创建后，可以使用下列属性访问序列中的值：
（1）NEXTVAL：返回序列中下一个有效的值，任何用户都可以引用。
（2）CURRVAL：存放序列的当前值。

【示例 8-27】使用序列

在 study 用户中，使用序列 seq_test 为数据库表 test 主键赋值。

```
SQL>CREATE TABLE test
2   (id NUMBER NOT NULL,
3   name VARCHAR2(25),
4   CONSTRAINT "id_pk" PRIMARY KEY(id)
5   )
6   TABLESPACE USERS;
SQL> COMMIT;
SQL> SELECT seq_test.CURRVAL FROM dual;
SELECT SEQ_TEST.CURRVAL FROM DUAL
       *
第 1 行出现错误:
ORA-08002: 序列 SEQ_TEST.CURRVAL 尚未在此会话中定义
SQL> INSERT INTO test VALUES(seq_test.NEXTVAL,'软件工程1');
已创建 1 行。
SQL> COMMIT;
SQL> SELECT * FROM test;
ID        NAME
--------  ----------------
1         软件工程1
SQL> SELECT seq_test.CURRVAL FROM dual;
  CURRVAL
----------
   1
SQL> INSERT INTO test VALUES(seq_test.NEXTVAL,'软件工程2');
已创建 1 行。
SQL> COMMIT;
SQL> SELECT * FROM test;
ID        NAME
--------  ----------------
```

1	软件工程 1
2	软件工程 2

注意：在序列操作中，只有当用户调用 NEXTVAL 之后，才真正创建了这个序列，而此后用户才可以调用 CURRVAL 属性获取当前序列值。

8.6.3 修改、删除序列

可以使用 ALTER SEQUENCE 语句修改序列，其语法格式如下：

```
ALTER SEQUENCE <sequence_name>
[INCREMENT BY n]
[START WITH n]
[{MAXVALUE n|NOMAXVALUE}]
[{MINVALUE n|NOMINVALUE}]
[{CYCLE|NOCYCLE}]
[{CACHE n|NOCACHE}];
```

其中，参数与序列创建时类似。修改序列需注意：
(1) 必须是序列的拥有者或对序列有 ALTER 权限者。
(2) 只有将来的序列值会被改变。
(3) 改变序列的初始值只能通过删除序列之后重建序列的方法实现。

可以使用 DROP SEQUENCE 删除序列，其语法格式如下：

```
DROP SEQUENCE <sequence_name>
```

8.6.4 自动序列

Oracle 12c 提供了类似于 MySQL 和 SQL Server 数据库那样的自动增长列，在数据库表创建时指定。

创建自动序列 GENERATED BY DEFAULTS AS IDENTITY 语句实现，其语法格式如下：

```
CREATE TABLE <table_name>
(
  column_name column_type GENERATED BY DEFAULTS AS IDENTITY(
[INCREMENT BY n]
[START WITH n]
[{MAXVALUE n|NOMAXVALUE}]
[{MINVALUE n|NOMINVALUE}]
[{CYCLE|NOCYCLE}]
[{CACHE n|NOCACHE}]),
column_name column_type, …
);
```

【示例 8-28】创建自动序列

在 study 用户中，为数据库表 test2 主键创建自动序列。

```
SQL>CREATE TABLE test2
2  (id NUMBER GENERATED BY DEFAULT AS IDENTITY(START WITH 10),
3   name VARCHAR2(25),
4   CONSTRAINT "id_pk" PRIMARY KEY(id)
5  )
6  TABLESPACE USERS;
SQL> INSERT INTO test(name)VALUES('软件工程1');
SQL> INSERT INTO test(name)VALUES('软件工程2');
SQL> COMMIT;
SQL> SELECT * FROM test2;
ID        NAME
--------  ---------------
10        软件工程1
11        软件工程2
```

注意：不能用 DROP SEQUENCE 删除自动序列。

8.7 同 义 词

同义词是数据方案对象的一个别名，常用于简化对象访问和提高对象访问的安全性，与视图类似，就是一种映射关系，不占用实际存储空间。Oracle 的同义词有两类：

1）公有同义词

由 Public 用户组拥有，数据库中所有的用户都可以使用公有同义词。

2）私有同义词

由创建该同义词的用户拥有。拥有者可以授权其他用户使用自己的私有同义词。

8.7.1 创建同义词

可以使用 CREATE SYNONYM 创建同义词，其语法格式如下：

```
CREATE [PUBLIC] SYNONYM <synonym_name> for [schema.]<obj_name>;
```

参数说明：

synonym_name：同义词名称。

obj_name：数据库对象名。

【示例 8-29】创建同义词

在 hr 用户中，为数据库表 departments 创建同义词 dept。

```
SQL> CREATE SYNONYM dept FOR hr.departments;
同义词已创建。
```

```
SQL> COMMIT;
SQL> SELECT * FROM dept;
DEPARTMENT_ID   DEPARTMENT_NAME         MANAGER_ID  LOCATION_ID
-------------   --------------------    ----------  -----------
10              Administration          200         1700
20              Marketing               201         1800
30              Purchasing              114         1700
40              Human Resources         203         2400
50              Shipping                121         1500
60              IT                      103         1400
...
已选择 27 行。
```

8.7.2 删除同义词

使用 DROP SYNONYM 可以删除同义词，其语法格式如下：

```
DROP [PUBLIC] SYNONYM <synonym_name>;
```

8.8 XML 和 Oracle 数据库

XML（Extensibe Markup Language，可扩展标记语言）是一种通用标记语言，特别在数据的传递和共享上有其重要的作用。

8.8.1 从关系数据生成 XML

Oracle 数据库包含了很多 SQL 函数，使用这些函数可以很方便地从关系数据生成 XML。下面介绍几个常用的函数：

1）XMLELEMENT()函数

使用 XMLELEMENT()函数可以从关系数据生成 XML 元素。该函数将返回一个 XMLTYPE 对象，XMLTYPE 是内置的 Oracle 数据库类型，用来表示 XML 数据。其语法格式如下：

```
XMLELEMENT(identifier[, xml_attribute_clause][, value_expr])
```

参数说明：

Identifier：用于指定元素名（xml 的标签名）。

xml_attribute_clause：用于指定元素属性子句。

value_expr：用于指定元素值。

【示例 8-30】导出数据生成 XML

在 hr 用户中，将数据库表 employees 中的数据生成 XML。

```
SQL> SELECT XMLELEMENT(
2  "employee",
3  XMLELEMENT("id", employee_id),
4  XMLELEMENT("name", first_name||' '||last_name)
5  )AS xml_employee
6  FROM employees WHERE department_id=60;
XML_EMPLOYEE
------------------------
<employee><id>103</id><name>Alexander Hunold</name></employee>
<employee><id>104</id><name>Bruce Ernst</name></employee>
<employee><id>105</id><name>David Austin</name></employee>
<employee><id>106</id><name>Valli Pataballa</name></employee>
<employee><id>107</id><name>Diana Lorentz</name></employee>
```

2) XMLATTRIBUTES()函数

XMLATTRIBUTES()函数一般与 XMLELEMENT()函数联合使用，用于设置元素的属性，其语法格式如下：

```
XMLATTRIBUTES(value_expr AS attr_name [, …])
```

参数说明：

value_expr：用于指定属性值。

attr_name：用于指定属性名。

【示例 8-31】导出数据生成带属性的 XML

在 hr 用户中，将数据库表 employees 中的数据生成带属性的 XML。

```
SQL> SELECT XMLELEMENT(
2  "employee",
3  XMLATTRIBUTES(
4  employee_id AS "id",
5  first_name||' '||last_name AS "name"
6  )
7  )AS xml_employee
8  FROM employees WHERE department_id=60;
XML_EMPLOYEE
----------------------------------
<employee id="103" name="Alexander Hunold"></employee>
<employee id="104" name="Bruce Ernst"></employee>
<employee id="105" name="David Austin"></employee>
<employee id="106" name="Valli Pataballa"></employee>
<employee id="107" name="Diana Lorentz"></employee>
```

8.8.2 XML DB 数据处理

Oracle XML DB 是一组专门为 XML 开发的内置高性能存储和检索技术，可用于存储、查询、更新、转换或处理 XML。

1) 创建能存储 XML 数据的 XML 表

能存储 XML 数据的表，一般有创建两种方式：

(1) 创建一个有 XMLTYPE 类型字段的表，其语法格式如下：

```
CREATE TABLE table_name (
col_name data_type, [, …, xml_col XMLTYPE]);
```

参数说明：

table_name：表名。

col_name data_type：普通列名及类型。

xml_col：存储 XML 数据列名。

XMLTYPE：XML 数据类型。

(2) 创建一个 XMLType 类型表，其语法格式如下：

```
CREATE TABLE table_name OF XMLTYPE;
```

参数说明：

table_name：表名。

xmltype：XML 数据类型。

【示例 8-32】创建 XML 表

在 study 用户中，创建 2 个 XML 表 xmlcontent 和 xmltable，xmlcontent 是一个有 XMLTYPE 类型字段的表，xmltable 是一个 XMLTYPE 类型的表。

```
SQL> CREATE TABLE xmlcontent(
2   keyvalue VARCHAR2(10)PRIMARY KEY,
3   xmlvalue XMLTYPE);
```

表已创建。

```
SQL> CREATE TABLE xmltable OF XMLTYPE;
```

表已创建。

```
SQL> COMMIT;
```

2) 直接插入 XML 数据

使用 INSERT INTO 语句，调用 sys.XMLType.createXML() 函数，可直接将 XML 数据插入到 XML 表中。

【示例 8-33】在 study 用户中，直接插入 XML 数据到表 xmlcontent。

```
SQL> INSERT INTO xmlcontent VALUES(1,
2   SYS.XMLTYPE.CREATEXML(
3   '<?xml version="1.0" encoding="UTF-8"?>
4     <shiporder orderid="000002"
5     xsi: noNamespaceSchemaLocation="shiporder.xsd"
```

```
 6        xmlns: xsi="http: //www.w3.org/2001/XMLSchema-instance">
 7          <orderperson>String1</orderperson>
 8          <shipto>
 9            <name>张三</name>
10            <address>成华区</address>
11            <city>成都市</city>
12            <country>中国</country>
13          </shipto>
14          <item>
15            <title>电脑1</title>
16            <note>小心轻放</note>
17            <quantity>32</quantity>
18            <price>1000.0</price>
19          </item>
20          <item>
21            <title>笔记本1</title>
22            <quantity>212</quantity>
23            <price>100.30</price>
24          </item>
25        </shiporder>
26  '));
```

1 行已插入。

SQL> COMMIT;
SQL> SELECT x.XMLVALUE.GETSTRINGVAL() FROM xmlcontent x;
X.XMLVALUE.GETSTRINGVAL()

<?xml version="1.0" encoding="GBK"?>
<shiporder orderid="000002" xsi:noNamespaceSchemaLocation="shiporder.xsd" xmlns:
xsi="http: //www.w3.org/2001/XMLSchema-instance">
  <orderperson>String1</orderperson>
  <shipto>
    <name>张三</name>
    <address>成华区</address>
    <city>成都市</city>
    <country>中国</country>
  </shipto>
  <item>

```
      <title>电脑 1</title>
      <note>小心轻放</note>
      <quantity>32</quantity>
      <price>1000.0</price>
   </item>
   <item>
      <title>笔记本 1</title>
      <quantity>212</quantity>
      <price>100.30</price>
   </item>
</shiporder>
```

如果往 xmltable 表中直接插入 XML 数据,可以使用下面方式:

```
INSERT INTO xmltable VALUES(SYS.XMLTYPE.CREATEXML(XML 数据));
```

3) 从 XML 文件导入数据

从 XML 文件中导入数据到 XML 表中,首先要创建一个目录(Directory)指向存放 XML 文件的路径(创建目录需要给用户 CREATE ANY DIRECTORY),再把要导入的 XML 文件存放于该目录中。

创建目录的语法如下:

```
CREATE DIRECTORY xmldir AS path_to_folder_containing_XML_file;
```

参数说明:

xmldir:要创建目录名称。

path_to_folder_containing_XML_file:目录指向的路径。

> 注意:创建目录需要给用户赋予 CREATE ANY DIRECTORY 权限。

为了演示从文件导入 XML 数据到 XML 表中,首先准备一个文件"PurchaseOrder.xml",该文件存放于 XMLDIR 目录中,内容如下:

```
<PurchaseOrder
   xmlns: xsi="http: //www.w3.org/2001/XMLSchema-instance"
   xsi: noNamespaceSchemaLocation="http: //www.oracle.com/xdb/po.xsd">
   <Reference>ADAMS-20011127121040988PST</Reference>
   <Actions>
      <Action>
         <User>SCOTT</User>
         <Date>2002-03-31</Date>
      </Action>
   </Actions>
   <Reject/>
   <Requestor>Julie P. Adams</Requestor>
   <User>ADAMS</User>
```

```
    <CostCenter>R20</CostCenter>
    <ShippingInstructions>
      <name>Julie P. Adams</name>
      <address>Redwood Shores, CA 94065</address>
      <telephone>650 506 7300</telephone>
    </ShippingInstructions>
    <SpecialInstructions>Ground</SpecialInstructions>
    <LineItems>
      <LineItem ItemNumber="1">
        <Description>The Ruling Class</Description>
        <Part Id="715515012423" UnitPrice="39.95" Quantity="2"/>
      </LineItem>
      <LineItem ItemNumber="2">
        <Description>Diabolique</Description>
        <Part Id="037429135020" UnitPrice="29.95" Quantity="3"/>
      </LineItem>
      <LineItem ItemNumber="3">
        <Description>8 1/2</Description>
        <Part Id="037429135624" UnitPrice="39.95" Quantity="4"/>
      </LineItem>
    </LineItems>
  </PurchaseOrder>
```

【示例 8-34】导入数据到 XML 表

在 study 用户中，从文件 PurchaseOrder.xml 导入数据到表 xmltable 中。

```
SQL> INSERT INTO xmltable VALUES(
2   XMLTYPE(bfilename('XMLDIR', 'PurchaseOrder.xml'),
3   nls_charset_id('AL32UTF8')));
1 行已插入。
SQL> SELECT x.OBJECT_VALUE.getSTRINGVal()FROM xmltable x;
X.OBJECT_VALUE.GETSTRINGVAL()
----------------------------------------------------
<PurchaseOrder xmlns:xsi="http://www.w3.org/2001/XMLSchema-instance" xsi:noNames
  paceSchemaLocation="http://www.oracle.com/xdb/po.xsd">
  <Reference>ADAMS-20011127121040988PST</Reference>
  <Actions>
    <Action>
      <User>SCOTT</User>
```

```
      <Date>2002-03-31</Date>
    </Action>
  </Actions>
…
</PurchaseOrder>
```

说明：XMLType()构造函数接受两个参数。第一个参数是 BFILE，它指向外部文件的指针。第二个参数是外部文件中的 XML 文本的字符集。在上面的 INSERT 语句中，BFILE 指向 PurchaseOrder.xml 文件，字符集是 AL32UTF8，这是标准的 UTF-8 编码。

4）从 XML 表中检索信息

XMLTYPE 对象有很多函数支持从 XML 表中检索信息，下面介绍几个主要操作(基于 PurchaseOrder.xml 文件的内容)：

◆ 查询整个 XML 文档信息

查询 XML 表中存储的整个 XML 文档信息，可以使用如下操作：

```
SELECT x.OBJECT_VALUE.getCLOBVal() FROM xmltable x;
SELECT x.OBJECT_VALUE.getSTRINGVal() FROM xmltable x;
--或者
SELECT value(x).getSTRINGVal() FROM xmltable x;
```

◆ 判断 XPath 中节点是否存在

【示例 8-35】判断 XPath 中节点是否存在

在表 xmltable 中，判断 XPath 中节点"/PurchaseOrder/Reference"是否存在，存在返回值 1 不存在返回 0。

```
SQL> SELECT existsNode(OBJECT_VALUE, '/PurchaseOrder/Reference') A
  2  FROM xmltable;

A
----------------------------------------------------------
1
```

◆ 提取 XPath 节点值

【示例 8-36】提取 XPath 节点值

在表 xmltable 中，提取 XPath 节点"/PurchaseOrder/Reference"值。

```
SQL> SELECT extractValue(OBJECT_VALUE, '/PurchaseOrder/Reference') A
  2  FROM xmltable
  3  WHERE existsNode(OBJECT_VALUE, '/PurchaseOrder/Reference') = 1;

A
----------------------------------------------------------
ADAMS-20011127121040988PST
```

◆ 提取 XPath 节点

【示例 8-37】提取 XPath 节点

在表 xmltable 中，提取 XPath 节点 "/PurchaseOrder/Reference"。

```
SQL> SELECT extract(OBJECT_VALUE, '/PurchaseOrder/Reference')
  2  "REFERENCE" FROM xmltable;
REFERENCE
--------------------------------------------------------
<Reference>ADAMS-20011127121040988PST</Reference>
```

◆ 解析 XML

【示例 8-38】使用 SQL 解析 XML 数据

在表 xmltable 中，使用 SQL 解析 XML 数据，获取相应信息。

```
SQL> SELECT extractValue(OBJECT_VALUE, '/PurchaseOrder/Reference')
  2  REFERENCE, extractValue(OBJECT_VALUE, '/PurchaseOrder/*//User'')
  3  USERID, CASE
  4  WHEN existsNode(OBJECT_VALUE, '/PurchaseOrder/Reject')= 1
  5  THEN 'Rejected'
  6  ELSE 'Accepted'
  7  END "STATUS",
  8  extractValue(OBJECT_VALUE, '//CostCenter')CostCenter
  9  FROM xmltable
 10 WHERE existsNode(OBJECT_VALUE, '//Reject')= 1;
REFERENCE                      USERID   STATUS      COSTCENTER
------------------------       ------   ---------   ----------
ADAMS-20011127121040988PST     SCOTT    Rejected    R20
```

5) 更新 XML 表数据

可以使用 UPDATE 语句更新 XML 表中的数据，主要有以下几种更新方式：

◆ 更新整个 XML 文档数据

【示例 8-39】更新整个 XML 文档数据

使用 PurchaseOrder2.xml 文件更新表 xmltable 中数据。

```
SQL> UPDATE xmltable x SET
  2  VALUE(x)=XMLTYPE(bfilename('XMLDIR', 'PurchaseOrder2.xml'),
  3  nls_charset_id('AL32UTF8'));
已更新 1 行。
```

◆ 更新一个节点值

【示例 8-40】更新一个节点值

更新 /PurchaseOrder/Reference 节点的值。

```
SQL> UPDATE xmltable x
  2  SET value(x)=updateXML(value(x),
```

```
3   '/PurchaseOrder/Reference/text()',
4   'MILLER-200203311200000000PST')
5   WHERE existsNode(value(x),
6   '/PurchaseOrder[Reference="ADAMS-20011127121040988PST"]')=1;
已更新 1 行。
SQL> COMMIT;
SQL> SELECT extract(OBJECT_VALUE, '/PurchaseOrder/Reference')
2   "REFERENCE" FROM xmltable;
REFERENCE
--------------------------------------------------------
<Reference>MILLER-200203311200000000PST</Reference>
```

◆ 更新一个节点树

【示例 8-41】更新一个节点树

更新/PurchaseOrders/LineItems/LineItem[2]节点树中数据。

```
SQL> UPDATE xmltable x
2   SET value(x)=
3   updateXML(value(x),
4   '/PurchaseOrder/LineItems/LineItem[2]',
5   xmltype('<LineItem ItemNumber="4">
6     <Description>Andrei Rublev</Description>
7     <Part Id="715515009928" UnitPrice="39.95" Quantity="2"/>
8   </LineItem>'
9   ))
10  WHERE existsNode(value(x),
11  '/PurchaseOrder[Reference="MILLER-200203311200000000PST"]')=1;
已更新 1 行。
SQL> COMMIT;
SQL> SELECT extract(OBJECT_VALUE,
2   '/PurchaseOrder/LineItems/LineItem[2]')"LineItem"
3   FROM xmltable;
LineItem
--------------------------------------------------------
<LineItem ItemNumber="4">
  <Description>Andrei Rublev</Description>
  <Part Id="715515009928" UnitPrice="39.95" Quantity="2"/>
</LineItem>
```

练习

一、判断题

1. Oracle 的非物化视图需要占用物理存储空间。（ ）
2. 分区表使用范围分区时，每一个分区都必须有一个 VALUES LESS THEN 子句，它指定了一个不包括在该分区中的上限值。（ ）
3. 视图修改后，不会保留视图上的权限。（ ）
4. 反向索引在索引列值递增且批量插入数据时，可使索引分布均匀。（ ）
5. 列表分区允许用户明确地控制无序行到分区的映射。（ ）

二、单项选择题

1. SQL 语句中的 UNIQUE 关键词表示（ ）
   A. 唯一约束   B. 特别约束
   C. 检查约束   D. 附加约束
2. Oracle 中，有一个名为 seq 的序列对象，以下语句能返回序列值但不会引起序列值增加的是（ ）
   A. SELECT seq.ROWNUM from dual   B. SELECT seq.ROWID from dual
   C. SELECT seq.CURRVAL from dual   D. SELECT seq.NEXTVAL from dual
3. 执行（ ）语句时，即使基表不存在也可以创建视图。
   A. CREATE FORCE VIEW   B. CREATE NOFORCE VIEW
   C. CREATE OR REPLACE VIEW   D. CREATE VIEW
4. 执行（ ）语句表示创建位图索引
   A. CREATE UNIQUE INDEX   B. CREATE BITMAP INDEX
   C. CREATE BTREE INDEX   D. CREATE INDEX
5. 在 Oracle 数据库中创建序列时，使用（ ）指定递增序列的初值。
   A. CYCLE   B. CACHE
   C. INCREMENT   D. START WITH

三、填空题

1. 在 CREATE TABLE 语句中，定义非空的关键字是_____。
2. 在 CREATE TABLE 语句中，定义主键的关键字是_____。
3. SQL 语句中修改表结构的命令是_____。
4. 在 ALTER TABLE 语句中，使用_____子句可修改列名。
5. 普通视图通常分为 _____、_____ 和 _____。

四、简答题

1. 阐述视图与基表的关系？
2. 什么是簇表？簇表与普通表有什么区别？
3. 为什么要使用分区表？简述范围分区的特点。
4. 索引的分类有哪些？在 Oracle 中如何创建和删除索引？
5. 简述在 Oracle 数据库中如何保存 XML 数据。

训练任务

1. 创建表和管理表

培养能力	工程能力、问题分析、设计/开发解决方案等		
掌握程度	★★★★★	难度	中
结束条件	表创建成功		

训练内容：
创建表和管理表是数据库操作的基础。训练要求：
(1) 创建 employees 表和 departments 表，表结构如下：
employees(employee_id number not null, name varchar2(40) not null, email varchar2(40), phone_number varchar2(40),
hire_date date not null, salary number(8, 2),
manager_id number, department_id number, photo blob)
departments(department_id number not null,
department_name varchar2(40) not null)
employees 表的主键是 employee_id；departments 表的主键是 department_id。
(2) 在 employees 表中，删除列 photo 列，为 email 列增加唯一约束，为 department_id 列增加外键约束（对应 departments 表中的主键），为 manager_id 列增加外键约束（对应 employees 表中的主键），为 salary 列增加检查约束（值大于 0），为 manager_id 列增加检查约束（不等于当前的 employee_id 值）。

2. 创建分区表

培养能力	工程能力、设计/开发解决方案		
掌握程度	★★★★	难度	中
结束条件	分区表创建成功		

训练内容：
当表中的数据量不断增大，查询数据的速度就会变慢，这时就应该考虑对表进行分区。按订单日期创建范围分区表 orders，将 2016 年以前的数据存储在 USERS 表空间中，将 2106 年的数据存储在 USERS02 表空间中。表的结构参见【示例 8-7】。

3. 创建索引

培养能力	工程能力、设计/开发解决方案		
掌握程度	★★★★	难度	中
结束条件	索引创建成功		

训练内容：
在 study 用户的 employees 表中按 name 创建非唯一索引，在 orders 表中，按 order_date 创建本地分区索引，按 order_id 创建全局索引。

第 9 章 表数据维护

本章目标

知识点	理解	掌握	应用
1.表数据的增加、修改、删除	✓	✓	✓
2.事务提交与回滚	✓	✓	✓
3.事务保存点	✓	✓	
4.事务 ACID 特性	✓	✓	
5.事务锁	✓	✓	

训练任务

- 完成表数据的增加、修改、删除
- 熟悉事务的 ACID 特性
- 熟悉事务锁

知识能力点

能力点 \ 知识点	知识点 1	知识点 2	知识点 3	知识点 4	知识点 5
工程知识	✓	✓			✓
问题分析			✓	✓	✓
设计/开发解决方案	✓		✓	✓	✓
研究			✓	✓	
使用现代工具	✓			✓	
工程与社会				✓	
环境和可持续发展			✓		✓
职业规范	✓	✓		✓	✓
个人和团队					✓
沟通					✓
项目管理				✓	✓
终身学习	✓			✓	✓

9.1 使用 INSERT INTO 语句添加行

当创建一个表之后,可以使用 INSERT 语句向指定的表中插入数据。INSERT 语句的

格式如下:

```
INSERT INTO <表名>(列名1,列名2,…,列名n)VALUES(值1,值2,…,值n);
```

列名是指表中定义的列,VALUES 子句的值一定要与列一一对应,且数据类型相同。

【示例 9-1】使用标准格式插入记录。

向 hr 用户的员工表 employees 插入 2 条记录。

```
SQL>INSERT INTO employees(
2  employee_id, first_name, last_name, email, phone_number, hire_date,
3  job_id, salary, manager_id, department_id )
4  VALUES( 400, '张三', '李', 'zs@qq.com', '123456789',
5  TO_DATE('2016-5-25', 'YYYY-MM-DD'), 'IT_PROG', 6000, 101, 60);
SQL>INSERT INTO employees(
2  employee_id, first_name, last_name, email, phone_number, hire_date,
3  job_id, salary, manager_id, department_id )
4  VALUES( 401, '李四', '李', 'ls@qq.com', '123456789',
5  TO_DATE('2016-7-25', 'YYYY-MM-DD'), 'IT_PROG', 5000, 101, 60);
SQL>COMMIT;
SQL>SELECT employee_id, first_name, salary, manager_id, department_id
2  FROM employees WHERE employee_id>=400;
EMPLOYEE_ID FIRST_NAME  SALARY      MANAGER_ID  DEPARTMENT_ID
----------- ----------  ----------  ----------  -------------
    400     张三          6000        101            60
    401     李四          5000        101            60
```

9.1.1 省略列的列表,默认值

如果插入语句中给出了表所有列的值,插入语句可以省略列名序列。

```
INSERT INTO <表名> VALUES(值1,值2,…,值n);
```

【示例 9-2】使用缺省列名插入记录。

向 hr 用户的工作表 jobs 插入 1 条记录。

```
SQL>INSERT INTO jobs VALUES('IT_MAN', 'IT Manager', 8000, 15000);
SQL>COMMIT;
SQL> SELECT * FROM jobs WHERE job_id='IT_MAN';
JOB_ID     JOB_TITLE              MIN_SALARY  MAX_SALARY
---------- ---------------------- ----------  ----------
IT_MAN     IT Manager              8000        15000
```

如果在表列上设置了 DEFAULT 属性(默认值),当插入数据时,可以不指定该列的值,Oracle 会自动为该列赋默认值。

【示例 9-3】使用默认值插入记录。

在 hr 用户的工作表 jobs 中，设置列 min_salary 的默认值为 4000，并采用默认值插入 1 条记录。

```
--设置列 MIN_SALARY 的默认值
SQL>ALTER TABLE jobs MODIFY min_salary DEFAULT 4000;
--使用默认值插入记录
SQL>INSERT INTO jobs(job_id, job_title, max_salary)
  2  VALUES('IT_PROG1', 'Programmer', 12000);
SQL>COMMIT;
SQL> SELECT * FROM jobs WHERE job_id='IT_PROG1';
JOB_ID      JOB_TITLE                    MIN_SALARY  MAX_SALARY
----------  ---------------------------  ----------  ----------
IT_PROG1    Programmer                       4000       12000
```

9.1.2 为列指定空值

插入数据时，未给出表中的某列，Oracle 将按以下两种情况处理：

(1) 当该列有默认值时，将以默认值插入。

(2) 如果该列没有设置默认值，将以空值(NULL)插入，如果该列声明了 NOT NULL，插入将导致错误。如果在 VALUES 子句中的列表中指定了 NULL，即使该列设置了默认值，也会以 NULL 插入。

【示例 9-4】为列指定空值插入记录。

在 hr 用户的工作表 jobs 中，插入一条记录，其中 min_salary 的值指定为 NULL，max_salary 未在插入列中给出。

```
SQL>INSERT INTO jobs(job_id, job_title, min_salary)
  2  VALUES('IT_PROG2', 'Programmer', null);
SQL>COMMIT;
SQL>SELECT * FROM jobs WHERE job_id LIKE 'IT%';
JOB_ID      JOB_TITLE                    MIN_SALARY  MAX_SALARY
----------  ---------------------------  ----------  ----------
IT_MAN      IT Manager                       8000       15000
IT_PROG     Programmer                       4000       10000
IT_PROG1    Programmer                       4000       12000
IT_PROG2    Programmer
```

注意：上面的插入操作结束后，新记录的 MIN_SALARY(列默认值：4000)和 MAX_SALARY 的值都为 NULL。

9.1.3 从一个表向另一个表复制行

从一个表向另一个表复制行，可以使用如下格式：

INSERT INTO <表名>(列名1, 列名2, …, 列名n)SELECT 列名1, 列名2, …, 列名n FROM <表名>;

> 注意：SELECT 子句中查询的列数与 INSERT INTO 语句中的列数要相同，且类型一致。

【示例9-5】从一个表向另一个表复制数据。

在 hr 用户中新建一个工作表 jobs1，其结构与 jobs 相同，并把 jobs 中 job_id 含有"IT"的记录，插入到 jobs1 中。

```
SQL>CREATE TABLE hr.jobs(
2  job_id VARCHAR2(10 BYTE),
3  job_title VARCHAR2(35 BYTE)NOT NULL,
4  min_salary NUMBER DEFAULT 4000,
5  max_salary NUMBER,
6  CONSTRAINT "job1_id_pk" PRIMARY KEY (job_id)
7  );
SQL>INSERT INTO jobs1 SELECT * FROM jobs WHERE job_id LIKE 'IT%';
SQL>COMMIT;
SQL>SELECT * FROM jobs1;
JOB_ID          JOB_TITLE            MIN_SALARY      MAX_SALARY
--------        ------------------   ----------      ----------
IT_MAN          IT Manager           8000            15000
IT_PROG1        Programmer           4000            12000
IT_PROG         Programmer           4000            10000
```

9.2 使用 UPDATE 语句修改行

可以使用 UPDATE 语句来修改表中的数据，其格式如下：

UPDATE <表名> SET 列名1=值1, 列名2=值2, …, 列名n=值n WHERE 条件表达式；

【示例9-6】修改表数据

修改 hr 用户中工作表 jobs，为 IT 类工种的 max_salary 增加 2000 元。

```
SQL> UPDATE jobs SET max_salary=max_salary+2000
2  WHERE job_id LIKE 'IT%';
SQL>COMMIT;
SQL>SELECT * FROM jobs WHERE job_id LIKE 'IT%';
JOB_ID      JOB_TITLE         MIN_SALARY      MAX_SALARY
```

```
----------      ---------------   ----------   ----------
IT_MAN          IT Manager        8000         17000
IT_PROG2        Programmer
IT_PROG1        Programmer        4000         14000
IT_PROG         Programmer        4000         12000
```

9.3 使用 DELETE 语句删除行

使用 DELETE 语句可以删除数据表中的记录，其格式如下：

```
DELETE FROM <表名> [WHERE 删除条件表达式]；
```

> 注意：如果被删除的表中有外键关系，需要先删除外键表中的数据，才能删除该表中的数据，否则将出现删除异常。

【示例 9-7】删除记录。

在 hr 用户工作表 jobs 中，删除一条记录。

```
SQL>DELETE FROM jobs WHERE job_id='IT_PROG2';
SQL>COMMIT;
SQL>SELECT * FROM jobs WHERE job_id LIKE 'IT%';
JOB_ID          JOB_TITLE         MIN_SALARY   MAX_SALARY
----------      ---------------   ----------   ----------
IT_MAN          IT Manager        8001         17000
IT_PROG         Programmer        4001         12000
IT_PROG1        Programmer        4001         14000
```

9.4 使用 MERGE 合并行

MERGE 语句是从 Oracle 9i R2 版本新增的 DML 语句，该语句可以合并数据表中的数据，简化一些数据处理的操作，比如：在一个步骤中同时完成更新和插入数据的操作。

MERGE 语句的格式如下：

```
MERGE INTO <表1> USING <表2> ON 条件表达式
WHEN MATCHED THEN UPDATE…
WHEN NOT MATCHED THEN INSERT…;
```

参数说明：

ON 子句：通常是表 1 和表 2 关联的条件；

UPDATE 和 INSERT 子句：是可选的，可以加 WHERE 子句进行过滤。

【示例 9-8】合并数据行。

使用 MERGE 语句，把 hr 用户工作表 jobs 中的 IT 工种的数据合并到工作表 jobs1 中，如果 jobs1 中有对应数据，就修改其 max_salary 的值；如果 jobs1 中没有对应数据，就插入新数据到 jobs1 中。

jobs 表与 jobs1 表中的原始数据如下：

```
SQL>SELECT * FROM jobs WHERE job_id LIKE 'IT%';
JOB_ID      JOB_TITLE         MIN_SALARY  MAX_SALARY
---------   ----------------  ----------  ----------
IT_MAN      IT Manager        8000        17000
IT_PROG     Programmer        4000        12000
IT_PROG1    Programmer        4000        14000
IT_TEST     Programmer test   2000        10000
SQL>SELECT * FROM jobs1;
JOB_ID      JOB_TITLE         MIN_SALARY  MAX_SALARY
---------   ----------------  ----------  ----------
IT_MAN      IT Manager        8000        15000
IT_PROG     Programmer        4000        10000
IT_PROG1    Programmer        4000        12000
```

使用 MERGE 语句操作如下：

```
SQL>MERGE INTO jobs1 j1 USING jobs j2 ON (j1.job_id=j2.job_id)
  2   WHEN MATCHED THEN
  3     UPDATE SET j1.max_salary=j2.max_salary
  4     WHERE j2.job_id LIKE 'IT%'
  5   WHEN NOT MATCHED THEN
  6     INSERT VALUES(j2.job_id, j2.job_title, j2.min_salary,
  7     j2.max_salary)WHERE j2.job_id LIKE 'IT%';
SQL>COMMIT;
SQL>SELECT * FROM jobs1;
JOB_ID      JOB_TITLE         MIN_SALARY  MAX_SALARY
---------   ----------------  ----------  ----------
IT_TEST     Programmer test   2000        10000
IT_MAN      IT Manager        8000        17000
IT_PROG     Programmer        4000        12000
IT_PROG1    Programmer        4000        14000
```

9.5　数据库事务

事务用于保证数据的一致性，它由一组相关的数据操作语句组成，该组操作语句要么全部成功，要么全部失败。对一组 SQL 语句操作构成事务，数据库操作系统必须确保这些操作的原子性、一致性、隔离性、持久性，即事务的 ACID 特性。事务的 ACID 特性有：

1) 原子性(Atomicity)

事务的原子性是指事务中包含的所有操作要么全做,要么不做,也就是说所有的活动在数据库中要么全部反映,要么全部不反映,以保证数据库的一致性。

2) 一致性(Consistency)

事务的一致性是指数据库在事务操作前和事务处理后,其中数据必须满足业务的规则约束。

3) 隔离性(Isolation)

隔离性是指数据库允许多个并发的事务同时对其中的数据进行读写或修改的能力,隔离性可以防止多个事务的并发执行时,由于它们的操作命令交叉执行而导致数据的不一致性。

4) 持久性(Durability)

事务的持久性是指在事务处理结束后,它对数据的修改应该是永久的。即便是系统在遇到故障的情况下也不会丢失,这是数据的重要性决定的。

在 Oracle 数据库中,没有提供开始事务处理语句,所有的事务都是隐式开始的,用户不能显示开始一个事务处理。当执行第一条修改数据库的语句,或者一些要求事务处理的场合都是事务的隐式开始。但是当用户想要终止一个事务处理时,必须显示使用 COMMIT 和 ROLLBACK 语句结束。

Oracle 提供了如下的事务控制语句:

1) SET TRANSACTION

设置事务属性。包括指定事务的隔离,指定回滚事务时所用的存储空间,命名事务等。

2) SET CONSTRAINS

设置事务的约束模式,即在事务中修改数据时,数据库中的约束是立即应用于数据,还是将约束推迟到当前事务结束后应用。

3) SAVEPOINT

在事务中建立一个存储的点。当事务处理发生异常而回滚事务时,可指定事务回滚到某存储点,然后从该存储点重新执行。

4) RELEASE SAVEPOINT

删除存储点。

5) ROLLBACK

回滚事务,取消对数据库所作的任何操作。

6) COMMIT

提交事务,对数据库的操作做持久的保存。

9.5.1 事务的提交和回滚

Oracle 数据库事务结束有两种方式:事务的提交(COMMIT)与事务回滚(ROLLBACK)。ROLLBACK 是取消数据的修改。

COMMIT 是确认数据的修改,过程是:重做记录中的事务被标记上所提交事务的唯一 SCN，日志写入进程将事务重做日志信息和事务 SCN，从重做日志缓冲区写到磁盘上的重做日志文件，释放 Oracle 持有的锁，标记事务为完成。

【示例 9-9】 事务的提交与回滚。

操作 hr 用户中工作表 jobs，插入记录过程中使用 COMMIT 和 ROLLBACK。

```
SQL>SELECT * FROM jobs WHERE job_id LIKE'IT%';
JOB_ID     JOB_TITLE           MIN_SALARY  MAX_SALARY
---------- ------------------- ----------- -----------
IT_MAN     IT Manager          8000        17000
IT_PROG    Programmer          4000        12000
IT_PROG1   Programmer          4000        14000
IT_TEST    Programmer test     2000        10000
SQL>INSERT INTO jobs VALUES('IT_TEST0', 'PROGRAMMER0', 4000, 10000);
SQL>INSERT INTO jobs VALUES('IT_TEST1', 'PROGRAMMER1', 4000, 10000);
SQL>INSERT INTO jobs VALUES('IT_TEST2', 'PROGRAMMER2', 4000, 10000);
SQL>ROLLBACK；  --回滚
回退已完成。
SQL>INSERT INTO jobs VALUES('IT_TEST3', 'PROGRAMMER3', 4000, 10000);
SQL>INSERT INTO jobs VALUES('IT_TEST4', 'PROGRAMMER4', 4000, 10000);
SQL>COMMIT；  --提交
提交完成。
SQL>SELECT * FROM jobs WHERE job_id LIKE 'IT%';
JOB_ID     JOB_TITLE           MIN_SALARY  MAX_SALARY
---------- ------------------- ----------- -----------
IT_MAN     IT Manager          8000        17000
IT_PROG    Programmer          4000        12000
IT_PROG1   Programmer          4000        14000
IT_TEST    Programmer test     2000        10000
IT_TEST3   PROGRAMMER3         4000        10000
IT_TEST4   PROGRAMMER4         4000        10000
```

说明：首先在 jobs 表中插入了三条记录，由于没有执行 COMMIT 操作，所以这三条记录并没有真正保存到 jobs 表中，当执行 ROLLBACK 时，将取消这三条记录的插入操作。而后面插入的两条记录，由于执行了 COMMIT 操作，这两条记录将永久保存到 jobs 表中。

注意：执行数据定义语句结束时，会默认执行 COMMIT；用户断开连接时，事务操作会自动 COMMIT；进程意外中止时，事务操作会自动 ROLLBACK。

9.5.2 事务的开始与结束

1) 事务开始

当第一个 DML（数据操纵语言：INSERT、DELETE、SELECT、UPDATE）语句执行时，将开始一个新的事务。当一个事务结束后，下一个可执行的 SQL 语句自动开始下一个事务。

2) 事务结束

事务结束主要有以下几种情况：

- ◆ 当执行 COMMIT 和 ROLLBACK 语句时，事务显示结束。
- ◆ 当执行完 DDL（数据定义语言：CREATE、ALTER、DROP、TRUNCATE）语句时，事务隐式结束。
- ◆ 当执行完 DCL（数据控制语言：GRANT、REVOKE、SET ROLE）语句时，事务隐式结束。
- ◆ 用户正常退出时，事务隐式结束，自动执行 COMMIT。
- ◆ 机器失效或者崩溃，事务隐式结束，自动执行 ROLLBACK。

9.5.3 保存点

使用 SAVEPOINT 语句在当前事务中创建一个标记，用 ROLLBACK TO SAVEPOINT 语句回退到标记处，这样可以把事务分为较小的部分来控制。

【示例 9-10】使用保存点。

操作 hr 用户中工作表 jobs，事务过程中使用 SAVEPOINT。

```
SQL>SELECT * FROM jobs WHERE job_id LIKE 'IT%';
JOB_ID              JOB_TITLE          MIN_SALARY         MAX_SALARY
-------------       -----------        ----------------   -----------
IT_MAN              T Manager          8000               17000
IT_PROG             Programmer         4000               12000
IT_PROG1            Programmer         4000               14000
IT_TEST             Programmer test    2000               10000
IT_TEST3            PROGRAMMER3        4000               10000
IT_TEST4            PROGRAMMER4        4000               10000
SQL> DELETE FROM jobs WHERE job_id='IT_TEST';
SQL> SAVEPOINT Del_OK;
保存点已创建。
SQL> UPDATE jobs SET min_salary=6000 WHERE job_id='IT_TEST3';
SQL> ROLLBACK TO Del_OK;
回退已完成。
SQL> COMMIT;
```

提交完成。

```
SQL> SELECT * FROM jobs WHERE job_id LIKE 'IT%';
JOB_ID      JOB_TITLE          MIN_SALARY  MAX_SALARY
----------  ----------------   ----------  ----------
IT_MAN      IT Manager         8000        17000
IT_PROG     Programmer         4000        12000
IT_PROG1    Programmer         4000        14000
IT_TEST3    PROGRAMMER3        4000        10000
IT_TEST4    PROGRAMMER4        4000        10000
```

说明：首先删除 jobs 表中一条记录，然后设置保存点，之后再修改一条记录，回滚到保存点，最后提交操作，查看结果可以看到，保存点前的删除操作执行成功，而保存点后的修改操作被取消。

> 注意：当使用多个相同的标记名设置保存点时，前面设置的保存点将被删除。

9.5.4 事务的 ACID 特性

1）事务的原子性（Atomicity）

是指事务中包含的所有操作要么全做，要么都不做。包括：语句级原子性、过程级原子性和事务级原子性。

【示例 9-11】事务的原子性操作。

```
--在 STUDY 用户中做如下操作
SQL> CREATE TABLE account(
2   id number(4)CONSTRAINT pk_id PRIMARY KEY,
3   name varchar2(30),
4   balance number(10, 2)
5   );
Table ACCOUNT 已创建。
SQL> INSERT INTO account VALUES(1001, '张三', 10000);
SQL> INSERT INTO account VALUES(1001, '李四', 10000);
第 1 行出现错误:
ORA-00001: 违反唯一约束条件 (STUDY.PK_ID)
SQL> SELECT * FROM account;
 ID    NAME                BALANCE
------ ------------------- -----------
 1001  张三                10000
SQL> BEGIN
2   INSERT INTO account VALUES(1002, '李四', 10000);
3   INSERT INTO account VALUES(1002, '高七', 10000);
```

```
  4  END;
  5  /
BEGIN
*
第 1 行出现错误：
ORA-00001：违反唯一约束条件 (STUDY.PK_ID)
ORA-06512：在 line 3
SQL> SELECT * FROM account;
  ID    NAME              BALANCE
------  --------------   ----------
 1001   张三              10000
SQL> INSERT INTO account VALUES(1003,'高七',10000);
SQL> COMMIT;
提交完成。
SQL> SELECT * FROM account;
  ID    NAME              BALANCE
------  --------------   ----------
 1001   张三              10000
 1003   高七              10000
```

说明：

◇ 语句级原子性

上面操作的每条语句就是最小级别的事务,该语句要么完全执行成功,要么完全失败,并且它不会影响其他语句的执行。

◇ 过程级原子性

过程中的所有代码要么都执行成功,要么都执行失败,并且不影响过程外的其他语句。

```
SQL> BEGIN
  2  INSERT INTO account VALUES(1002,'李四',10000);
  3  INSERT INTO account VALUES(1002,'高七',10000);
  4  END;
  5  /
```

上面的这个匿名块执行时,被看成一个整体,由于在执行第二条插入语句时报错,插入失败,则两条插入语句都会失败。

◇ 事务级原子性

示例代码中的所有语句和匿名块都当作一个整体,一个事务。用户在提交或回滚事务时,要么所有语句都执行,要么都失败。当最后执行 COMMT 时,"张三"和"高七"两条记录插入到了 ACCOUNT 表中。

2)事务的一致性(Consistency)

事务的一致性是指数据库在事务操作前和事务处理后,数据必须满足业务的规则约束。

【示例 9-12】 事务的一致性操作。

```
SQL> BEGIN
2    UPDATE account SET balance=balance-500 WHERE id=1001;
3    UPDATE account SET balance=balance+500 WHERE cid=1003;
4    END;
5    /
UPDATE account SET balance=balance+500 WHERE cid=1003;
                                             *
第 3 行出现错误:
ORA-06550: 第 3 行, 第 46 列:
PL/SQL: ORA-00904: "CID": 标识符无效
ORA-06550: 第 3 行, 第 1 列:
PL/SQL: SQL Statement ignored
SQL> SELECT * FROM account;
  ID   NAME                   BALANCE
------ ---------------------- ----------
 1001  张三                     10000
 1003  高七                     10000
```

说明：当从账户 1001 向账户 1003 转账 500 元时，由于修改账户 1003 时失败，为了保证事务的一致性(银行要求：转账前后 1001 账户和 1003 账的总金额必须一致)，修改 1001 的操作将被取消。

```
SQL> BEGIN
2    UPDATE account SET balance=balance-500 WHERE id=1001;
3    UPDATE account SET balance=balance+500 WHERE id=1002;
4    END;
5    /
PL/SQL 过程已成功完成。
SQL> COMMIT;
提交完成。
SQL> SELECT * FROM account;
  ID   NAME                   BALANCE
------ ---------------------- ----------
 1001  张三                     9500
 1003  高七                     10500
```

说明：当账户 1001 和账户 1003 都修改成功时，转账成功，事务可正常提交结束。

3) 事务的隔离性(Isolation)

事务的隔离性是指数据库允许多个并发的事务同时对其中的数据进行读写或修改的能力。隔离性可以防止多个事务的并发执行时，由于它们的操作命令交叉执行而导致数据的不一致性。并发事务对同一个数据库进行访问时，可能发生数据异常，常发生如下 3 种情况：

✧ 错读(或脏读)：当一个事务修改数据时，另一事务读取了该数据，但是第一事务由于某种原因取消对数据修改，使数据返回了原状态，这时第二个事务读取的数据与数据库中数据不一致，这就叫错读。

✧ 非重复读取(或不可重复读)：是指一个事务读取数据库中的数据后，另一个事务则更新了数据，当第一个事务再次读取其中的数据时，就会发现数据已经发生了改变，这就是非重复读取。非重复读取所导致的结果就是一个事务前后两次读取的数据不相同。

✧ 假读(或幻读)：如果一个事务基于某个条件读取数据后，另一个事务则更新了同一个表中的数据，这时第一个事务再次读取数据时，根据搜索的条件返回了不同的行，这就是假读。

事务中遇到的这些异常与事务的隔离性设置有关，事务的隔离性设置越多，异常就出现的越少，但并发效果就越低，事务的隔离性设置越少，异常出现的越多，并发效果越高。

针对上面 3 种读取数据时产生的不一致现象，在 ANSI SQL 标准 92 中定义了 4 个事务的隔离级别：

✧ READ UNCOMMITTED(非提交读)：并发事务时，可以读取未提交数据，不能避免上面 3 种数据异常；

✧ READ COMMITTED(提交读)：并发事务时，只能读取提交后的数据，可以避免错读，但不能避免非重复读取和假读；

✧ REPEATABLE READ(可重复读)：并发事务时，可以避免错读和非重复读取，但不能避免假读；

✧ SERIALIZALBE(串行读)：并发事务时，事务顺序执行，可以避免上面 3 种数据异常，但代价花费最高，性能很低。

Oracle 支持上述四种隔离层中的两种：READ COMMITTED 和 SERIALIZALBE。除此之外 Oracle 中还定义 READ ONLY 和 READ WRITE 隔离层。

✧ READ COMMITTED：是 oracle 默认的隔离层。

✧ SERIALIZALBE：事务与事务之间完全隔开，事务以串行的方式执行。

✧ READ ONLY：当使用 READ ONLY 时，事务中不能有任何修改数据库中数据的操作语句，这包括 INSERT、UPDATE、DELETE、CREATE 语句。

✧ READ WRITE：是默认设置，该选项表示在事务中可以有访问语句、修改语句。

使用 SET TRANSACTION 语句建立隔离级别：

✧ SET TRANSACTION READ ONLY

✧ SET TRANSACTION READ WRITE

✧ SET TRANSACTION ISOLATION LEVEL READ COMMITTED

✧ SET TRANSACTION ISOLATION LEVEL SERIALIZABLE

【示例 9-13】Oracle 默认隔离级 READ COMMITTED 并发事务操作。

打开 sqlplus 窗口 1，执行如下操作：

```
SQL> SELECT * FROM account;  --原始数据
```

```
ID         NAME                    BALANCE
------     --------------------    ----------
1001       张三                    9500
1003       高七                    10500
```

```
SQL> INSERT INTO account VALUES(1002,'李四',10000);
SQL> UPDATE account SET NAME='张山' WHERE ID=1001;
SQL> SELECT * FROM account;

ID         NAME                    BALANCE
------     --------------------    ----------
1001       张山                    9500
1003       高七                    10500
1002       李四                    10000
```

打开 sqlplus 窗口 2，查看 account 表中的数据：

```
SQL> SELECT * FROM account;
ID         NAME                    BALANCE
------     --------------------    ----------
1001       张三                    9500
1003       高七                    10500
```

从上面的操作可以看出，由于在窗口 1 中进行的事务操没有 COMMIT，所以在窗口 2 中是看不到操作结果的，这就是 READ COMMITTED 隔离级，避免了并发事务的错读。

如果在窗口 1 中执行了 COMMIT 操作：

```
SQL> COMMIT;
```
提交完成。

再到窗口 2 中去查看 account 表中的数据，数据已更新：

```
SQL> SELECT * FROM account;
ID         NAME                    BALANCE
------     --------------------    ----------
1001       张山                    9500
1003       高七                    10500
1002       李四                    10000
```

【示例 9-14】隔离级 READ ONLY 和 READ WRITE 并发事务操作。

打开 3 个 sqlplus 窗口，在窗口 1 中向 account 表中插入一行新数据：

```
SQL> SELECT * FROM account;   --查看原始数据
ID         NAME                    BALANCE
------     --------------------    ----------
1001       张山                    9500
1003       高七                    10500
1002       李四                    10000
```

第 9 章 表数据维护

```
SQL> INSERT INTO account VALUES(1004,'赵兵',12000);
SQL> SELECT * FROM account;
  ID      NAME                  BALANCE
  ------  --------------------  -----------
  1001    张山                   9500
  1003    高七                   10500
  1002    李四                   10000
  1004    赵兵                   12000
```

在窗口 2 中设置事务的隔离属性是 READ ONLY：

```
SQL> SET TRANSACTION READ ONLY;
事务处理集。
```

在窗口 3 中设置事务的隔离属性是 READ WRITE：

```
SQL> SET TRANSACTION READ WRITE;
事务处理集。
```

窗口 1 提交前，观察窗口 2 和窗口 3 数据，都不能看到在窗口 1 中插入的新数据：

```
SQL> SELECT * FROM account;   --在窗口 2 中查看数据
  ID      NAME                  BALANCE
  ------  --------------------  -----------
  1001    张山                   9500
  1003    高七                   10500
  1002    李四                   10000
SQL> SELECT * FROM account;   --在窗口 3 中查看数据
  ID      NAME                  BALANCE
  ------  --------------------  -----------
  1001    张山                   9500
  1003    高七                   10500
  1002    李四                   10000
```

在窗口 1 提交（COMMIT）后，观察窗口 2 和窗口 3 中的数据：

```
SQL> SELECT * FROM account;   --在窗口 2 中查看数据
  ID      NAME                  BALANCE
  ------  --------------------  -----------
  1001    张山                   9500
  1003    高七                   10500
  1002    李四                   10000
SQL> SELECT * FROM account;   --在窗口 3 中查看数据
  ID      NAME                  BALANCE
  ------  --------------------  -----------
  1001    张山                   9500
```

1003	高七	10500
1002	李四	10000
1004	赵兵	12000

从上面的执行结果可以看出，由于窗口 2 设置了 READ ONLY，所以当在窗口 1 提交数据后，在窗口 2 中仍然不能看到新插入的数据；而窗口 3 设置的是 READ WRITE，则可以看到新插入的数据。

在窗口 2 中向 account 表中插入数据（插入数据失败）：

```
SQL>INSERT INTO account VALUES(1005,'杨雨',10000);
INSERT INTO account VALUES(1005,'杨雨',10000)
            *
第 1 行出现错误:
ORA-01456：不能在 READ ONLY 事务处理中执行插入/删除/更新操作
```

在窗口 3 中向 account 表中插入数据（可以插入数据）：

```
SQL>INSERT INTO account VALUES(1005,'杨雨',10000);
SQL>SELECT * FROM account;
    ID   NAME                    BALANCE
------  ----------------------  ------------
    1001 张山                     9500
    1005 杨雨                     10000
    1003 高七                     10500
    1002 李四                     10000
    1004 赵兵                     12000
```

4）持久性（Durability）

事务的持久性是指在事务处理结束后，它对数据的修改应该是永久的。

9.5.5 锁

Oracle 数据库是一个多用户使用的共享资源，多个用户可同时访问相同的数据。当多个用户同时存取相同数据时，如果对并发操作不加以控制，就可能造成存取不正确的数据，破坏数据的一致性。

锁就是数据库用来控制共享资源并发访问的一种机制。Oracle 的锁按处理对象不同，可分为：

- ❖ 数据锁（Data Locks）：也称 DML 锁，用于保护数据的完整性；
- ❖ 字典锁（Dictionary Locks）：可称 DDL 锁，用于保护数据库对象的结构（如视图、表和索引的结构定义等）；
- ❖ 内部锁与闩（Internal Locks 和 Latches）：保护内部数据库结构。
- ❖ 分布式锁（Distributed Locks）：用于 OPS（并行服务器）中。
- ❖ 并行高速缓存管理锁（PCM Locks）：用于 OPS（并行服务器）中。

在 Oracle 中最主要的锁是数据锁(也称 DML 锁)，DML 锁主要用于保证并发情况下数据的完整性。一个 DML 操作，会对表或行加锁，即 TM 锁(表锁)和 TX 锁(行锁)。Oracle 数据锁的类型主要有 5 种：

- 行级共享锁(Row Share，RS 锁)
- 行级排他锁(Row Exclusive Table Lock，RX 锁)
- 共享锁(Share Table Lock，S 锁)
- 共享行级排他锁(Share Row Exclusive Table Lock，SRX 锁)
- 排他锁(Exclusive Table Lock，X 锁)

在 Oracle 数据库中，常使用视图 v$session 和 v$lock 查询锁的信息，其中 v$lock.LMODE 字段中的数字对应锁的模式为：

- 0：none
- 1：空(NULL)
- 2：行级共享锁(RS)
- 3：行级排他锁(RX)
- 4：共享锁(S)
- 5：共享行级排他锁(SRX)
- 6：排他锁(X)

下面通过示例对 Oracle 数据锁进行详细介绍。

1) 行级共享锁(Row Share，RS 锁)

RS 锁在锁类型中是限制最少的，在表的并发操作中使用率最高。当一个事务持有行级共享锁时，其他事务可以同时持有除排他锁之外的所有锁，允许其他事务进行查询、插入、更新、删除等操作。

加锁语法：

```
LOCK TABLE <表名> IN ROW SHARE MODE;
```

【示例 9-15】并发事务行级共享锁操作。

打开 3 个 Sql*Plus 窗口，窗口 1(hr 用户操作)，窗口 2(hr 用户操作)，窗口 3(system 用户操作)，首先在窗口 3 中查看锁信息：

```
SQL>SELECT a.sid, a.serial#, a.username, b.lmode, b.block, b.type
  2  FROM v$session a, v$lock b WHERE a.sid=b.sid AND USERNAME='HR';
SID            SERIAL#      USERNAME     LMODE    BLOCK    TY
-------------  -----------  -----------  -------  -------  -------
45             8584         HR           4        0        AE
226            10556        HR           4        0        AE
```

说明：从上面的信息可以看到，在窗口 1 中打开一个会话，SID=45，在窗口 2 中打开了一个会话，SID=226，其中 LMODE 字段是锁的模式(取值：0~6)，BLOCK 字段表示事务是否被阻塞(0：未阻塞；1：已阻塞)，TY 字段表示锁的类型(TM：表锁；TX 行锁)。

在窗口 1 中对 account 表设置行级共享锁(RS)：

```
SQL>LOCK TABLE account IN ROW SHARE MODE;
```

表已锁定。

窗口 3 中查看锁信息：

```
SQL>SELECT a.sid, a.serial#, a.username, b.lmode, b.block, b.type
2  FROM v$session a, v$lock b WHERE a.sid=b.sid AND USERNAME='HR';
   SID        SERIAL# USERNAME     LMODE    BLOCK TY
---------- ---------- ---------- ------- -------- -------
    45         8584   HR            2        0    TM
    45         8584   HR            4        0    AE
   226        10556   HR            4        0    AE
```

说明：SID=45 的事务持有了一个 account 表的 RS 锁（LMODE=2）。

在窗口 2 中对 account 表设置行级共享锁（RS）：

```
SQL>LOCK TABLE account IN ROW SHARE MODE;
```

表已锁定。

窗口 3 中查看锁信息：

```
SQL>SELECT a.sid, a.serial#, a.username, b.lmode, b.block, b.type
2  FROM v$session a, v$lock b WHERE a.sid=b.sid AND USERNAME='HR';
   SID        SERIAL#    USERNAME     LMODE   BLOCK    TY
---------- ----------  ----------    ------  -------  -------
   226        10556      HR             2       0      TM
    45         8584      HR             2       0      TM
    45         8584      HR             4       0      AE
   226        10556      HR             4       0      AE
```

说明：此时 SID=45 和 SID=226 的事务都持有了一个 account 表的 RS 锁（LMODE=2）。

在窗口 2 中对 account 表设置行级排他锁（RX）：

```
SQL>COMMIT;  --结束前面的事务，开启一个新的事务
```

提交完成。

```
SQL>LOCK TABLE account IN ROW EXCLUSIVE MODE;
```

表已锁定。

窗口 3 中查看锁信息：

```
SQL>SELECT a.sid, a.serial#, a.username, b.lmode, b.block, b.type
2  FROM v$session a, v$lock b WHERE a.sid=b.sid AND USERNAME='HR';
   SID        SERIAL#    USERNAME     LMODE   BLOCK    TY
---------- ----------  ----------    ------  -------  -------
   226        10556      HR             3       0      TM
    45         8584      HR             2       0      TM
    45         8584      HR             4       0      AE
   226        10556      HR             4       0      AE
```

说明：此时 SID=45 的事务仍然持有 account 表的 RS 锁（LMODE=2），而 SID=226 的

事务持有了一个 account 表的 RX 锁(LMODE=3)。

在窗口 2 中对 account 表设置共享锁(S)：

```
SQL>COMMIT;  --结束前面的事务,开启一个新的事务
提交完成。
SQL>LOCK TABLE account IN SHARE MODE;
表已锁定。
```

窗口 3 中查看锁信息：

```
SQL>SELECT a.sid, a.serial#, a.username, b.lmode, b.block, b.type
  2  FROM v$session a, v$lock b WHERE a.sid=b.sid AND USERNAME='HR';
SID         SERIAL#      USERNAME      LMODE    BLOCK    TY
-------     --------     ----------    ------   ------   -----
226         10556        HR            4        0        TM
45          8584         HR            2        0        TM
45          8584         HR            4        0        AE
226         10556        HR            4        0        AE
```

说明：此时 SID=45 的事务仍然持有 account 表的 RS 锁(LMODE=2)，而 SID=226 的事务持有了一个 account 表的 S 锁(LMODE=4)。

在窗口 2 中对 account 表设置共享行级排他锁(SRX)：

```
SQL>COMMIT;   --结束前面的事务,开启一个新的事务
提交完成。
SQL>LOCK TABLE account IN SHARE ROW EXCLUSIVE MODE;
表已锁定。
```

窗口 3 中查看锁信息：

```
SQL>SELECT a.sid, a.serial#, a.username, b.lmode, b.block, b.type
FROM v$session a, v$lock b WHERE a.sid=b.sid AND USERNAME='HR';
SID         SERIAL#      USERNAME      LMODE    BLOCK    TY
---------   ----------   -----------   ------   ------   ------
226         10556        HR            5        0        TM
45          8584         HR            2        0        TM
45          8584         HR            4        0        AE
226         10556        HR            4        0        AE
```

说明：此时 SID=45 的事务仍然持有 account 表的 RS 锁(LMODE=2)，而 SID=226 的事务持有了一个 account 表的 SRX 锁(LMODE=5)。

在窗口 2 中对 account 表设置排他锁(X)：

```
SQL>COMMIT;   --结束前面的事务,开启一个新的事务
提交完成。
SQL>LOCK TABLE account IN EXCLUSIVE MODE;
操作处于等待状态(被阻塞)…
```

窗口3中查看锁信息：

```
SQL>SELECT a.sid, a.serial#, a.username, b.lmode, b.block, b.type
FROM v$session a, v$lock b WHERE a.sid=b.sid AND USERNAME='HR';
   SID    SERIAL#    USERNAME        LMODE      BLOCK      TY
   ----   --------   -----------     -------    -------    -------
   226    10556      HR              0          0          TM
   45     8584       HR              2          1          TM
   45     8584       HR              4          0          AE
   226    10556      HR              4          0          AE
```

说明：此时SID=45的事务仍然持有account表的RS锁（LMODE=2），并设置了阻塞标志（BLOCK=1），而SID=226的事务对account表加X锁未成功（LMODE=0），处于等待状态，只有当窗口1中的事务提交（COMMIT）或回滚（ROLLBACK）后，才能继续执行。

在窗口1中提交事务：

```
SQL>COMMIT;
提交完成。
```

窗口3中查看锁信息：

```
SQL>SELECT a.sid, a.serial#, a.username, b.lmode, b.block, b.type
FROM v$session a, v$lock b WHERE a.sid=b.sid AND USERNAME='HR';
   SID    SERIAL#    USERNAME        LMODE      BLOCK      TY
   ----   --------   -----------     -------    -------    -------
   226    10556      HR              6          0          TM
   45     8584       HR              4          0          AE
   226    10556      HR              4          0          AE
```

说明：此时SID=45的事务持有的RS锁已解锁，SID=226的事务对account表加上了X锁（LMODE=6）。

当三个会话窗口中的事务都提交或回滚后，在窗口1中对account表重新设置行级共享锁（RS）：

```
SQL>LOCK TABLE account IN ROW SHARE MODE;
表已锁定。
```

窗口3中查看锁信息：

```
SQL>SELECT a.sid, a.serial#, a.username, b.lmode, b.block, b.type
FROM v$session a, v$lock b WHERE a.sid=b.sid AND USERNAME='HR';
   SID    SERIAL#    USERNAME        LMODE      BLOCK      TY
   ----   --------   -----------     -------    --------   -------
   45     8584       HR              2          0          TM
   45     8584       HR              4          0          AE
   226    10556      HR              4          0          AE
```

在窗口2中做如下操作：

第 9 章 表数据维护

```
SQL> UPDATE account SET NAME='杨小雨' WHERE ID=1005;
已更新 1 行。
SQL> SELECT * FROM account;
    ID  NAME                          BALANCE
------  --------------------      -----------
  1001  张山                             9500
  1005  杨小雨                           10000
  1003  高七                            10500
  1002  李四                            10000
  1004  赵兵                            12000
```

窗口 3 中查看锁信息：

```
SQL>SELECT a.sid, a.serial#, a.username, b.lmode, b.block, b.type
FROM v$session a, v$lock b WHERE a.sid=b.sid AND USERNAME='HR';
   SID  SERIAL#          USERNAME         LMODE    BLOCK    TY
------  ---------------  -----------      ------   -------  -------
   226  10556            HR               3        0        TM
    45  8584             HR               2        0        TM
   226  10556            HR               6        0        TX
    45  8584             HR               4        0        AE
   226  10556            HR               4        0        AE
```

说明：此时 SID=45 的事务持有一个行级共享锁 RS(TM，LMODE=2)，但仍然可以在窗口 2 中执行更新操作，更新完成后，SID=226 的事务将持有一个行级排他锁 RX(TM，LMODE=3)和一个排他锁(TX，LMODE=6)。

2) 行级排他锁(Row Exclusive Table Lock，RX 锁)

RX 锁比行级共享锁更多一些限制。当一个事务对数据资源加上行级排他锁(RX)后，其他事务仍可以对该资源再加行级共享锁(RS)和行级排他锁(RX)，但不能加共享锁(S)、共享行级排他锁(SRX)及排他锁(X)，允许其他事务对未加锁的数据行进行插入、更新、删除等操作，查询不受限制。

加锁语法：

```
LOCK TABLE <表名> IN ROW EXCLUSIVE MODE;
SELECT ... FROM <表名> ... FOR UPDATE OF...
INSERT INTO <表名>...
DELETE FROM <表名>...
UPDATE <表名>...
```

【示例 9-16】并发事务行级排他锁操作。

打开 3 个 sqlplus 窗口，窗口 1(HR 用户操作)，窗口 2(HR 用户操作)，窗口 3(SYSTEM 用户操作)，首先在窗口 3 中查看锁信息：

```
SQL>SELECT a.sid, a.serial#, a.username, b.lmode, b.block, b.type
```

```
FROM v$session a, v$lock b WHERE a.sid=b.sid AND USERNAME='HR';
   SID     SERIAL#         USERNAME      LMODE    BLOCK     TY
-------  --------------  -----------   -------  -------   -------
    45     8584           HR             4        0         AE
   226    10556           HR             4        0         AE
```

在窗口 1 中对 account 表设置行级排他锁（RX）：

```
SQL>LOCK TABLE account IN ROW EXCLUSIVE MODE;
表已锁定。
```

窗口 3 中查看锁信息：

```
SQL>SELECT a.sid, a.serial#, a.username, b.lmode, b.block, b.type
FROM v$session a, v$lock b WHERE a.sid=b.sid AND USERNAME='HR';
   SID     SERIAL#         USERNAME      LMODE    BLOCK     TY
-------  --------------  -----------   -------  -------   -------
    45     8584           HR             3        0         TM
    45     8584           HR             4        0         AE
   226    10556           HR             4        0         AE
```

说明：SID=45 的事务持有了一个 account 表的 RX 锁（LMODE=3）。

在窗口 2 中对 account 表做如下操作：

```
SQL>LOCK TABLE account IN ROW SHARE MODE;
表已锁定。
SQL>COMMIT;
提交完成。
SQL>LOCK TABLE account IN ROW EXCLUSIVE MODE;
表已锁定。
SQL>COMMIT;
提交完成。
SQL>LOCK TABLE account IN SHARE MODE;
操作处于等待状态(被阻塞)...
SQL>LOCK TABLE account IN SHARE ROW EXCLUSIVE MODE;
操作处于等待状态(被阻塞)...
SQL>LOCK TABLE account IN EXCLUSIVE MODE;
操作处于等待状态(被阻塞)...
```

说明：RX 锁与 RS 锁、RX 锁可以并发操作，而与 S 锁、SRX 锁、X 锁无法并发。

当三个会话窗口中的事务都提交或回滚后，在窗口 1 中做如下操作：

```
SQL>SELECT * FROM account WHERE id<1003 FOR UPDATE;
    ID     NAME           BALANCE
------  ------------   ------------
  1001   张山            9500
```

第 9 章 表数据维护

| 1002 | 李四 | 10000 |

窗口 3 中查看锁信息：

```
SQL>SELECT a.sid, a.serial#, a.username, b.lmode, b.block, b.type
FROM v$session a, v$lock b WHERE a.sid=b.sid AND USERNAME='HR';
    SID    SERIAL#         USERNAME      LMODE    BLOCK    TY
 -------  ------------   ------------   -------  -------  -------
    45     8584            HR              3        0      TM
    45     8584            HR              6        0      TX
    45     8584            HR              4        0      AE
    226    10556           HR              4        0      AE
```

说明：此时 SID=45 的事务持有一个行级排他锁 RX（LMODE=3），加锁对象是表（TM），还持有一个排他锁 X（LMODE=6）加锁对象是数据行（TX），而且只对 account 表中的 ID=1001 和 ID=1002 这两条数据行加锁，其他的数据未加锁。

如果在窗口 2 中做以下修改：

```
SQL> UPDATE account SET name='杨小雨' WHERE id=1005;
已更新 1 行。
SQL> SELECT * FROM account;
    ID     NAME                        BALANCE
 ------  ---------------------      -----------
   1001   张山                          9500
   1005   杨小雨                        10000
   1003   高七                          10500
   1002   李四                          10000
   1004   赵兵                          12000
```

窗口 3 中查看锁信息：

```
SQL>SELECT a.sid, a.serial#, a.username, b.lmode, b.block, b.type
FROM v$session a, v$lock b WHERE a.sid=b.sid AND USERNAME='HR';
    SID    SERIAL#         USERNAME      LMODE    BLOCK    TY
 -------  ------------   ------------   -------  -------  -------
    226    10556           HR              3        0      TM
    45     8584            HR              3        0      TM
    45     8584            HR              6        0      TX
    226    10556           HR              6        0      TX
    45     8584            HR              4        0      AE
    226    10556           HR              4        0      AE
```

说明：SID=45 的事务持有一个行级排他锁 RX（LMODE=3），加锁对象是表（TM），还持有一个排他锁 X（LMODE=6）加锁对象是数据行（account 表中的 ID=1001 和 ID=1002 两条记录）。当 SID=226 的事务修改 account 表中 ID=1005 的记录时，由于未加锁，所以修改成功，并对 account 表加上一个行级排他锁(TM，LMODE=3)和一个排他锁(TX，LMODE=6，该锁只对 ID＝1005 的记录生效)。

在窗口 2 中继续做修改：

```
SQL> UPDATE account SET name='李方方' WHERE id=1002;
操作处于等待状态(被阻塞)…
```

窗口 3 中查看锁信息：

```
SQL>SELECT a.sid, a.serial#, a.username, b.lmode, b.block, b.type
FROM v$session a, v$lock b WHERE a.sid=b.sid AND USERNAME='HR';
   SID    SERIAL#        USERNAME     LMODE    BLOCK    TY
------- -------------- ------------ -------- -------- -------
   226   10556          HR           3        0        TM
    45    8584          HR           3        0        TM
    45    8584          HR           6        1        TX
   226   10556          HR           0        0        TX
   226   10556          HR           6        0        TX
    45    8584          HR           4        0        AE
   226   10556          HR           4        0        AE
```

说明：SID=226 的事务增加了一个 LMODE=0 的数据行锁(TX)，该加锁过程未完成，被 SID=45 的事务锁阻塞，此时 SID=45 的排他锁(TX)的 BLOCK 已变成 1。

3) 共享锁(Share Locks，即 S 锁)：

当一个事务对数据资源加上共享锁(S)后，其他事务仍可以对该资源再加行级共享锁(RS)和共享锁(S)，但不能加行级排他锁(RX)、共享行级排他锁(SRX)及排他锁(X)，允许其他事务查询被锁定的表，不允许其他事务做插入、更新、删除操作。

加锁语法：

```
LOCK TABLE <表名> IN SHARE MODE;
```

【示例 9-17】并发事务共享锁操作。

打开 3 个 sqlplus 窗口，窗口 1(hr 用户操作)，窗口 2(hr 用户操作)，窗口 3(system 用户操作)，首先在窗口 3 中查看锁信息：

```
SQL>SELECT a.sid, a.serial#, a.username, b.lmode, b.block, b.type
FROM v$session a, v$lock b WHERE a.sid=b.sid AND USERNAME='HR';
   SID    SERIAL#        USERNAME     LMODE    BLOCK    TY
------- -------------- ------------ -------- -------- -------
    45    8584          HR           4        0        AE
   226   10556          HR           4        0        AE
```

在窗口 1 中对 account 表设置共享锁(S)：

```
SQL>LOCK TABLE account IN SHARE MODE;
表已锁定。
```

窗口 3 中查看锁信息：

第 9 章 表数据维护

```
SQL>SELECT a.sid, a.serial#, a.username, b.lmode, b.block, b.type
FROM v$session a, v$lock b WHERE a.sid=b.sid AND USERNAME='HR';
   SID    SERIAL#        USERNAME       LMODE    BLOCK    TY
-------  -------------  ------------   -------  -------  -------
   45      584            HR              4        0      TM
   45      8584           HR              4        0      AE
   226     10556          HR              4        0      AE
```

说明：SID=45 的事务持有了一个 account 表的 S 锁（LMODE=4）。

在窗口 2 中对 account 表做如下操作：

```
SQL>LOCK TABLE account IN ROW SHARE MODE;
表已锁定。
SQL>COMMIT;
提交完成。
SQL> LOCK TABLE account IN SHARE MODE;
表已锁定。
SQL>COMMIT;
提交完成。
SQL> LOCK TABLE account IN ROW EXCLUSIVE MODE;
操作处于等待状态(被阻塞)…
SQL>LOCK TABLE account IN SHARE ROW EXCLUSIVE MODE;
操作处于等待状态(被阻塞)…
SQL>LOCK TABLE account IN EXCLUSIVE MODE;
操作处于等待状态(被阻塞)…
```

说明：S 锁与 RS 锁、S 锁可以并发操作，而与 RX 锁、SRX 锁、X 锁无法并发。当三个会话窗口中的事务都提交或回滚后，在窗口 1 中重新对 account 表设置共享锁：

```
SQL > LOCK TABLE account IN SHARE MODE;
表已锁定。
```

窗口 3 中查看锁信息：

```
SQL>SELECT a.sid, a.serial#, a.username, b.lmode, b.block, b.type
FROM v$session a, v$lock b WHERE a.sid=b.sid AND USERNAME='HR';
   SID    SERIAL#        USERNAME       LMODE    BLOCK    TY
-------  -------------  ------------   -------  -------  -------
   45      8584           HR              4        0      TM
   45      8584           HR              4        0      AE
   226     10556          HR              4        0      AE
```

说明：SID=45 的事务持有了一个 account 表的 S 锁（LMODE=4）。

在窗口 2 中对 account 表进行查询和修改：

```
SQL>SELECT * FROM account;
```

```
    ID  NAME                  BALANCE
    --- --------------------- ----------
    1001 张山                  9500
    1005 杨雨                  10000
    1003 高七                  10500
    1002 李四                  10000
    1004 赵兵                  12000
SQL > UPDATE account SET NAME='杨小雨' WHERE ID='1005';
操作处于等待状态(被阻塞)...
```

说明：窗口 2 中的查询操作可正常执行，而修改操作无法完成，将等待窗口 1 中的事务解锁。在窗口 3 中查看锁信息：

```
SQL>SELECT a.sid, a.serial#, a.username, b.lmode, b.block, b.type
FROM v$session a, v$lock b WHERE a.sid=b.sid AND USERNAME='HR';
   SID    SERIAL#        USERNAME     LMODE    BLOCK     TY
   ----   -------------  -----------  ------   -------   -------
   226    10556          HR           0        0         TM
   45     8584           HR           4        1         TM
   45     8584           HR           4        0         AE
   226    10556          HR           4        0         AE
```

说明：SID=226 的事务加锁未完成(TM，LMODE=0)，这是因为被 SID=45 的事务锁阻塞(BLOCK=1)。

在窗口 1 中提交事务(或回滚事务)，将解除窗口 1 中事务的共享锁 S：

```
SQL>COMMIT;
提交完成。
```

当窗口 1 中的共享锁被解除后，窗口 2 中的更新操作将继续进行：

```
SQL> UPDATE account SET NAME='杨小雨' WHERE ID=1005;
已更新 1 行。
```

在窗口 3 中查看锁信息：

```
SQL>SELECT a.sid, a.serial#, a.username, b.lmode, b.block, b.type
FROM v$session a, v$lock b WHERE a.sid=b.sid AND USERNAME='HR';
   SID    SERIAL#        USERNAME     LMODE    BLOCK     TY
   ----   -------------  -----------  ------   -------   -------
   226    10556          HR           3        0         TM
   226    10556          HR           6        0         TX
   45     8584           HR           4        0         AE
   226    10556          HR           4        0         AE
```

说明：窗口 1 中事务已提交，其所加 S 锁被解除；而窗口 2 中的修改操作还未提交，故 UPDATE 操作为 account 表加上了一个行级排他锁 RX(TM，LMODE=3)和一个排他锁

X(TX,LMODE=6)。

4）共享行级排他锁（Share Row Exclusive Table Lock，SRX 锁）

当一个事务对数据资源加上共享行级排他锁（SRX）后，其他事务就只能对该资源再加行级共享锁（RS），不能加行级排他锁（RX）、共享锁（S）、共享行级排他锁（SRX）及排他锁（X），允许其他事务查询被锁定的表，不允许其他事务做插入、更新、删除操作。

加锁语法：

```
LOCK TABLE <表名> IN SHARE ROW EXCLUSIVE MODE;
```

【示例 9-18】并发事务共享行级排他锁操作。

打开 3 个 sqlplus 窗口，窗口 1（hr 用户操作），窗口 2（hr 用户操作），窗口 3（system 用户操作），首先在窗口 3 中查看锁信息：

```
SQL>SELECT a.sid, a.serial#, a.username, b.lmode, b.block, b.type
FROM v$session a, v$lock b WHERE a.sid=b.sid AND USERNAME='HR';
SID      SERIAL#        USERNAME      LMODE   BLOCK   TY
-------  -------------- ------------  ------  ------- ------
45       8584           HR            4       0       AE
226      10556          HR            4       0       AE
```

在窗口 1 中对 account 表设置共享行级排他锁（SRX）：

```
SQL>LOCK TABLE account IN SHARE ROW EXCLUSIVE MODE;
表已锁定。
```

窗口 3 中查看锁信息：

```
SQL>SELECT a.sid, a.serial#, a.username, b.lmode, b.block, b.type
FROM v$session a, v$lock b WHERE a.sid=b.sid AND USERNAME='HR';
SID      SERIAL#        USERNAME      LMODE   BLOCK   TY
-------  -------------- ------------  ------  ------- ------
45       8584           HR            5       0       TM
45       8584           HR            4       0       AE
226      10556          HR            4       0       AE
```

说明：SID=45 的事务持有了一个 account 表的 SRX 锁（LMODE=5）。

在窗口 2 中对 account 表做如下操作：

```
SQL>LOCK TABLE account IN ROW SHARE MODE;
表已锁定。
SQL>COMMIT;
提交完成。
SQL> LOCK TABLE account IN SHARE MODE;
操作处于等待状态（被阻塞）...
SQL> LOCK TABLE account IN ROW EXCLUSIVE MODE;
操作处于等待状态（被阻塞）...
SQL>LOCK TABLE account IN SHARE ROW EXCLUSIVE MODE;
```

操作处于等待状态(被阻塞)…

SQL>LOCK TABLE account IN EXCLUSIVE MODE;

操作处于等待状态(被阻塞)…

说明:SRX 锁与 RS 锁可以并发操作,而与 S 锁、RX 锁、SRX 锁、X 锁无法并发。当然,如果在窗口 2 中进行插入、更新、删除操作,也会被阻塞。

5)排它锁(Exclusive Locks,即 X 锁)

X 锁是在锁机制中限制最多的一种锁类型。当一个事务对数据资源加上排他锁(X)后,其他事务就无法再加其他任何类型的锁,不允许其他事务做插入、更新、删除操作(查询除外)。

加锁语法:

LOCK TABLE <表名> IN EXCLUSIVE MODE;

【示例 9-19】并发事务排它锁操作。

打开 3 个 sqlplus 窗口,窗口 1(hr 用户操作),窗口 2(hr 用户操作),窗口 3(system 用户操作),首先在窗口 3 中查看锁信息:

```
SQL>SELECT a.sid, a.serial#, a.username, b.lmode, b.block, b.type
FROM v$session a, v$lock b WHERE a.sid=b.sid AND USERNAME='HR';
   SID    SERIAL#         USERNAME         LMODE    BLOCK    TY
------- --------------- ---------------- -------- -------- ------
   45     8584            HR                 4        0      AE
   226    10556           HR                 4        0      AE
```

在窗口 1 中对 account 表设置排他锁(X):

SQL>LOCK TABLE account IN SHARE ROW EXCLUSIVE MODE;

表已锁定。

窗口 3 中查看锁信息:

```
SQL>SELECT a.sid, a.serial#, a.username, b.lmode, b.block, b.type
FROM v$session a, v$lock b WHERE a.sid=b.sid AND USERNAME='HR';
   SID    SERIAL#         USERNAME         LMODE    BLOCK    TY
------- --------------- ---------------- -------- -------- ------
   45     8584            HR                 6        0      TM
   45     8584            HR                 4        0      AE
   226    10556           HR                 4        0      AE
```

说明:SID=45 的事务持有了一个 account 表的 X 锁(LMODE=6)。

在窗口 2 中对 account 表做如下操作:

SQL>LOCK TABLE account IN ROW SHARE MODE;

操作处于等待状态(被阻塞)…

SQL> LOCK TABLE account IN SHARE MODE;

操作处于等待状态(被阻塞)…

SQL> LOCK TABLE account IN ROW EXCLUSIVE MODE;

```
操作处于等待状态(被阻塞)...
SQL>LOCK TABLE account IN SHARE ROW EXCLUSIVE MODE;
操作处于等待状态(被阻塞)...
SQL>LOCK TABLE account IN EXCLUSIVE MODE;
操作处于等待状态(被阻塞)...
```

说明：X 锁与其他任何锁都无法并发，在窗口 2 中进行插入、更新、删除操作，也会被阻塞。

当查出锁的信息后，还可以"杀掉"对应的 SESSION 以解除加锁：

```
SQL>ALTER SYSTEM KILL SESSION '45,8584';   --杀掉锁对应的Session
系统已更改。
```

练习

一、判断题

1. 如果在表列上设置了默认值，当插入数据时，可以不指定该列的值，Oracle 会自动以默认值填充该列。（　　）

2. 使用 INSERT…SELECT 语句时，SELECT 子句中查询的列数与 INSERT INTO 语句中的列数要相同，且类型一致。（　　）

3. 删除表中数据时，如果该表中有外键关系，需要先删除外键表中的数据，才能删除该表中的数据，否则将出现删除异常。（　　）

4. 用户断开连接时，事务操作会自动提交。（　　）

5. 不能使用多个相同的标记名设置保存点。（　　）

二、单项选择题

1. 在 Oracle 中，通过命令（　　）可以释放锁
   A. INSER   B. DELETE
   C. ROLLBACK   D. UNLOCK

2. 在 MERGE 语句中使用（　　）语句指定匹配时的操作。
   A. MATCHED   B. NOT MATCHED
   C. UPDATE   D. WHERE

3. 用于修改表中数据的语句是（　　）
   A.MODIFY   B.EDIT
   C.ALTER   D.UPDATE

4. 事务提交的命令是（　　）
   A.ROLLBACK   B.COMMIT
   C.SUBMIT   D.SAVEPOINT

5. 能与 RX 锁并发操作的锁是（　　）
   A.RS 锁   B. S 锁
   C.SRX 锁   D. X 锁

6. 在 Oracle 中，事务中使用下列 SQL 语句不会引起锁定（　　）
A. SELECT　　　　　　　　　　B. INSERT
C. UPDATE　　　　　　　　　　D. DELETE

三、填空题

1. 事务的 ACID 特性是指_____、_____、_____和_____。
2. 可使用_____子句返回聚合函数的计算结果。
3. 执行数据定义语句事务结束时，会默认执行_____语句。
4. 在事务中创建一个保存点后，用_____语句可以回退到标记处。
5. 当一个事务修改数据时，另一事务读取了该数据，但是第一事务由于某种原因取消对数据修改，使数据返回了原状态，这时第二个事务读取的数据与数据库中数据不一致，这就叫_____。

四、简答题

1. 简述事务的开始与结束方式。
2. 简述事务的 ACID。
3. 简述事务锁的类型。

训练任务

1. 表数据的增加、修改、删除

培养能力	工程能力、设计/开发解决方案		
掌握程度	★★★★★	难度	中
结束条件	编写对表数据进行增删改操作的 SQL 语句		

训练内容：
创建 employees 表和 departments 表，表结构如下：
employees（employee_id number not null, name varchar2(40) not null,
 email varchar2(40), phone_number varchar2(40),
 hire_date date not null, salary number(8, 2), manager_id number,
 department_id number, photo blob）
departments（department_id number not null,
 department_name varchar2(40) not null）
其中 employees 表的主键是 employee_id；departments 表的主键是 department_id；
employees.department_id 是外键，外键表是 departments；employees.salary 默认值是 2000。
训练要求：
(1) 使用 INSERT INTO 语句插入 2 条数据，给出完整的列名和列值；
(2) 使用 INSERT INTO 语句插入数据，salary 填充默认值，photo 为空值；
(3) 创建一个与 employees 结构相同的表 employees1，使用 INSERT…SELECT 语句把 employees 中的数据复制到 employees1 中；
(4) 使用 UPDATE 语句更新一条记录的 SALARY 值；
(5) 使用 DELETE 语句删除一条记录。
以上操作的每一步使用 SELECT 语句查看操作结果

2. 熟悉事务的 ACID 特性

培养能力	工程能力、问题分析、设计/开发解决方案等		
掌握程度	★★★	难度	中
结束条件	编写测试事务 ACID 特性的 SQL 语句		
训练内容： 验证事务的原子性、一致性和隔离性操作，参考教材 9.5.4 完成训练任务			

3. 熟悉事务锁

培养能力	工程能力、设计/开发解决方案		
掌握程度	★★★	难度	中
结束条件	编写测试事务锁的 SQL 语句		
训练内容：按照"9.5.5 锁"的示例进行实验，要求： (1) 给数据表加行级共享锁(RS)； (2) 给数据表加行级排他锁(RX)； (3) 给数据表加共享锁(S)； (4) 给数据表加共享行级排他锁(SRX)； (5) 给数据表加排他锁(X)			

第10章 SQL 语言基础

本章目标

知识点	理解	掌握	应用
1.基本 SQL 查询	✓	✓	✓
2.子查询	✓	✓	
3.递归查询	✓	✓	

训练任务

- 熟悉分页查询
- 熟悉递归查询

知识能力点

能力点 \ 知识点	知识点 1	知识点 2	知识点 3
工程知识	✓		
问题分析		✓	✓
设计/开发解决方案	✓	✓	✓
研究			✓
使用现代工具	✓		
工程与社会	✓		
环境和可持续发展			✓
职业规范	✓	✓	
个人和团队	✓		
沟通		✓	
项目管理	✓		
终身学习	✓		

10.1 SQL 语言概述

结构化查询语言 SQL(Structured Query Language)是操作关系型数据库的一种通用语言。SQL 语言具有使用方式灵活、功能强大、语言简捷等优点，许多关系数据库如 DB2、ORACLE、SQL Server 等都实现了 SQL 语言。

SQL 语言的分类：
(1) 数据定义语言(DDL)：主要包括 CREATE、ALTER、DROP、DECLARE 等语句。
(2) 数据操纵语言(DML)：主要包括 SELECT、DELETE、UPDATE、INSERT 等语句。
(3) 数据控制语言(DCL)：主要包括 GRANT、REVOKE、COMMIT、ROLLBACK 等语句。

本章主要介绍标准 SQL 语言的查询语句，ORACLE 数据库使用的 PL/SQL 语言将在第 12 章作详细介绍。

数据库查询语句的基本格式如下：
SELECT [ALL｜DISTINCT] <*｜目标列> FROM <表名｜视图>
[WHERE <条件表达式>]
[GROUP BY <列名 1> [HAVING <内部函数表达式>]]
[ORDER BY <列名 2> <ASC｜DESC>]

说明：
- "[ ]"表示可选项，"< >"表示必选项，"｜"表示或者。
- ALL：查询所有符合条件的记录，缺省为 ALL。
- DISTINCT：查询去掉重复的记录。
- "*"表示所有列。

10.2 选择部分列

【示例 10-1】选择部分或者全部列，行操作。

选择全部列和行：

```
SQL> SELECT * FROM employees;
EMPLOYEE_ID FIRST_NAME          ...    MANAGER_ID     DEPARTMENT_ID
----------- ----------------    ----   -------------  -----------------
    400     张三                ...        101              60
    401     李四                ...        101              60
...
已选择 109 行。
```

选择部分列和行：

```
SQL> SELECT employee_id, first_name, job_id, salary, manager_id,
```

```
  2  department_id FROM employees WHERE department_id=60;
EMPLOYEE_ID FIRST_NAME    JOB_ID      SALARY  MANAGER_ID  DEPARTMENT_ID
----------- ----------    -------     ------  ----------  -------------
     400    张三          IT_PROG      6000       101           60
     401    李四          IT_PROG      5000       101           60
     103    Alexander     IT_PROG      9000       102           60
     104    Bruce         IT_PROG      6000       103           60
     105    David         IT_PROG      4800       103           60
     106    Valli         IT_PROG      4800       103           60
     107    Diana         IT_PROG      4200       103           60

已选择 7 行。
```

10.3　WHERE 子句

在 SQL 查询中，可以使用 WHERE 子句来筛选查询结果。WHERE 子句中的条件表达式主要有以下几种形式：

算术运算符：+、一、*、/。
逻辑运算符：AND、OR、NOT。
比较运算符：=、>、>=、<、<=、<>。
判断列值是否为空：<列名> IS [NOT] NULL；
区间值判断：<表达式 1> [NOT]　BETWEEN <表达式 2> AND <表达式 3>。
值是否在一集合中：<表达式> [NOT] IN（目标值表列）。
样式匹配：<列名> [NOT] LIKE <'模式串'>；（'模式串'中的%表示一个或多个字符，_表示一个字符。）

【示例 10-2】WHERE 子句操作。
查询指定行（manager_id=101）的记录：

```
SQL> SELECT employee_id, first_name, job_id, manager_id, department_id
  2  FROM employees WHERE manager_id=101;
EMPLOYEE_ID FIRST_NAME        JOB_ID      MANAGER_ID  DEPARTMENT_ID
----------- ----------------  ----------  ----------  -------------
     400    张三              IT_PROG        101           60
     401    李四              IT_PROG        101           60
     108    Nancy             FI_MGR         101          100
     200    Jennifer          AD_ASST        101           10
     203    Susan             HR_REP         101           40
     204    Hermann           PR_REP         101           70
     205    Shelley           AC_MGR         101          110
```

第 10 章 SQL 语言基础

查询 manager_id 值为空的记录：

```
SQL> SELECT employee_id, first_name, job_id, manager_id, department_id
  2  FROM employees WHERE manager_id IS NULL;
```

EMPLOYEE_ID	FIRST_NAME	JOB_ID	MANAGER_ID	DEPARTMENT_ID
100	Steven	AD_PRES		90

查询 manager_id 值在 102 到 105 之间的记录：

```
SQL> SELECT employee_id, first_name, job_id, manager_id, department_id
  2  FROM employees WHERE manager_id BETWEEN 102 AND 105;
```

EMPLOYEE_ID	FIRST_NAME	JOB_ID	MANAGER_ID	DEPARTMENT_ID
103	Alexander	IT_PROG	102	60
104	Bruce	IT_PROG	103	60
105	David	IT_PROG	103	60
106	Valli	IT_PROG	103	60
107	Diana	IT_PROG	103	60

查询 manager_id 值为 103 和 123 的记录：

```
SQL> SELECT employee_id, first_name, job_id, manager_id, department_id
  2  FROM employees WHERE manager_id IN (103, 123);
```

EMPLOYEE_ID	FIRST_NAME	JOB_ID	MANAGER_ID	DEPARTMENT_ID
104	Bruce	IT_PROG	103	60
105	David	IT_PROG	103	60
106	Valli	IT_PROG	103	60
107	Diana	IT_PROG	103	60
137	Renske	ST_CLERK	123	50
138	Stephen	ST_CLERK	123	50
139	John	ST_CLERK	123	50
140	Joshua	ST_CLERK	123	50
192	Sarah	SH_CLERK	123	50
193	Britney	SH_CLERK	123	50
194	Samuel	SH_CLERK	123	50
195	Vance	SH_CLERK	123	50

查询 first_name 第 1 个字母为 'J'，第 2 个字母任意，第 3 个字母为 's' 的所有记录：

```
SQL> SELECT employee_id, first_name, job_id, manager_id, department_id
  2  FROM employees WHERE first_name LIKE 'J_s%';
```

EMPLOYEE_ID	FIRST_NAME	JOB_ID	MANAGER_ID	DEPARTMENT_ID

133	Jason	ST_CLERK	122	50
140	Joshua	ST_CLERK	123	50
112	Jose Manuel	FI_ACCOUNT	108	100

10.4 列算术运算

在 SQL 查询中常使用算术表达式来进行列算术运算，包括加、减、乘、除四则运算以及函数的引用。

【示例 10-3】 列算术运算查询。

查询（department_id=100）人员的薪水，并以"万"为单位显示查询结果：

```
SQL> SELECT employee_id, first_name, ROUND(salary/10000, 2)
  2  AS "薪水(万)", department_id FROM employees
  3  WHERE department_id='100';
```

EMPLOYEE_ID	FIRST_NAME	薪水(万)	DEPARTMENT_ID
108	Nancy	1.2	100
109	Daniel	.9	100
110	John	.82	100
111	Ismael	.77	100
112	Jose Manuel	.78	100
113	Luis	.69	100

查询（department_id=100）人员的薪水，并以"万"为单位显示查询结果：

```
SQL> SELECT employee_id, first_name, ROUND(salary/10000, 2)
  2  AS "薪水(万)", department_id FROM employees
  3  WHERE department_id='100';
```

EMPLOYEE_ID	FIRST_NAME	薪水(万)	DEPARTMENT_ID
108	Nancy	1.2	100
109	Daniel	.9	100
110	John	.82	100
111	Ismael	.77	100
112	Jose Manuel	.78	100
113	Luis	.69	100

查询管理员管理部门 ID 为 20（Marketing）、80（Sales）、60（IT）的人数：

```
SQL> SELECT DISTINCT manager_id AS 管理员ID, department_id AS 部门ID,
  2  SUM(CASE WHEN department_id=20 THEN 1 ELSE 0 END)AS Marketing 人数,
  3  SUM(CASE WHEN department_id=80 THEN 1 ELSE 0 END)AS Sales 人数,
```

```
4 SUM(CASE WHEN department_id=60 THEN 1 ELSE 0 END)as IT人数
5 FROM employees  WHERE department_id IN(20, 80, 60)
6 GROUP BY manager_id, department_id;
```

管理员ID	部门ID	MARKETING人数	SALES人数	IT人数
147	80	0	6	0
102	60	0	0	1
103	60	0	0	4
148	80	0	6	0
146	80	0	6	0
149	80	0	5	0
100	80	0	5	0
145	80	0	6	0
100	20	1	0	0
201	20	1	0	0

10.5　禁止重复行

在 SQL 查询中，可以使用 DISTINCT 关键字来禁止重复行的出现。

【示例 10-4】 禁止重复行查询。

在 employees 表中查询所有不同的部门 ID：

```
SQL> SELECT DISTINCT department_id FROM employees
  2  WHERE department_id IS NOT NULL;
DEPARTMENT_ID
-------------
    10
    20
    30
    40
    50
    60
    70
    80
    90
   100
```

10.6 排 序

在 SQL 查询中，可以使用 ORDER BY 子句对查询结果排序。

【示例 10-5】排序查询。

在 employees 表中查询(department_id=60)所有人员的薪水，并按降序输出结果：

```
SQL> SELECT employee_id, first_name, salary from employees
  2  WHERE department_id=60 ORDER BY salary DESC;
EMPLOYEE_ID FIRST_NAME       SALARY
----------- ------------ ----------
        103 Alexander          9000
        104 Bruce              6000
        105 David              4800
        106 Valli              4800
        107 Diana              4200
```

注意：默认排序是升序(或使用 ASC 关键字)。

10.7 表别名及多表查询

在 SQL 查询中，如果一个查询语句需要显示多张表的数据，则必须应用到多表查询的操作，多表查询的种类包括：

- ◇ 笛卡尔积连接
- ◇ 等值连接(或不等值连接)
- ◇ 外连接(包括：左连接、右连接、全连接)
- ◇ 自连接

多表查询的基本语法如下：

SELECT [DISTINCT] * | <字段> [别名] [，字段 [别名] ，…]
FROM <表名> [别名]，[<表名> [别名] ，…]
[WHERE <条件>]
[ORDER BY <排序字段> [ASC|DESC] [，<排序字段> [ASC|DESC] ，…]];

【示例 10-6】多表查询示例一。

查询 employees 表和 departments 表的笛卡尔积(employees 中的每一条记录与 departments 表中的每一条记录连接成一条新记录)：

```
SQL> SELECT COUNT(*)FROM employees, departments;
  COUNT(*)
----------
      2889
```

第 10 章 SQL 语言基础

说明：在表 employees 中有 107 条记录，表 departments 中有 27 条记录，两个表的笛卡尔查询结果有 107*27=2889 条记录。（笛卡积查询一般不常用）

等值查询，查询所有员工的部门名称(员工所在部门在 departments 表中有的，才连接成一条新的记录，即员工有部门的才连接)：

```
SQL> SELECT emp.employee_id, emp.first_name, dep.department_name
  2  FROM employees emp, departments dep
  3  WHERE emp.department_id=dep.department_id;
EMPLOYEE_ID FIRST_NAME       DEPARTMENT_NAME
----------- ---------------- ---------------
    200     Jennifer         Administration
    201     Michael          Marketing
    202     Pat              Marketing
    114     Den              Purchasing
...
已选择 106 行。
```

说明：在 employees 表中有 107 个员工，其中有一个员工的部门为空，故等值连接查询的结果就只有 106 条记录。

左连接查询，查询所有员工的部门名称(如果员工所在部门在 departments 表中没有，也将连接成一条新的记录，门部名称为 NULL，即左表记录将全部出现在查询结果中)：

```
SQL> SELECT emp.employee_id, emp.first_name, dep.department_name
  2  FROM employees emp, departments dep
  3  WHERE emp.department_id=dep.department_id(+);
EMPLOYEE_ID FIRST_NAME       DEPARTMENT_NAME
----------- ---------------- ---------------
    200     Jennifer         Administration
    201     Michael          Marketing
    202     Pat              Marketing
...
已选择 107 行。
```

说明：

(1)"(+)"放在连接条件等号的右边，表示左连接查询。

(2)在 employees 表中，员工(employee_id=178)的门部在 departments 中没有对应的值，故其部门为空。

右连接查询，查询所有部门对应的员工信息(如果某部门没有员工，将以空值显示员工信息，即右表记录将全部出现在查询结果中)：

```
SQL> SELECT emp.employee_id, emp.first_name, dep.department_name
  2  FROM employees emp, departments dep
  3  WHERE emp.department_id(+)=dep.department_id;
```

```
EMPLOYEE_ID    FIRST_NAME        DEPARTMENT_NAME
-----------    ---------------   ---------------
200            Jennifer          Administration
201            Michael           Marketing
202            Pat               Marketing
114            Den               Purchasing
...
Null           null              Recruiting
Null           null              Payroll
...
已选择 122 行。
```

说明：

(1)"(+)"放在连接条件等号的左边，表示右连接查询；

(2)没有员工的部门也会连接成一条记录，其员工信息为空值。

自连接查询(指同一张表的连接查询)，查询所有员工的管理员姓名：

```
SQL> SELECT e1.employee_id 员工编号, e1.first_name 员工姓名,
  2  e2.first_name 管理员姓名 FROM employees e1, employees e2
  3  WHERE e1.manager_id=e2.employee_id;
员工编号     员工姓名         管理员姓名
---------   ---------------  --------------------
   173      Sundita          Gerald
   172      Elizabeth        Gerald
   171      William          Gerald
   170      Tayler           Gerald
   169      Harrison         Gerald
   168      Lisa             Gerald
   103      Alexander        Lex
...
已选择 106 行。
```

Oracle 还支持以下连接方式：

(1)交叉连接(CROSS JOIN)：用于产生笛卡尔积。

(2)自然连接(NATURAL JOIN)：自动找到匹配的关联字段，消除掉笛卡尔积。

(3)使用 JOIN…USING 子句建立连接：用户自己指定一个消除笛卡尔积的关联字段。

(4)使用 JOIN…ON 子句建立连接：用户自己指定一个可以消除笛卡尔积的关联条件。

(5)左(外)连接：LEFT OUTER JOIN…ON。

(6)右(外)连接：RIGHT OUTER JOIN…ON。

(7)全(外)连接：FULL OUTER JOIN…ON。

【示例 10-7】多表查询示例二。

```
--交叉连接
SQL> SELECT * FROM employees CROSS JOIN departments;
--自然连接
SQL> SELECT * FROM employees NATURAL JOIN departments;
--使用 JOIN...USING 子句
SQL> SELECT * FROM employees JOIN departments USING(department_id);
--使用 JOIN...ON 子句
SQL> SELECT * FROM employees e JOIN departments d
2  ON e.department_id = d.department_id;
--左(外)连接
SQL> SELECT * FROM employees e LEFT OUTER JOIN departments d
2  ON e.department_id= d.department_id;
--右(外)连接
SQL> SELECT * FROM employees e RIGHT OUTER JOIN departments d
2  ON e.department_id =d.department_id;
--全(外)连接
SQL> SELECT * FROM employees e FULL OUTER JOIN departments d
2  ON e.department_id= d.department_id;
```

10.8 子 查 询

子查询，也叫嵌套查询，允许 SELECT 语句嵌入到其他 SQL 语句中，最常用的格式如下：

```
SELECT <字段列表> FROM <表名1>
WHERE <表达式> 运算符 (SELECT <字段列表> FROM <表名2>);
```

注意：对单行子查询使用单行运算符，对多行子查询使用多行运算符。

10.8.1 单行子查询

如果子查询返回单行结果，则为单行子查询，可以在主查询中对其使用相应的单行记录比较运算符，常用的运算符有：=、>、>=、<、<=、<>。

【示例 10-8】单行子查询。

查询所有薪水高于员工(employee_id=108)的员工信息：

```
SQL> SELECT employee_id, first_name, salary from employees WHERE
2  salary > (SELECT salary FROM employees WHERE employee_id=108);
EMPLOYEE_ID FIRST_NAME         SALARY
----------- ----------------   ----------
100         Steven             24000
```

```
    101          Neena              17000
    102          Lex                17000
    145          John               14000
    146          Karen              13500
    201          Michael            13000
已选择 6 行。
```

> 注意：如果子查询未返回任何行，则主查询不会返回任何结果。

10.8.2 多行子查询

如果子查询返回多行结果，则为多行子查询，此时不允许对其使用单行记录比较运算符，而使用多行记录比较运算符，常用的运算符有：

- in：等于列表中的任何一个
- any：和子查询返回的任意一个值比较
- all：和子查询返回的所有值比较

【示例 10-9】 多行子查询。

查询在 job_history 表中有过记载的员工信息：

```
SQL> SELECT employee_id, first_name, job_id from employees WHERE
  2  employee_id IN(SELECT DISTINCT employee_id FROM job_history);
EMPLOYEE_ID FIRST_NAME        JOB_ID
----------- ----------------  ----------
    101     Neena             AD_VP
    102     Lex               AD_VP
    114     Den               PU_MAN
    122     Payam             ST_MAN
    176     Jonathon          SA_REP
    200     Jennifer          AD_ASST
    201     Michael           MK_MAN
已选择 7 行。
```

查询员工薪水低于所有部门平均工资的员工信息：

```
SQL> SELECT employee_id, first_name, salary from employees WHERE
  2  salary < ALL(SELECT AVG(salary)FROM employees
  3  GROUP BY department_id);
EMPLOYEE_ID FIRST_NAME        SALARY
----------- ----------------  ----------
    186     Julia             3400
    133     Jason             33200
    129     Laura             3300
```

125	Julia	3200
194	Samuel	3200
180	Winston	3200
138	Stephen	3200
...		

已选择 37 行。

查询员工薪水高于任意部门平均工资的员工信息：

```
SQL> SELECT employee_id, first_name, salary from employees WHERE
  2  salary > ANY(SELECT avg(salary)FROM employees
  3  GROUP BY department_id);
EMPLOYEE_ID FIRST_NAME       SALARY
----------- ---------------- ----------
        100 Steven            24000
        101 Neena             17000
        102 Lex               17000
        145 John              14000
        146 Karen             13500
...
```

已选择 70 行。

10.8.3 Top N 查询

在 Oracle 中通常采用子查询的方式来实现 Top N 查询，其格式如下：

✧ 格式 1：

```
SELECT <字段列表>
FROM(SELECT <字段列表> FROM <表名> ORDER BY <排序字段>)
WHERE ROWNUM <=n;
```

格式 2（Oracle 12c 适用）：

```
SELECT <字段列表>
FROM <表名> ORDER BY <排序字段> FETCH FIRST n ROWS ONLY;
```

注意：ROWNUM 是 oracle 系统顺序分配从查询返回行的编号，返回的第一行编号是 1，第二行是 2，依此类推。

【示例 10-10】Top N 查询。

查询薪水排在前 5 名的员工信息：

✧ 格式 1：

```
SQL> SELECT employee_id,first_name,salary FROM(SELECT *
  2  FROM employees ORDER BY salary DESC) WHERE ROWNUM<=5;
EMPLOYEE_ID      FIRST_NAME           SALARY
```

```
-----------  --------------------  ----------
        100  Steven                     24000
        101  Neena                      17000
        102  Lex                        17000
        145  John                       14000
        146  Karen                      13500
```

✧ 格式 2：

```
SQL> SELECT employee_id,first_name,salary FROM employees
  2  ORDER BY salary DESC FETCH FIRST 5 ROWS ONLY;
EMPLOYEE_ID  FIRST_NAME                SALARY
-----------  --------------------  ----------
        100  Steven                     24000
        101  Neena                      17000
        102  Lex                        17000
        145  John                       14000
        146  Karen                      13500
```

10.8.4 分页查询

可以利用 ROWNUM 和子查询实现分页查询，主要有以下几种方式：

✧ 格式 1：

```
SELECT * FROM (SELECT a.*, ROWNUM rn FROM (SELECT * FROM <表名>)a)
WHERE rn BETWEEN n AND m
```

✧ 格式 2：

```
SELECT * FROM (SELECT a.*, ROWNUM rn FROM(SELECT * FROM <表名>)a
WHERE ROWNUM <= m)WHERE RN>=n
```

✧ 格式 3（使用 HINT）：

```
SELECT /*+ FIRST_ROWS */ * FROM (SELECT a.*, ROWNUM rn
FROM (SELECT * FROM <表名>)a WHERE ROWNUM <= m)WHERE RN >= n
```

✧ 格式 4（Oracle 12c 适用）：

```
SELECT * FROM <表名> OFFSET n ROWS FETCH NEXT m ROWS ONLY;
```

> 注意：在格式 1-3 中，a 是表的别名，n 是本页起始记录编号，m 是本页末记录编号；在格式 4 中，n 是本页起始记录偏移量，m 是本页记录数大小。

【示例 10-11】分页查询。

如果每页 10 条记录，查询第 3 页（记录编号：21~30）的员工信息：

✧ 格式 1：

```
SQL> SELECT employee_id, first_name, salary, rn FROM (
  2  ELECT a.*, ROWNUM rn  FROM (SELECT * FROM employees)a )
```

第 10 章　SQL 语言基础

```
3  WHERE rn BETWEEN 21 AND 30;
```

EMPLOYEE_ID	FIRST_NAME	SALARY	RN
120	Matthew	8000	21
121	Adam	8200	22
122	Payam	7900	23
123	Shanta	6500	24
124	Kevin	5800	25
125	Julia	3200	26
126	Irene	2700	27
127	James	2400	28
128	Steven	2200	29
129	Laura	3300	30

已选择 10 行。

- ◆ 格式 2：

```
SQL> SELECT employee_id, first_name, salary, rn FROM (
2  SELECT a.*, ROWNUM rn  FROM (SELECT * FROM employees)a
3  WHERE ROWNUM<=30)WHERE rn >= 21;
```

EMPLOYEE_ID	FIRST_NAME	SALARY	RN
120	Matthew	8000	21
121	Adam	8200	22
122	Payam	7900	23
123	Shanta	6500	24
124	Kevin	5800	25
125	Julia	3200	26
126	Irene	2700	27
127	James	2400	28
128	Steven	2200	29
129	Laura	3300	30

已选择 10 行。

- ◆ 格式 3：

```
SQL> SELECT /*+ FIRST_ROWS */employee_id, first_name, salary, rn FROM (
2  SELECT a.*, ROWNUM rn  FROM (SELECT * FROM employees)a
3  WHERE ROWNUM<=30)WHERE rn >= 21;
```

EMPLOYEE_ID	FIRST_NAME	SALARY	RN
120	Matthew	8000	21

121	Adam	8200	22
122	Payam	7900	23
123	Shanta	6500	24
124	Kevin	5800	25
125	Julia	3200	26
126	Irene	2700	27
127	James	2400	28
128	Steven	2200	29
129	Laura	3300	30

已选择 10 行。

◇ 格式 4（推荐使用）：

```
SQL> SELECT employee_id,first_name,salary FROM employees
  2  OFFSET 10*(3-1)+1 ROWS FETCH NEXT 10 ROWS ONLY;
EMPLOYEE_ID FIRST_NAME          SALARY
----------- ------------------- --------
        120 Matthew             8000
        121 Adam                8200
        122 Payam               7900
        123 Shanta              6500
        124 Kevin               5800
        125 Julia               3200
        126 Irene               2700
        127 James               2400
        128 Steven              2200
        129 Laura               3300
```

已选择 10 行。

10.9 递归查询

Oracle 的递归查询可以将一个记录之间有父子关系的表，以树的顺序列出来，也称树查询。其基本的语法格式如下：

```
SELECT ... FROM <表名> START WITH 条件1 CONNECT BY 条件2
WHERE 条件3;
```

说明：

(1) 是根结点的限定语句，如果有多个根结点，查询结果将对应多棵树。

(2) 是连接条件，常使用"PRIOR 字段1=字段2"或"字段1= PRIOR 字段2"的形式。运算符 PRIOR 被放置于等号前后的位置，决定着查询时的检索顺序。PRIOR 被置于

CONNECT BY 子句等号的前面时,则强制从根节点到叶节点的顺序检索,为自顶向下的方式查询。PRIOR 运算符被置于 CONNECT BY 子句等号的后面时,则强制从叶节点到根节点的顺序检索,为自底向上的方式查询。PRIOR 表示上一条记录,比如:CONNECT BY PRIOR employee_id = manager_id 表示上一条记录的 employee_id 是本条记录的 manager_id,即本记录的父亲是上一条记录。

(3)是过滤条件,用于对返回的所有记录进行过滤。

【示例 10-12】递归查询。

查询管理员编号为 108 和 205 管理员所管理的员工:

```
SQL> SELECT employee_id, manager_id, first_name from employees
  2  START WITH manager_id = 108 or manager_id = 205
  3  CONNECT BY PRIOR employee_id = manager_id;
EMPLOYEE_ID MANAGER_ID FIRST_NAME
----------- ---------- --------------------
        109        108 Daniel
        110        108 John
        111        108 Ismael
        112        108 Jose Manuel
        113        108 Luis
        206        205 William
```

已选择 6 行。

查询员工编号为 101 管理员所管理的所有员工(树的顺序排列):

```
SQL> SELECT employee_id, manager_id, first_name, level, CONNECT_BY_ISLEAF
  2  FROM employees START WITH employee_id = 101
  3  CONNECT BY PRIOR employee_id = manager_id;
EMPLOYEE_ID MANAGER_ID FIRST_NAME      LEVEL CONNECT_BY_ISLEAF
----------- ---------- --------------- ----- -----------------
        101        100 Neena               1                 0
        108        101 Nancy               2                 0
        109        108 Daniel              3                 1
        110        108 John                3                 1
        111        108 Ismael              3                 1
        112        108 Jose Manuel         3                 1
        113        108 Luis                3                 1
        200        101 Jennifer            2                 1
        203        101 Susan               2                 1
        204        101 Hermann             2                 1
        205        101 Shelley             2                 0
        206        205 William             3                 1
```

已选择 12 行。level 是节点所处树的层号,层号根据节点与根节点的距离确定。根节点的层号始终为 1,根节点的子节点为 2,依此类推。CONNECT_BY_ISLEAF 判断是否为叶子节点,1 表示是叶子节点,0 表示非叶子节点。查询结果树形结构如图 10-1 所示。

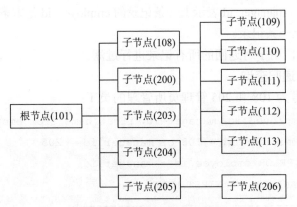

图 10-1　递归查询结果树形结构图

查询员工编号为 101 和 102 管理的所有员工,并显示根节点的信息:

```
SQL> SELECT employee_id, manager_id, first_name,
  2  CONNECT_BY_ROOT(employee_id)root_id ,
  3  CONNECT_BY_ROOT(first_name)root_name From employees
  4  START WITH employee_id = 101 or employee_id = 102
  5  CONNECT BY PRIOR employee_id = manager_id;
EMPLOYEE_ID MANAGER_ID FIRST_NAME       ROOT_ID   ROOT_NAME
----------- ---------- ---------------- --------- --------------
    101        100     Neena              101     Neena
    108        101     Nancy              101     Neena
    109        108     Daniel             101     Neena
    110        108     John               101     Neena
    111        108     Ismael             101     Neena
    112        108     Jose Manuel        101     Neena
    113        108     Luis               101     Neena
    200        101     Jennifer           101     Neena
    203        101     Susan              101     Neena
    204        101     Hermann            101     Neena
    205        101     Shelley            101     Neena
    206        205     William            101     Neena
    102        100     Lex                102     Lex
    103        102     Alexander          102     Lex
    104        103     Bruce              102     Lex
    105        103     David              102     Lex
```

106	103	Valli	102	Lex
107	103	Diana	102	Lex

已选择 18 行。

练习

一、判断题

1. ROWNUM 是 Oracle 系统顺序分配从查询返回行的编号，返回的第一行编号是 1，第二行是 2，依此类推。（　　）
2. SQL 中可以使用 DISTINCT 关键字来禁止重复行的出现。（　　）
3. SQL 中可以使用 ORDER BY 子句对查询结果排序，默认是升序。（　　）
4. Oracle 不允许表的自我连接查询。（　　）
5. 左连接查询，左表记录将全部出现在查询结果中。（　　）

二、单项选择题

1. 在子查询中可以使用（　　）关键字，该关键字只注重子查询是否返回行，如果返回一行或多行，那么它将返回 TRUE，否则返回 FALSE。

　　A. EXISTS　　　　　　　　　　B. IN
　　C. AND　　　　　　　　　　　D. BETWEEN AND

2. 在 Oracle 中下列哪个数据库对象不能直接从 SELECT 语句中引用？（　　）

　　A. 表　　　　　　　　　　　　B. 序列
　　C. 索引　　　　　　　　　　　D. 视图

3. 定义外连接时，下列哪个选项正确描述了外连接语句？（　　）

　　A. 由于外连接操作允许一个表中有 NULL 值，因此连接这些表时不必指定相等性比较。

　　B. 在外连接语句中，如果不管右表有无相应记录，都要显示左表的所有行，则可以使用右外连接。

　　C. 在外连接语句中，如果不管左表有无相应记录，都要显示右表的所有行，则可以使用左外连接。

　　D. 尽管外连接操作允许一个表中有 NULL 值，但连接这些表时仍要指定相等性比较。

4. 根据下面代码块回答问题：

```
SQL> SELECT department_id, avg(salary)FROM employees
2   GROUP BY department_id HAVING avg(salary)>
3   ( SELECT salary FROM employees WHERE employee_id=108 );
```

上述语句使用下面哪种子查询？（　　）

　　A. 单行子查询　　　　　　　　B. 多行子查询
　　C. FROM 子句子查询　　　　　D. 多列子查询

5. JOB 表有三个列 JOB_NAME、JOB_DESC 和 JOB_WAGE。根据下面代码块回答问题：

```
SQL> INSERT INTO job(job_name, job_desc)
2   VALUES ('Programmer', 'Write code');
```

```
SQL> SELECT * FROM job WHERE job_name = 'Programmer';
JOB_NAME        JOB_DESC        JOB_WAGE
-----------     ------------    --------
Programmer      Write code      3500
```

数据是如何填充 JOB_WAGE 列的？（ ）

A. JOB_WAGE 列的默认值设置为 3500。

B. 生成表时 JOB_WAGE 列定义的 DEFAULT 子句指定了默认值。

C. INSERT 语句的 VALUES 子句包含隐藏值，在插入行时加入。

D. 随后的一个 UPDATE 语句增加了 JOB_WAGE 值。

三、填空题

1. 外连接可以分为左连接、_____和_____。

2. 连接可以使用一个表同其自身进行连接，这种连接称为_____。

3. 允许将 SELECT 语句嵌入到其他 SQL 语句进行操作，这种查询称为_____。

4. 如果子查询返回单行结果，则称为_____。

5. Oracle 的递归查询可以将一个记录之间有父子关系的表，以树的顺序列出来，也称_____。

四、简答题

1. 在 Oracle 中如何实现多表查询？

2. 什么是子查询？如何实现子查询？

3. 什么是递归查询？如何实现递归查询？

训练任务

1. 分页查询

培养能力	工程能力、设计/开发解决方案		
掌握程度	★★★★	难度	中
结束条件	编写分页查询 SQL 语句		
训练内容： 对 hr 用户中的 employees 表进行分页查询，要求：每页 10 条记录，查询前 3 页的记录			

2. 递归查询

培养能力	工程能力、设计/开发解决方案		
掌握程度	★★★★	难度	中
结束条件	编写递归查询 SQL 语句		
训练内容： 对 hr 用户中的 employees 表进行递归查询，训练要求： (1) 查询管理员编号为 108 和 205 管理员所管理的员工； (2) 查询员工编号为 101 的管理员所管理的所有员工(树的顺序排列)； (3) 查询员工编号为 101 和 102 的管理员所管理的所有员工，并显示根节点的信息			

第 11 章 使 用 函 数

本章目标

知识点	理解	掌握	应用
1.常用单行函数的用法	✓	✓	✓
2.分组查询函数的用法	✓	✓	✓
3.聚合函数的用法	✓	✓	✓
4.SQL 语句的优化	✓	✓	

项目任务

- 能够使用单行函数
- 能够使用分组查询及聚合函数
- 掌握查询优化，改进 SQL 语句

知识能力点

能力点 \ 知识点	知识点 1	知识点 2	知识点 3	知识点 4
工程知识	✓	✓	✓	✓
问题分析		✓	✓	✓
设计/开发解决方案				✓
研究				✓
使用现代工具	✓	✓	✓	✓
工程与社会	✓	✓	✓	
环境和可持续发展				✓
职业规范	✓	✓	✓	✓
个人和团队				✓
沟通		✓	✓	
项目管理	✓	✓	✓	✓
终身学习	✓	✓	✓	✓

11.1 单行函数

单行函数格式也称标题函数，只对表中的一行数据进行操作，并且对每一行数据只产生一个输出结果。单行函数可用在 SELECT、WHERE 和 ORDER BY 的子句中，而且单行函数可以嵌套。单行函数有以下几种：

(1) 字符函数：接受字符输入并且返回字符或数值。
(2) 数值函数：接受数值输入并返回数值。
(3) 日期函数：对日期型数据进行操作。
(4) 转换函数：从一种数据类型转换为另一种数据类型。
(5) 通用函数：Oracle 自己提供的函数。

11.1.1 字符处理函数

字符函数接收的是字符类型的参数，这些字符可以来自于一个表中的列或者任意表达式。字符函数会按照某种方式处理输入参数，并返回一个结果。本节简单介绍 Oracle 数据库中常用的字符函数。

1) 字母大小写转换函数

UPPER(str)：将指定的将字符串全部转为大写；

LOWER(str)：将指定的字符串转换成小写；

INITCAP(str)：把每个字符串的首字符转换成大写。

【示例 11-1】UPPER()、LOWER()、INITCAP() 函数的应用。

本示例将字符串 "cddx" 转换为全部大写和首字符大写，将 "CDDX" 转换为全部小写。

```
SQL> SELECT UPPER('cddx')大写, LOWER('CDDX')小写,
  2  INITCAP('cddx')首字符大写 FROM dual;
大写              小写             首字符大写
------------------------   ------------
CDDX            cddx            Cddx
```

上述代码中的 dual 表是系统的一张虚表(伪表)。Oracle 中所有的查询都必须符合标准的 SQL 语句，因此，在 FROM 子句后面必须有一张表的名称。Oracle 保证 dual 里面永远只有一条记录。

2) 替换字符串

REPLACE() 函数是用另外一个值来替代字符串中的某个值。例如，可以用一个匹配数字来替代字母的每一次出现。REPLACE() 函数的基本语法如下：

```
REPLACE (string, search_string [, replace_string])
```

在上述语法中，REPLACE() 函数把 string 中的所有子字符串 search_string 使用可选的 replace_string 替换。如果没有指定 replace_string 参数的值，所有的 string 中的子字符串 search_string 都将被删除。string 可以为任何字符数据类型，如 char、varchar2、nchar、 nvarchar2、clob 或 nclob 等。

【示例 11-2】 REPLACE()函数的应用。

本示例将字符串"Hello cddx"中的字母"l"进行替换。

```
SQL> SELECT REPLACE('hello cddx ', 'l')替换1,
2  REPLACE('hello cddx', 'l', '5')替换2 FROM dual;
替换1                    替换2
---------               ---------
heo cddx                he55o cddx
```

REGEXP_REPLACE()函数也是字符串替换函数,它相当于增强的 REPLACE()函数。基本语法如下:

```
REGEXP_REPLACE(source_string, pattern[, replace_string[, position[, occurrence[, match_parameter ]]]])
```

其中,source_string 指定源字符串;pattern 指定正则表达式;replace_string 指定用于替换的字符串;position 指定起始搜索位置;occurrence 指定替换出现的第 n 个字符串;match_parameter 指定默认匹配操作的文本字符串。当 match_parameter 参数的取值为 i 时表示大小写不敏感;取值为 c 时表示大小写敏感,这是默认取值;取值为 n 时表示不匹配换行符号;取值为 m 表示多行模式;取值为 x 表示扩展模式,忽略正则表达式中的空白字符。

【示例 11-3】 REGEXP_REPLACE()函数的应用。

本例使用 REGEXP_REPLACE()函数替换 x 后面带数字的字符串为 Love,并且一个区分大小写,一个不区分大小写。执行语句和输出结果如下。

```
SQL> SELECT REGEXP_REPLACE('cddx! x12', 'x[0-9]+', 'love')v
2  FROM dual;
V
-----------
cddx! love
```

3) 截取字符串

截取字符串可以使用 SUBSTR()函数,该函数有两种使用方法。当从指定位置截取到结尾时,可以使用以下语法:

```
SUBSTR(string, position);
```

如果只截取部分的字符串,可以使用以下语法:

```
SUBSTR(string, position, length);
```

在前面的语法中,string 表示源字符串,position 表示截取字符串的开始位置;length 表示截取的长度。当 position 取值为 0 或者 1 时,它们的运行结果是一样的,都表示从第一个字符开始截取。

【示例 11-4】 SUBSTR()函数的第一种使用方法。

分别从字符串"成都大学"的第 0 个和第 1 个位置处进行截取,不指定截取长度,这时会截取整个字符串。语句和执行结果如下。

```
SQL> SELECT SUBSTR('成都大学', 0)s1, SUBSTR('成都大学', 1)s2
  2  FROM dual;
S1                      S2
--------------------    ----------------------
成都大学                 成都大学
```

【示例 11-5】截取字符串"成都大学"的内容,起始位置是 2,截取的长度为 3。

```
SQL>SELECT SUBSTR('成都大学', 2, 3)s1 FROM dual;
S1
----------
都大学
```

从上述结果可以看出,当指定起始位置为 2 时,会从字符串的正数第二个位置开始截取,由于该位置之后只有 3 个长度,因此截取的子字符串为"都大学"。

> 注意:使用 SUBSTR()函数截取字符串时,指定的截取长度小于 1,那么将返回一个空值。

4)连接字符串

连接字符串可以使用 CONCAT()函数来实现,CONCAT()只能连接两个字符串。语法如下:

```
CONCAT(n, m);
```

其中,n 和 m 两个参数既可以是字符,也可以是字符串。

【示例 11-6】CONCAT()函数的应用。

```
SQL> SELECT CONCAT('成都大学', '信息科学与工程学院.')concat
  2  FROM dual;
CONCAT
----------------------------------------------------------------
成都大学信息科学与工程学院.
```

5)获取字符串长度

Oracle 数据库提供多个用于获取字符串长度的函数,如表 11-1 所示,其中 UCS2 和 UCS4 是 Unicode 类别的一种编码。

表 11-1 获取字符串长度函数使用

函数	说明
LENGTH(x)	返回以字符为单位的长度
LENGTHB(string)	返回以字节为单位的长度
LENGTHC(string)	返回以 Unicode 为单位的长度
LENGTH2(string)	返回以 UCS2 代码点为单位的长度
LENGTH4(string)	返回以 UCS4 代码点为单位的长度

第 11 章 使用函数

【示例 11-7】获取字符串长度函数。

使用表列出的 LENGTH()函数查询字符串 "Hello Cddx" 的长度。

```
SQL>SELECT LENGTH('Hello Cddx')len FROM dual;
len
----------
10
```

6）其他字符函数

除了前面介绍的几种字符函数外，Oracle 还有其他的字符函数，下面再简单介绍几种。

❖ INSTR()函数

INSTR()函数返回要截取的字符串在源字符串中的位置。基本语法如下：

```
INSTR(string1, string2[, start_position, nth_appearance]);
```

其中，string1 表示源字符串，string2 表示要在 string1 中查找的字符串；start_position 是可选参数，表示从 string1 的哪个位置开始查找；nth_appearance 也是可选参数，表示要查找第几次出现的 string2。

当省略 start_position 参数时，默认值为 1，字符串索引从 1 开始。如果该参数为正，从左到右开始查找；如果该参数为负，从右到左查找，返回要查找的字符串在源字符串中的开始索引。当省略 nth_appearance 参数时，默认值为 1，如果为负数系统会报错。

【示例 11-8】INSTR()函数的应用。

返回截取的字符串'hello cddx'在源字符串中的位置

```
SQL>SELECT INSTR('hello cddx', 'l')i1, INSTR('hello cddx', 'cd')
2   i2, INSTR('hello cddx', 'l', 1, 2)i3 FROM dual;
I1         I2         I3
---------- ---------- ----------
3          7          4
```

如果要查找的字符串在源字符串中没有找到时，INSTR()函数将返回 0。

❖ LTRIM()、RTRIM()和 TRIM()函数

LTRIM()函数和 RTRIM()函数的语法相似，基本语法如下。

LTRIM(string1，string2);

RTRIM(string1，string2);

LTRIM()返回从 string1 左边删除 string2 后的字符串，RTRIM()返回从 string1 右边删除 string2 后的字符串，当遇到非 string2 的第一个字符时，结果将被返回。string2 默认设置为单个的空格串。

【示例 11-9】LTRIM()、RTRIM()函数的应用

分别使用 LTRIM()函数和 RTRIM()函数删除字符串 " cddx " 中的左边空格和右边空格，为了明显地观察效果，需要将它们与其他字符连接起来。

```
SQL> SELECT '1'||LTRIM(' cddx ')||'2' l1, '1'||RTRIM(' cddx ')
2  ||'2' r1 FROM dual;
L1                  R1
```

```
--------------------- ---------------------
1cddx 2                1 cddx2
```

TRIM()函数同时实现了 LTRIM()和 RTRIM()的功能,完整语法如下:

TRIM([LEADING|TRAILING|BOTH] [trimchar FROM] string);

其中,LEADING 表示只将字符串的头部分字符删除;TRAILING 表示只将字符串的尾部字符删除;BOTH 是默认值,既可以删除头部字符,也可以删除尾部字符;trimchar 是可选参数,表示试图删除什么字符,默认被删除的是空格,string 表示任意一个等待被处理的字符串。

【示例 11-10】TRIM()函数的使用。

本示例使用 TRIM()函数对字符串进行删除操作,删除了字母 c。

```
SQL> SELECT TRIM('c' from 'ccddxc')ltr FROM dual;
LTR
----
ddx
```

❖ ASCII()和 CHR()函数

ASCII()函数将字符转换为 ASCII 码值,CHR()函数将 ASCII 码值转换为字符。

【示例 11-11】ASCII()函数和 CHR()函数的应用。

```
SQL>SELECT ASCII('C')C, ASCII('成')成, chr(70)Z FROM dual;
C          成         Z
---------- ---------- ----------
65         46025      F
```

注意:CHR()函数与 ASCII()函数功能相反。

11.1.2 数值函数

数值函数主要用来处理数值数据,主要的数值函数有:绝对值函数、三角函数、对数函数、随机数函数等。在有错误产生时,数值函数将会返回 NULL。

1) 绝对值函数 ABS()

ABS()函数用来返回一个数的绝对值

【示例 11-12】ABS()函数的应用。

本示例求 2,-3.3 和-33 的绝对值。

```
SQL> SELECT ABS(2),ABS(-3.3),ABS(-33)FROM dual;
ABS(2)     ABS(-3.3)  ABS(-33)
---------- ---------- ----------
2          3.3        33
```

2) 精度函数

❖ ROUND(x)函数返回最接近于参数 x 的整数,对 x 值进行四舍五入。

❖ ROUND(x,y)返回最接近于参数 x 的数,其值保留到小数点后面 y 位,若 y 为负值，则将保留 x 值到小数点左边 y 位。

【示例 11-13】 ROUND(x) 函数和 ROUND(x, y) 函数的应用。

本示例使用 ROUND(x) 函数和 ROUND(x, y) 函数对操作数进行四舍五入操作。

```
SQL> SELECT ROUND(-1.13), ROUND(-1.68), ROUND(1.39, 1),
2        ROUND(232.38, -2)FROM dual;
ROUND(-1.13) ROUND(-1.68)    ROUND(1.39,1)   ROUND(232.38,-2)
------------ ------------    -------------   ----------------
-1           -2              1.4             200
```

注意：ROUND(1.39，1) 保留小数点后面 1 位，四舍五入的结果为 1.4；ROUND(232.38, -2) 保留小数点左边 2 位。y 值为负数时，保留的小数点左边的相应位数直接保存为 0，不进行四舍五入。

◆ TRUNCATE(x, y) 返回被舍去至小数点后 y 位的数字 x。若 y 的值为 0，则结果不带有小数点或不带有小数部分。若 y 设为负数，则截去(归零)x 小数点左起第 y 位开始后面所有低位的值。

【示例 11-14】 使用 TRUNC(x, y) 函数。

本示例对操作数进行四舍五入操作，结果保留小数点后面指定 y 位。

```
SQL>SELECT TRUNC(1.31, 1), TRUNC (1.99, 1), TRUNC (1.99, 0),
2   TRUNC (19.99, -1)FROM dual;
TRUNC(1.31, 1)   TRUNC(1.99, 1)   TRUNC(1.99, 0)   TRUNC(19.99, -1)
--------------   --------------   --------------   ----------------
1.3              1.9              1                10
```

TRUNC(1.31，1) 和 TRUNC(1.99，1) 都保留小数点后 1 位数字，返回值分别为 1.3 和 1.9；TRUNC(1.99，0) 返回整数部分值 1；TRUNC(19.99，-1) 截去小数点左边第 1 位后面的值，并将整数部分的 1 位数字置 0，结果为 10。

注意：ROUND(x, y) 函数在截取值的时候会四舍五入，而 TRUNC(x, y) 直接截取值，并不进行四舍五入。

3) 求余函数

MOD(x, y) 返回 x 被 y 除后的余数，MOD() 对于带有小数部分的数值也起作用，它返回除法运算后的精确余数。

【示例 11-15】 求余函数的应用。

```
SQL> SELECT MOD(31, 8), MOD(234, 10), MOD(45.5, 6)FROM dual;
MOD(31, 8)       MOD(234, 10)     MOD(45.5, 6)
----------       ------------     ------------
7                4                3.5
```

4) 三角函数

Oracle 提供了一些与求正弦值、求余弦值有关的三角函数

◆ SIN(x) 是正弦函数，其中 x 为弧度值。

◆ ASIN(x)是反正弦函数，若 x 不在-1 到 1 的范围内，则返回 NULL。

【示例 11-16】SIN(x)和 ASIN(x)函数使用。

```
SQL> SELECT SIN(1),ASIN(0.841470985)FROM dual;
SIN(1)       ASIN(0.841470985)
----------   -----------------
0.841470985  1
```

由结果可以看到，函数 ASIN 和 SIN 互为反函数。

相似地，Oracle 还提供了余弦函数 COS(x)和反余弦函数 ACOS(x)以及正切函数 TAN(x)和反正切函数 ATAN(x)。

5) 其他数值函数

◆ CEIL(x)返回不小于 x 的最小整数值。
◆ EXP(x)返回 e 的 x 乘方后的值。
◆ FLOOR(x)返回不大于 x 的最大整数值，返回值转化为一个 BIGINT。
◆ LN(x)返回 x 的自然对数，x 相对于基数 e 的对数，参数 n 要求大于 0。
◆ LOG(x, y)返回以 x 为底 y 的对数。
◆ POWER(x, y)函数返回 x 的 y 次乘方的结果值。
◆ SQRT(x)返回非负数 x 的二次方根。
◆ SIGN(x)函数返回参数 x 的符号。正数返回 1，0 返回 0，负数返回-1。

【示例 11-17】其他常用数值函数的应用。

```
SQL> SELECT SQRT(9),SQRT(40)FROM dual;
SQRT(9)      SQRT(40)
----------   ----------
3            6.32455532
SQL> SELECT CEIL(-3.35),CEIL (3.35)FROM dual;
CEIL(-3.35)  CEIL(3.35)
----------   ----------
-3           4
SQL> SELECT FLOOR(-3.35),FLOOR(3.35)FROM dual;
FLOOR(-3.35)    FLOOR(3.35)
---------------  -----------
-4              3
SQL> SELECT SIGN (3),SIGN (-10),SIGN (0)FROM dual;
 SIGN(3      SIGN(-10)   SIGN(0)
----------   ----------  ----------
1            -1          0
SQL> SELECT POWER(2,2),POWER(2,-2)FROM dual;
POWER(2,2)   POWER(2,-2)
----------   -----------
```

第 11 章 使用函数

```
4                0.25
SQL> SELECT EXP(3),EXP(-3),EXP(0)FROM dual;
EXP(3)         EXP(-3)              EXP(0)
----------     ---------------      ----------
20.08553692    0.04978706837        1
SQL>SELECT LOG(10,100),LOG(7,49)FROM dual;
LOG(10,100)    LOG(7,49)
----------     ----------
2              2
SQL> SELECT LN(2),LN(100)FROM dual;
LN(2)          LN(100)
----------     ----------
0.6931471806   4.605170186
```

11.1.3 类型转换函数

类型转换函数是将一种类型转换为另一种类型的函数,通常这类函数名称后面跟着待转换类型以及输出类型。

1) 字符串转 ASCII 类型字符串函数

ASCIISTR(string)函数可以将任意字符串转换为数据库字符集对应的 ASCII 字符串。

【示例 11-18】使用 ASCIISTR 函数。

```
SQL> SELECT ASCIISTR('从零开始学')FROM dual;
ASCIISTR('从零开始学')
------------------------
\4ECE\96F6\5F00\59CB\5B66
```

2) 二进制转十进制函数

BIN_TO_NUM()函数可以实现将二进制转换成对应的十进制。

【示例 11-19】使用 BIN_TO_NUM 函数。

```
SQL> SELECT BIN_TO_NUM (1,1,0) FROM dual;
BIN_TO_NUM(1,1,0)
-----------------
6
```

3) 数据类型转换函数

在 Oracle 中,用户如果想把数字转化为字符或者字符转化为日期,通常使用 CAST(expr as type_name)函数来完成。

【示例 11-20】使用 CAST 函数。

```
SQL> SELECT CAST('4321' AS NUMBER)+1 as a FROM dual;
A
```

```
--------------------
4322
```

4)数值转换为字符串函数

TO_CHAR 函数将一个数值型参数转换成字符型数据。具体语法格式如下：

TO_CHAR(n，[fmt[nlsparam]])

其中参数 n 代表数值型数据；参数 ftm 代表要转换成字符的格式；nlsparam 参数代表指定 fmt 的特征，包括小数点字符、组分隔符和本地钱币符号。

【示例 11-21】 使用 TO_CHAR 函数。

```
SQL> SELECT TO_CHAR(10.13245,'99.999'), TO_CHAR (10.13245)
  2  FROM dual;
TO_CHAR(10.13245,'99.999')       TO_CHAR(10.13245)
--------------------------       -----------------
 10.132                          10.13245
SQL> SELECT TO_CHAR (SYSDATE, 'YYYY-MM-DD')A,
  2 TO_CHAR (SYSDATE, 'HH24-MI-SS')B FROM dual;
A                         B
------------------------  ------------------------
2017-04-20                23-44-27
```

由结果可知，如果不指定转换的格式，则数值直接转化为字符串，不做任何格式处理。另外，TO_CHAR 函数还可以将日期类型转换为字符串类型。上面示例中的 YYYY 表示 4 位年，MM 表示两位月，DD 表示两位日，HH24 表示 24 小时制的小时，MI 表示两位分钟，SS 表示两位秒。

5)字符转日期函数

TO_DATE 函数将一个字符型数据转换成日期型数据。具体语法格式如下：

```
TO_DATE(char[, fmt[, nlsparam]])
```

其中参数 char 代表需要转换的字符串。参数 ftm 代表要转换成字符的格式；nlsparam 参数控制格式化时使用的语言类型。

【示例 11-22】 使用 TO_DATE 函数。

```
SQL> SELECT TO_CHAR(TO_DATE('1998-11-26', 'YYYY-MM-DD'), 'month')A
  2 FROM dual;
A
-----------------------------------------------------
11月
```

6)字符串转数字函数

TO_NUMBER() 函数将一个字符型数据转换成数字型数据。具体语法格式如下：

```
TO_NUMBER (expr[, fmt[, nlsparam]])
```

其中参数 expr 代表需要转换的字符串。参数 ftm 代表要转换成数字的格式；nlsparam 参数指定 fmt 的特征。包括小数点字符、组分隔符和本地钱币符号。

【示例 11-23】使用 TO_NUMBER 函数。

```
SQL> SELECT TO_NUMBER('1999.123', '9999.999')FROM dual;
TO_NUMBER('1999.123','9999.999')
--------------------------------
1999.123
```

11.1.4 日期和时间函数

日期和时间函数主要用来处理日期和时间值，一般的日期函数除了使用 DATE 类型的参数外，也可以使用 TIMESTAMP 类型的参数，但会忽略这些值的时间部分。

1) 获取当前日期和时间的函数
 ✧ SYSDATE 函数获取当前系统日期。
 ✧ SYSTIMESTAMP 函数获取当前系统时间，该时间包含时区信息，精确到微秒。返回类型为带时区信息的 TIMESTAMP 类型。

【示例 11-24】获取系统当前日期和当前时间。

```
SQL> SELECT TO_CHAR(SYSDATE,'YYYY-MM-DD HH24:MI:SS')v1,
  2   SYSTIMESTAMP FROM dual;
V1                      SYSTIMESTAMP
-------------------     ----------------------------------------
2017-04-20 23:57:44     20-4月 -17 11.59.31.571046 下午 +08:00
```

2) 获取时区的函数
 ✧ DBTIMEZONE 函数返回数据库所在的时区。
 ✧ SESSIONTIMEZONE 函数返回当前会话所在的时区。

【示例 11-25】使用时区函数。

```
SQL> SELECT DBTIMEZONE, SESSIONTIMEZONE FROM dual;
DBTIMEZONE  SESSIONTIMEZONE
----------  ---------------
+00:00      +08:00
```

> 注意，SYSDATE、SYSTIMESTAMP、DBTIMEZONE 和 SESSIONTIMEZONE 虽然叫函数，但是在使用的时候却规定不能加引号()。

3) 获取指定月份最后一天函数
 ✧ LAST_DAY(date)函数返回参数指定日期对应月份的最后一天。

【示例 11-26】使用 LAST_DAY 函数。

```
SQL> SELECT LAST_DAY(SYSDATE)A FROM dual;
A
-----------------
30-4月 -17
```

4) 获取指定日期后一周的日期函数

◆ NEXT_DAY(date，char)函数获取当前日期向后的一周对应日期，char 表示是星期几，全称和缩写都允许，但必须是有效值。

【示例 11-27】使用 NEXT_DAY 函数。

```
SQL> SELECT NEXT_DAY(SYSDATE, '星期五')A FROM dual;
A
------------------------------
28-4月 -17
```

NEXT_DAY(SYSDATE，'星期五')返回当前日期后第一个星期五的日期。

5）获取指定日期特定部分的函数

◆ EXTRACT(datetime)函数可以从指定的时间中提取特定部分。例如提取年份、月份或者时等。

【示例 11-28】使用 EXTRACT 函数。

```
SQL> SELECT EXTRACT (YEAR FROM SYSDATE)A FROM dual;
A
--------------
2017
```

6）获取两个日期之间的月份数

◆ MONTHS_BETWEEN(date1，date2)函数返回 date1 和 date2 之间的月份数。

【示例 11-29】使用 MONTHS_BETWEEN 函数。

```
SQL> SELECT MONTHS_BETWEEN(TO_DATE('20100228', 'YYYYMMDD'),
  2 TO_DATE('20110228', 'YYYYMMDD'))MONTHS FROM dual;
MONTHS
----------
    12
```

11.1.5 正则表达式函数

正则表达式通常用来检索或替换那些符合某个模式的文本内容,根据指定的匹配模式匹配文本中符合要求的特殊字符串。例如，从一个文本文件中提取电话号码，查找一篇文章中重复的单词或者替换用户输入的某些敏感词语等，这些地方都可以使用正则表达式。正则表达式强大而且灵活，可以用于非常复杂的查询。Oracle 中的支持正则表达式的函数主要有下面四个：

◆ REGEXP_LIKE：与 LIKE 的功能相似。
◆ REGEXP_INSTR：与 INSTR 的功能相似。
◆ REGEXP_SUBSTR：与 SUBSTR 的功能相似。
◆ REGEXP_REPLACE：与 REPLACE 的功能相似。

本节主要介绍常用的 REGEXP_LIKE。表 11-2 是 Oracle 中使用 REGEXP_LIKE 函数指定正则表达式的常用字符匹配模式。

表 11-2　正则表达式常用匹配列表

选项	说明	例子
^	匹配文本的开始字符	^a 匹配 arwe 但不匹配 barwe
$	匹配文本的结束字符	en$匹配 arwe 但不匹配 arwen
.	点号，匹配除 null，换行以外的任意单个字符	arw.n.可以匹配 arwen，arwin，但不能匹配 arween 或 arwn
*	匹配零个或多个在它前面的字符	a*rw 可以匹配 rw 或 aaarw
+	匹配前面的字符 1 次或多次	a+rwen 可以匹配 arwen 或 aarwen.但不能匹配 rwen
?	匹配前面的字符 0 次或 1 次	a?rwen 可以匹配 arwen 或 rwen.但不能匹配 aarwen.
<字符串>	匹配包含指定字符串的文本	fa 可以匹配 fan 或 afa 或 faad
[字符集合]	匹配字符集合中的任何一个字符	heelo[ab]可以匹配 helloa 或 hellob
[^]	匹配不在括号中的任何一个字符	[^abc] 匹配 desk 或 fox 等不包含 a、b、c 的字符串
字符串{n, }	匹配前面的字符串至少 n 次	b{2}可以匹配 Bbb 或 bbbb 或 bbbbbb.
字符串{n, m}	匹配前面的字符至少 n 次，最多 m 次.	b{2, 4}可以匹配 Bbb 或 bbb 或 bbbb.

1) 查询以特定字符或字符串开头的记录

字符"^"匹配以特定字符或者字符串开头的文本

【示例 11-30】查询 first_name 以字母 T 开头的记录。

```
SQL> SELECT first_name FROM employees WHERE
  2  REGEXP_LIKE(first_name,'^T');
FIRST_NAME
--------------------
Tayler
Timothy
TJ
Trenna
```

2) 查询以特定字符或字符串结尾的记录

字符"$"匹配以特定字符或者字符串结尾的文本

【示例 11-31】查询 first_name 以字母 h 结尾的记录。

```
SQL> SELECT first_name FROM employees
  2    WHERE REGEXP_LIKE(first_name,'h$');
FIRST_NAME
```

```
-----------
Elizabeth
Sarah
Sarath
```

3) 用符号"."来替代字符串中的任意一个字符

【示例 11-32】查询 first_name 包含字母 e 与 n 且两个字母之间只有一个字母的记录。

```
SQL> SELECT first_name FROM employees
2    WHERE REGEXP_LIKE(first_name , 'e.n');
FIRST_NAME
-----------
Jennifer
Jean
Neena
Trenna
Jennifer
```

4) 使用 "*" 和 "+" 来匹配多个字符

星号 '*' 可以匹配任意多个字符，包括 0 次。加号 '+' 匹配前面的字符至少一次。

【示例 11-33】查询 first_name 以字母 S 开头，且 S 后面出现 ar 的记录。

```
SQL> SELECT first_name FROM EMPLOYEES
2    WHERE REGEXP_LIKE(first_name, '^Sar*');
FIRST_NAME
-----------
Sarah
Sarath
```

【示例 11-34】查询 first_name 以字母 J 开头，且 J 后面出现字母 a 至少一次的记录。

```
SQL> SELECT first_name FROM employees
2    WHERE REGEXP_LIKE(first_name, '^Ja+');
FIRST_NAME
-----------
Janette
James
Jack
Jason
James
```

5) 匹配指定字符串

正则表达式可以匹配指定字符串，只要这个字符串在查询文本中即可，如要匹配多个字符串，多个字符串之间使用分隔符"|"隔开。

第 11 章 使用函数

【示例 11-35】 查询 first_name 包含字符串 on 或者 as 的记录。

```
SQL> SELECT first_name FROM employees
  2    WHERE REGEXP_LIKE(first_name , 'on|as');
FIRST_NAME
-----------
Harrison
Anthony
Jason
Donald
Jonathon
Winston
```

6）匹配指定字符中的任意一个

方括号"[]"指定一个字符集合，只匹配其中任何一个字符，即为所查找的文本。

【示例 11-36】 查找 first_name 包含字母 z 或者 w 的记录。

```
SQL> SELECT first_name FROM employees
  2    WHERE REGEXP_LIKE(first_name , '[zw]');
FIRST_NAME
-----------
Mozhe
Elizabeth
Hazel
Matthew
```

7）使用{n, }或者{n, m}来指定字符串连续出现的次数

字符串{n, }表示匹配前面的字符串至少 n 次；字符串{n, m}表示匹配前面的字符至少 n 次，最多 m 次。例如，a{2, }表示 a 连续出现至少 2 次，也可以大于 2 次；a{2, 4}表示 a 连续出现最少 2 次，最多 4 次。

【示例 11-37】 查询 first_name 出现 e 至少 2 次的记录。

```
SQL>SELECT first_name FROM employees
  2    WHERE REGEXP_LIKE(first_name , 'e{2, }');
FIRST_NAME
-----------
Neena
```

【示例 11-38】 查询 last_name 出现字符串 z 最少 1 次，最多 2 次的记录。

```
SQL>  SELECT last_name  FROM EMPLOYEES
  2    WHERE REGEXP_LIKE(last_name , 'z{1, 2}');
LAST_NAME
-----------
Errazuriz
```

```
Gietz
Lorentz
Ozer
```

11.2　分组查询及聚合函数

聚合函数也称统计函数或集合函数，返回基于多行的单一结果，行的准确数量并不确定，因此聚合函数并不是单行函数，而是多行函数。本节将介绍这些函数以及如何使用它们。这些聚合函数的名称和作用如表 11-3 所示。

表 11-3　Oracle 聚合函数

函数	作用
AVG()	返回某列的平均值
COUNT()	返回某列的行数
MAX()	返回某列的最大值
MIN()	返回某列的最小值
STDDEV()	返回某列的标准误差
SUM()	返回某列的和
VARIANCE()	返回某列的方差

1) AVG()

AVG() 函数通过计算返回行数和每一行数据的和，求得指定列数据的平均值。

【示例 11-39】查询职员工资的平均值。

```
SQL> SELECT AVG(salary)avg_salary FROM employees;
AVG_SALARY
-----------
6461.83178
```

2) COUNT()

COUNT() 函数统计数据表中包含的记录行的总数，或者根据查询结果返回列中包含的数据行数。其使用方法有两种：

◇ COUNT(*)计算表中总的行数，不管某列有数值或者为空值。
◇ COUNT(字段名)计算指定列下总的行数，计算时将忽略空值的行。

【示例 11-40】COUNT 函数示例。

本示例统计 EMPLOYEES 表中职工总人数 A 以及有销售提成的总人数 B。COUNT(*)计算所有人数，而 COUNT(commission_pct)只计算 commission_pct 不为空的人数。

```
SQL> select COUNT(*)A, COUNT(commission_pct)B from employees;
A        B
```

```
---------- --------------
    107           35
```

3) MAX()和 MIN()

- MAX()返回指定列中的最大值。
- MIN()函数返回查询列中的最小值。

【示例 11-41】在 employees 表中查找职员最高工资和最低工资。

```
SQL> SELECT MAX(salary)max, MIN(salary)min FROM employees;
   MAX          MIN
----------   ----------
  24000         2100
```

4) STDDEV()

STDDEV()返回某列的标准误差。

【示例 11-42】返回 EMPLOYEES 表中职员工资的标准偏差。

```
SQL>SELECT STDDEV(salary)"Deviation" FROM employees;
Deviation
----------
3909.36575
```

5) SUM()

SUM()函数是一个求总和的函数，返回指定列值的总和。

【示例 11-43】在 EMPLOYEES 表中查询各部门职工工资总和。

```
SQL> SELECT department_id, SUM(salary)AS depart_total
  2    FROM employees GROUP BY department_id;
DEPARTMENT_ID    DEPART_TOTAL
-------------    ------------
     100             51608
      30             24900
     Null             7000
      90             58000
      20             19000
...
```

注意：由查询结果可以看到，GROUP BY 按照 department_id 进行分组，SUM()函数计算每个分组中工资的总和。

6) VARIANCE()

VARIANCE()返回某列的方差

【示例 11-44】计算 EMPLOYEES 表中所有薪水的方差。

```
SQL>SELECT VARIANCE(salary)"Variance" FROM employees;
Variance
----------
```

```
15283140.5
```

7) GROUP BY

在实际应用中，经常需要进行分组聚合，而创建分组是通过 GROUP BY 子句实现的，即将查询对象按一定条件分组，然后对每一个组进行聚合分析。GROUP BY 通常和集合函数一起使用。

【示例 11-45】 根据 department_id 统计出各部门总人数。

```
SQL> SELECT department_id, COUNT(*)AS Total
2    FROM EMPLOYEES GROUP BY department_id;
DEPARTMENT_ID    TOTAL
-------------    ----------
100              6
30               6
...
```

8) HAVING 过滤分组

GROUP BY 子句分组，只是简单地依据所选列的数据进行分组，将该列具有相同值的行划为一组。而实际应用中，往往还需要过滤那些不能满足条件的行组，Oracle 提供了 HAVING 子句来实现。

【示例 11-46】 根据 department_id 进行分组，只显示平均工资大于 10000 的部门信息。

```
SQL> SELECT department_id, AVG(salary)avg_salary
2    FROM employees
3    GROUP BY DEPARTMENT_ID HAVING AVG(salary)>10000;

DEPARTMENT_ID    AVG_SALARY
-------------    ----------
90               19333.3333
110              10154
```

WHERE，GROUP BY，HAVING 之间的区别和用法如下：

（1）HAVING 通常与 GROUP BY 子句同时使用。GROUP BY 子句筛选数据行到各个组中，HAVING 只能用在 GROUP BY 之后，对分组后的结果进行筛选（即使用 HAVING 的前提条件是分组）。

（2）WHERE 子句在聚合前从数据源中去掉不符合筛选条件的数据。也就是说作用在 GROUP BY 子句和 HAVING 子句前。

（3）WHERE 后的条件表达式里不允许使用聚合函数，而 HAVING 可以。

【示例 11-47】 组合使用 WHERE，GROUP BY，HAVING

本示例根据 department_id 对 EMPLOYEES 表中的数据进行分组，并显示平均工资大于 10000 的部门名称、部门编号信息。

```
SQL>SELECT e.department_id, d.department_name, AVG(salary)avg_salary
2    FROM employees e, departments d
```

```
3  WHERE e.department_id=d.department_id
4  GROUP BY e.department_id,department_name
5  HAVING AVG(salary)>10000 ;
DEPARTMENT_ID     DEPARTMENT_NAME
-------------     ---------------
90                19333.3333
110               10154
```

11.3 SQL 语句优化

在 SQL 查询中，为了提高查询的效率，常常需要对 SQL 语句进行性能优化。

注意：由于篇幅有限，后面的示例不再赘述执行计划的对比分析，请读者自行练习操作，完成课后的训练任务。

（1）用 EXISTS 替换 DISTINCT，当提交一个包含一对多表信息的查询时，避免在 SELECT 子句中使用 DISTINCT，一般可以考虑用 EXISTS 替换。

【示例 11-48】显示部门编号和部门名称。

低效：

```
SQL> SELECT DISTINCT d.department_id, d.department_name
2    FROM departments d, employees e
3    WHERE d.department_id = e.department_id;
```

高效：

```
SQL> SELECT d.department_id, d.department_name
2    FROM  departments d
3    WHERE EXISTS (SELECT department_id  FROM employees e
4    WHERE d.department_id = e.department_id);
```

执行分析的过程如下：以 HR 身份登录到 PDBORCL，执行低效语句，输出查询结果，显示执行计划，如图 11-1 所示。

```
$ sqlplus hr/***@localhost/pdborcl
SQL> SET AUTOTRACE ON
SQL> SET LINESIZE 120
SQL> SELECT DISTINCT d.department_id, d.department_name
2    FROM departments d, employees e
3    WHERE d.department_id = e.department_id;
```

```
执行计划
----------------------------------------------------------
Plan hash value: 1467903405

---------------------------------------------------------------------------------------
| Id  | Operation           | Name              | Rows | Bytes | Cost (%CPU)| Time     |
---------------------------------------------------------------------------------------
|   0 | SELECT STATEMENT    |                   |   11 |   209 |     4  (25)| 00:00:01 |
|   1 |  HASH UNIQUE        |                   |   11 |   209 |     4  (25)| 00:00:01 |
|   2 |   NESTED LOOPS SEMI |                   |   11 |   209 |     3   (0)| 00:00:01 |
|   3 |    TABLE ACCESS FULL| DEPARTMENTS       |   27 |   432 |     3   (0)| 00:00:01 |
|*  4 |    INDEX RANGE SCAN | EMP_DEPARTMENT_IX |   44 |   132 |     0   (0)| 00:00:01 |
---------------------------------------------------------------------------------------

Predicate Information (identified by operation id):
---------------------------------------------------

   4 - access("D"."DEPARTMENT_ID"="E"."DEPARTMENT_ID")


统计信息
----------------------------------------------------------
        552  recursive calls
          0  db block gets
        743  consistent gets
         12  physical reads
          0  redo size
        868  bytes sent via SQL*Net to client
        552  bytes received via SQL*Net from client
          2  SQL*Net roundtrips to/from client
         39  sorts (memory)
          0  sorts (disk)
         11  rows processed
```

图 11-1　低效 SQL 语句执行计划

执行高效语句，输出结果，显示执行计划，如图 11-2 所示。

```
SQL>SET AUTOTRACE ON
SQL>SET LINESSIZE 120
SQL>SELECT d.department_id, d.department_name
  2  FROM  departments d
  3  WHERE EXISTS (SELECT department_id  FROM employees e
  4  WHERE d.department_id = e.department_id);
```

注意：上述语句需要 HR 用户具有 SELECT_CATALOG_ROLE、SELECT ANY DICTIONARY、ADVISOR 以及 ADMINISTER SQL TUNING SET 权限才可以进行优化指导。授权方式见：3.6.1　授予查询执行计划的权限。

表 11-4 对比展示了低效和高效语句的执行计划，效率差别是相当明显的。

表 11-4　执行计划对比分析

语句	Cost（%CPU）	Consistent Gets	Physical Reads
低效语句	4	743	12
高效语句	3	31	0

```
执行计划
----------------------------------------------------------
Plan hash value: 2605691773

--------------------------------------------------------------------------------
| Id  | Operation           | Name            | Rows  | Bytes | Cost (%CPU)| Time     |
--------------------------------------------------------------------------------
|   0 | SELECT STATEMENT    |                 |    11 |   209 |     3   (0)| 00:00:01 |
|   1 |  NESTED LOOPS SEMI  |                 |    11 |   209 |     3   (0)| 00:00:01 |
|   2 |   TABLE ACCESS FULL | DEPARTMENTS     |    27 |   432 |     3   (0)| 00:00:01 |
|*  3 |   INDEX RANGE SCAN  | EMP_DEPARTMENT_IX|   44 |   132 |     0   (0)| 00:00:01 |
--------------------------------------------------------------------------------

Predicate Information (identified by operation id):
---------------------------------------------------

   3 - access("D"."DEPARTMENT_ID"="E"."DEPARTMENT_ID")


统计信息
----------------------------------------------------------
         17  recursive calls
          0  db block gets
         31  consistent gets
          0  physical reads
          0  redo size
        868  bytes sent via SQL*Net to client
        552  bytes received via SQL*Net from client
          2  SQL*Net roundtrips to/from client
          0  sorts (memory)
          0  sorts (disk)
         11  rows processed
```

图 11-2 高效 SQL 语句执行计划

(2) 用表连接替换 EXISTS，采用表连接的方式比 EXISTS 更有效率。

【示例 11-49】查找部门名称包含 a 字母的职员姓名。

低效：

```
SQL>SELECT first_name, last_name
2    FROM employees e WHERE exists (SELECT department_id
3    FROM departments d WHERE d.department_id = e.department_id
4    AND d.department_name LIKE '%a%');
```

高效：

```
SQL> SELECT FIRST_NAME, LAST_NAME
2    FROM EMPLOYEES E , DEPARTMENTS D
3    WHERE D.DEPARTMENT_ID = E.DEPARTMENT_ID
4    AND d.department_name LIKE '%a%';
```

(3) 避免使用 HAVING 子句，HAVING 只会在检索出所有记录之后才对结果集进行过滤，这个处理需要排序，总计等操作。如果能通过 WHERE 子句限制记录的数目，那就能减少这方面的开销。

【示例 11-50】查找 job_id='ST_CLERK'，显示 job_id 和部门平均工资。

低效：

```
SQL> SELECT job_id, AVG(salary)FROM employees
2     GROUP BY job_id   HAVING job_id = 'ST_CLERK';
```
高效：
```
SQL> SELECT job_id, AVG(salary)FROM employees
2   WHERE job_id='ST_CLERK' GROUP BY job_id';
```

(4) 用 EXISTS 代替 IN，用 NOT EXISTS 代替 NOT IN 效率更高。

【示例 11-51】 显示部门名称，部门编号的信息。

低效：
```
SQL>SELECT department_name, department_id  FROM departments
2    WHERE department_id IN (SELECT department_id FROM employees);
```
高效：
```
SQL> SELECT department_name, department_id
2     FROM departments d
3     WHERE EXISTS (SELECT department_id FROM employees e
4     WHERE d.department_id = e.department_id);
```

(5) 减少访问数据库的次数。当执行每条 SQL 语句时，Oracle 在内部执行了许多工作：解析 SQL 语句、估算索引的利用率、绑定变量、读数据块等。由此可见，减少访问数据库的次数，就能实际上减少 Oracle 的工作量。

【示例 11-52】 查找部门编号为 100 和 90 的学生信息。

低效：
```
SQL> SELECT employee_id, first_name, last_name
2     FROM employees WHERE department_id=100;
SQL> SELECT employee_id, first_name, last_name
2     FROM employees WHERE department_id=90;
```
高效：
```
SQL>SELECT DISTINCT employee_id, first_name, last_name
2     FROM employees
3    WHERE department_id=100 OR department_id=90;
```

【示例 11-53】 查找 EMPLOYEES 表中部门编号=50 或部门编号=30 的职员数量和部门工资之和。

本示例使用 DECODE() 函数来减少处理时间，可以避免重复扫描相同记录或重复连接相同的表。

低效：
```
SQL> SELECT COUNT(*), SUM(salary)FROM employees
2     WHERE department_id='50';
SQL>SELECT COUNT(*), SUM(salary)FROM employees
2     WHERE department_id='30';
```

高效：

```
SQL> SELECT COUNT(DECODE(department_id,'50','XYZ',null))count_01,
  2    COUNT(DECODE(department_id,'30','XYZ',null))count_02,
  3    SUM(DECODE(department_id,'50',SALARY,null))sum_01,
  4    SUM(DECODE(department_id,'30',SALARY,null))sum_02
  5  FROM employees;
```

练习

一、判断题

1. Oralce 数据库中，TO_CHAR()也可用于数字的格式化。（　）
2. DECODE()的作用类似于一系列的 IF…ELSE 语句的组合。（　）
3. 单行函数分为字符函数、数字函数、日期函数、转换函数。（　）
4. 单行函数可用在 SELECT、WHERE 和 ORDER BY 中，而且可以嵌套。（　）
5. 单行函数只对表中的一行数据进行操作，对每一行数据只产生一个输出结果。
（　）

二、单项选择题

1. 哪个函数能去掉字符串右边的空格。（　）
   A. LTRIM()　　　　　　　　　　B. RTRIM()
   C. MOD()　　　　　　　　　　　D. INSERT()
2. 在 Oracle 中，执行下面的语句，哪个函数的返回值不等于-97（　）
   SELECT CEIL(-97.342)，FLOOR(-97.342)，ROUND(-97.342)，
   TRUNC(-97.342)FROM DUAL;
   A. CEIL()　　　　　　　　　　　B. FLOOR()
   C. ROUND()　　　　　　　　　　D. TRUNC()
3. 以下需求中哪个需要用分组函数来实现？（　）
   A. 把 ORDER 表中的订单时间显示成 'DD MM YYYY' 格式
   B. 把字符串 'JANUARY 28，2000' 转换成日期格式
   C. 显示 PRODUCT 表中的 COST 列值总量
   D. 把 PRODUCT 表中的 DESCRIPTION 列用小写形式显示
4. 在 Oracle 提供的日期函数中，（　）函数可以返回两个期间的月数。
   A. ADD_MONTHS()　　　　　　　B. MONTHS_BETWEEN()
   C. NEXT_DAY()　　　　　　　　D. LAST_DAY()
5. 使用语句查看公司员工工资的最大值、最小值和平均工资时，不涉及（　）
   A. MAX()　　　　　　　　　　　B. MIN()
   C. SUM()　　　　　　　　　　　D. AVG()

三、填空题

1. 执行下面的语句，替换后的返回结果是_____

SELECT REPLACE('accd'，'cd'，'ef') FROM dual；

2. 执行下面的语句，截取后的返回结果是_____

SELECT SUBSTR('HelloWorld'，6，5) value FROM dual；

3. Oracle 数据库提供的 ROUND() 函数和_____函数与精度有关。

4. Oracle 数据库提供_____函数连接两个字符串。

5. 获取系统的日期时需要借助_____伪列。

四、简答题

1. 简述 SUBSTR() 和 LENGTH() 的主要功能。

2. 分析以下的 SQL 命令：

SELECT * FROM product WHERE LOWER(description) = 'CABLE'；

命令能否执行？是否有结果返回？为什么？

3. 什么是多行函数？常用的多行函数有哪些？

训练任务

1. 使用单行函数

培养能力	工程能力、问题分析、使用现代工具、工程与社会、职业规范、项目管理、项目管理、终身学习		
掌握程度	★★★★★	难度	中
结束条件	能够创建和执行单行函数的 SQL 语句		
训练内容： 在 EMPLOYEES 表中查询出在(任何年份)2 月受聘的所有员工，显示满 10 年服务年限的员工的姓名和受雇日期，查询 50 部门的雇员工作的月数			

2. 使用分组查询及聚合函数

培养能力	工程能力、使用现代工具、工程与社会、职业规范、项目管理、沟通、项目管理、终身学习		
掌握程度	★★★★★	难度	中
结束条件	能够创建和执行分组查询及聚合函数的 SQL 语句		
训练内容： 使用分组查询及聚合函数，操作完成教材"11.2 分组查询及聚合函数"中的所有示例			

3. 掌握查询优化方法，改进 SQL 语句

培养能力	工程知识、问题分析、设计/开发解决方案研究、使用现代工具、工程与社会、环境和可持续发展、职业规范、个人和团队、项目管理、终身学习		
掌握程度	★★★★	难度	高
结束条件	找出最优的 SQL 语句		
训练内容： 掌握执行计划与 SQL 优化方法，请读者自行设计低效和高效的 SQL 语句，对比分析执行计划、执行效率，找出最优的 SQL 语句			

第 12 章　PL/SQL 语言

本章目标

知识点	理解	掌握	应用
1.PL/SQL 语言的基本语法结构	✓	✓	✓
2.PL/SQL 程序中的流程控制语句	✓	✓	✓
3.游标的基本操作	✓	✓	✓
4.过程和函数的创建	✓	✓	✓
5.包与触发器的创建	✓	✓	✓
6.正确使用异常处理	✓	✓	

项目任务

- 能够使用 PL/SQL 的基本语法和流程控制结构编写程序块。
- 能够利用游标逐行处理查询结果,以编程的方式访问数据。
- 能够将 PL/SQL 子程序写成函数和过程保存到数据库中。
- 能够利用 SQL 语句创建包和触发器。

知识能力点

能力点 \ 知识点	知识点 1	知识点 2	知识点 3	知识点 4	知识点 5
工程知识	✓	✓	✓	✓	✓
问题分析		✓	✓	✓	✓
设计/开发解决方案		✓	✓	✓	
研究		✓	✓	✓	
使用现代工具	✓				✓
工程与社会			✓	✓	
环境和可持续发展			✓	✓	
职业规范	✓	✓	✓	✓	
个人和团队			✓	✓	
沟通		✓	✓		
项目管理	✓	✓	✓	✓	✓
终身学习			✓		✓

12.1 PL/SQL 简介

在 SQL*Plus 环境下进行程序设计的语言称为 PL/SQL(PL 是 Procedural Language 的缩写)，是 Oracle 的一种开发工具，是对 SQL 语言的一种扩充。在在 PL/SQL 中可以使用查询语句和数据操纵语句，也可以编写过程、函数、包及数据库触发器。

PL/SQL 是一种块结构的语言,它将一组语句放在一个块中,一次性发送给服务器，PL/SQL 引擎分析收到 PL/SQL 语句块中的内容,把其中的过程控制语句由 PL/SQL 引擎自身去执行,把 PL/SQL 块中的 SQL 语句交给服务器的 SQL 语句执行器执行,如图 12-1 所示。

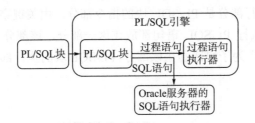

图 12-1 PL/SQL 体系结构

PL/SQL 块发送给服务器后，先被编译再执行，对于有名称的 PL/SQL 块(如子程序)可以单独编译，永久存储在数据库中，随时准备执行。

PL/SQL 的优点还有：

1) 支持 SQL 语言

SQL 是访问数据库的标准语言，通过 SQL 命令，用户可以操纵数据库中的数据。PL/SQL 支持所有的 SQL 数据操纵命令、游标控制命令、事务控制命令、SQL 函数、运算符和伪列。同时 PL/SQL 和 SQL 语言紧密集成，PL/SQL 支持所有的 SQL 数据类型和 NULL 值。

2) 支持面向对象编程

PL/SQL 支持面向对象的编程，在 PL/SQL 中可以创建类型，可以对类型进行继承，可以在子程序中重载方法等。

3) 更好的性能

SQL 是非过程语言，只能一条一条执行，而 PL/SQL 把一个 PL/SQL 块统一进行编译后执行，同时还可以把编译好的 PL/SQL 块存储起来，以备重用，减少了应用程序和服务器之间的通信时间，PL/SQL 是快速而高效的。

4) 可移植性

使用 PL/SQL 编写的应用程序，可以移植到任何操作系统平台上的 Oracle 服务器，同时还可以编写可移植程序库，在不同环境中重用。

5) 安全性

可以通过存储过程对客户机和服务器之间的应用程序逻辑进行分隔，这样可以限制对

Oracle 数据库的访问，数据库还可以授权和撤销其他用户访问的能力。

12.1.1 PL/SQL 基本结构

PL/SQL 是一种块结构的语言，一个 PL/SQL 程序包含了一个或者多个逻辑块，逻辑块中可以声明变量，变量在使用之前必须先声明。除了正常的执行程序外，PL/SQL 还提供了专门的异常处理部分进行异常处理。每个逻辑块分为三个部分，其语法结构如图 12-2 所示。

块结构说明：

◆ 声明部分：声明部分包含了变量和常量的定义。这个部分由关键字 DECLARE 开始，如果不声明变量或者常量，可以省略这部分。

◆ 执行部分：执行部分是 PL/SQL 块的指令部分，由关键字 BEGIN 开始，关键字 END 结尾。所有的可执行 PL/SQL 语句都放在这一部分，该部分执行命令并操作变量。其他的 PL/SQL 块可以作为子块嵌套在该部分。PL/SQL 块的执行部分是必选的。注意 END 关键字后面用分号结尾。

图 12-2　PL/SQL 块的结构

◆ 异常处理部分：该部分是可选的，该部分用 EXCEPTION 关键字把可执行部分分成两个小部分，一旦出现异常就跳转到异常部分执行。

> 注意：PL/SQL 是一种编程语言，与 Java 和 C#一样，除了有自身独有的数据类型、变量声明和赋值以及流程控制语句外，PL/SQL 还有自身的语言特性。
>
> PL/SQL 对大小写不敏感，为了良好的程序风格，开发团队都会选择一个合适的编码标准。比如有的团队规定：关键字全部大写，其余的部分小写。
>
> PL/SQL 块中的每一条语句都必须以分号结束，SQL 语句可以是多行的，但分号表示该语句结束。一行中可以有多条 SQL 语句，他们之间以分号分隔，但是不推荐一行中写多条语句。

12.1.2 变量和常量

1) 变量

PL/SQL 支持 SQL 中的数据类型，声明变量必须指明变量的数据类型，也可以声明变量时对变量初始化，变量声明必须在声明部分。声明变量的语法如下：

```
DECLARE
变量名 数据类型[:=初始值];
```

注意：DECLARE 声明数据类型如果需要长度，可以用括号指明长度，比如：varchar2(20)。

【示例 12-1】声明一个变量 sname，初始化值是 eric。

```
SET SERVEROUTPUT ON;
DECLARE
sname VARCHAR2(20):='eric';
BEGIN
sname:=sname||' and tom';
dbms_output.put_line(sname);
END;
/
eric and tom
PL/SQL 过程已成功完成。
```

注意：
(1) 声明一个变量 sname，初始化值是'eric'。字符串用单引号，若字符串中出现单引号可以使用两个单引号('')来表示，即单引号同时也具有转义的作用。
(2) 对变量 sname 重新赋值，赋值运算符是":="。
(3) dbms_output.put_line 是输出语句，可以把一个变量的值输出，在 SQL*Plus 中输出数据时，可能没有结果显示，可以使用命令：set serveroutput on 设置输出到 SQL*Plus 控制台上。

对变量赋值可以使用 SELECT…INTO 语句从数据库中查询数据对变量进行赋值。但是查询的结果只能是一行记录，不能是零行或者多行记录。

【示例 12-2】使用 SELECT…INTO 语句对变量 sname 赋值。

```
SET SERVEROUTPUT ON
DECLARE
sname VARCHAR2(20)DEFAULT 'John';
BEGIN
 SELECT first_name INTO sname FROM employees WHERE EMPLOYEE_ID =201;
 dbms_output.put_line(sname);
END;
```

/
Michael
PL/SQL 过程已成功完成。

> 注意：变量初始化时，可以使用 DEFAULT 关键字对变量进行初始化，也可以通过 RETURNING 子句返回聚合函数的计算结果，给变量赋值。

RETURNING 通常结合 DML(INSERT UPDATE DELETE)语句使用。常用的使用方法：

> UPDATE <表名> SET 列名=值 RETURNING 列名 INTO 变量名；
> INSERT INTO <表名>(列名1，列名2，…，列名n)VALUES(值1，值2，…，值n)
> RETURNING 列名 INTO 变量名；
> DELETE <表名> WHERE 条件表达式 RETURNING 列名 INTO 变量名；

> 注意：INSERT 返回的是添加后的值；UPDATE 返回更新后的值；DELETE 返回删除前的值。

【示例 12-3】 使用 RETURNING 子句。

操作 HR 用户中工作表 JOBS，使用 RETURNING 子句返操作后的数据。

```
SET SERVEROUTPUT ON
DECLARE  v_id VARCHAR2(10);
BEGIN
   INSERT INTO jobs VALUES ('IT_EXPL', 'Programmer Test', 5000, 12000)
     RETURNING job_id INTO v_id;
     DBMS_OUTPUT.PUT_LINE('INSERT: ' || v_id);
   v_id: = NULL;
   UPDATE jobs SET job_id='IT_TEST' WHERE job_id='IT_EXPL'
     RETURNING job_id INTO v_id;
     DBMS_OUTPUT.PUT_LINE('UPDATE: ' || v_id);
   v_id: = NULL;
   FROM jobs WHERE job_id='IT_TEST'
     RETURNING job_id INTO v_id;
     DBMS_OUTPUT.PUT_LINE('DELETE: ' || v_id);
END;
/
INSERT: IT_EXPL
UPDATE: IT_TEST
DELETE: IT_TEST
```
PL/SQL 过程已成功完成。

> 注意：INSERT INTO SELECT 和 MERGE 语句不支持 RETURNING

2) 常量

常量在声明时赋予初值，并且在运行时不允许重新赋值。使用 CONSTANT 关键字声明常量。常量初值可以使用赋值运算符":="赋值，也可以使用 DEFAULT 关键字赋值。声明常量的语法如下：

```
DECLARE
常量名 CONSTANT 数据类型：=常量值；
```

【示例 12-4】已知圆半径为 3，圆周率为 3.14，计算圆面积。

```
SET SERVEROUTPUT ON
DECLARE
  pi CONSTANT number:=3.14;     --圆周率长值
  r number DEFAULT 3;           --圆的半径默认值 3
  area number;                  --面积。
BEGIN
  area: =pi*r*r;                --计算面积
  dbms_output.put_line(area);   --输出圆的面积
END;
/
28.26
PL/SQL 过程已成功完成。
```

12.1.3 可变数组

可变数组(Varray)与嵌套表相似，也是一种集合。一个可变数组是对象的一个集合，其中每个对象都具有相同的数据类型。可变数组的大小由创建时决定。在表中建立可变数组后，可变数组在主表中作为一个列对待，从概念上讲，可变数组是一个限制了行集合的嵌套表。所有可变数组有连续的存储位置，最低的地址对应于第一个元素，最高地址对应于最后一个元素，如图 12-3 所示。

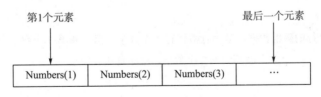

图 12-3 可变数组的存储结构

一个可变数组类型是用 CREATE TYPE 语句创建。它的基本语法如下：

```
CREATE OR REPLACE TYPE 类型名称 IS VARRAY(n)of 数据类型
```

PL/SQL 块创建 VARRAY 类型的基本语法是：

```
TYPE 类型名称 IS VARRAY(n)of 数据类型
```

其中 n 是 VARRAY 元素(最大值)的数目，并且以后可以使用 ALTER TYPE 语句来改变。可变数组是一维序列,起始索引值始终为 1。

【示例 12-5】用可变数组实现杨辉三角形的打印。

本示例输出杨辉三角,定义了数组类型 T_NUMBER,以及相应的数组变量 ROWARRAY,变量 N 表示输出的行数。本算法是从第一行开始,逐行打印,每一行的数字存储在数组变量 ROWARRAY 中,数组元素的值是从按右到左的顺序生成的。

```
SET SERVEROUTPUT ON;
DECLARE
TYPE T_NUMBER IS VARRAY (100) OF INTEGER NOT NULL;   --数组
I INTEGER;
J INTEGER;
SPACES VARCHAR2(30):='   ';   --三个空格,用于打印时分隔数字
N INTEGER:=9;   --一共打印9行数字
ROWARRAY T_NUMBER:=T_NUMBER();
BEGIN
   DBMS_OUTPUT.PUT_LINE('1');   --先打印第1行
   DBMS_OUTPUT.PUT(RPAD(1,9,' '));   --先打印第2行
   DBMS_OUTPUT.PUT(RPAD(1,9,' '));   --打印第一个1
   DBMS_OUTPUT.PUT_LINE('');   --打印换行
   --初始化数组数据
   FOR I IN 1..N LOOP
      ROWARRAY.EXTEND;
   END LOOP;
   ROWARRAY(1):=1;
   ROWARRAY(2):=1;
   FOR I IN 3..N  --打印每行,从第3行起
   LOOP
      ROWARRAY(I):=1;
      J:=I-1;
   --准备第J行的数组数据,第J行数据有J个数字,第1和第J个数字为1
   --这里从第J-1个数字循环到第2个数字,顺序是从右到左
      WHILE J>1
      LOOP
         ROWARRAY(J):=ROWARRAY(J)+ROWARRAY(J-1);
         J:=J-1;
      END LOOP;
      --打印第I行
      FOR J IN 1..I
      LOOP
```

```
            DBMS_OUTPUT.PUT(RPAD(ROWARRAY(J), 9, ' '));--打印第一个1
        END LOOP;
        DBMS_OUTPUT.PUT_LINE('');   --打印换行
    END LOOP;
END;
/
1
1    1
1    2    1
1    3    3    1
1    4    6    4    1
1    5    10   10   5    1
1    6    15   20   15   6    1
1    7    21   35   35   21   7    1
1    8    28   56   70   56   28   8    1
```

12.1.4 运算符

运算符是程序的重要部分，在 PL/SQL 程序中，可以将运算符分为多类，如算数运算符、连接运算符、比较(关系)运算符和逻辑运算符等。

1) 算数运算符

Oracle 中的算术运算符，只有+、-、*、/四种，其中除号(/)的结果是浮点数。求余运算只能借助函数 MOD(x, y)，返回 x 除以 y 的余数。

2) 连接运算符

连接运算符用双竖线‖表示，可以将两个字符串连接在一起。

【示例 12-6】表达式求值

本示例计算表达式的值为 6.5，并转化为整数(integer)输出，结果是 7。

```
set serveroutput on
DECLARE
    result integer;
BEGIN
    result:=1+3*5/2-MOD(12, 5);
    dbms_output.put_line('运算结果是：'||result);
END;
/
运算结果是：7
```

3) 比较(关系)运算符

用于将一个表达式与另一个表达式进行比较。常用的运算符如表 12-1 所示。

表 12-1　关系运算符

运算符	意义
<	小于
<=	小于等于
>	大于
>=	大于等于
=	等于(不是赋值运算符：=)
like	类似于 in 在……之中
!= 或<>	不等于
between	在……之间

4) 逻辑运算符

用于合并两个条件的结果以产生单个结果。逻辑运算符有 AND(逻辑与)、OR(逻辑或)和 NOT(逻辑非)，如表 12-2 所示。

表 12-2　逻辑运算符

运算符	意义
NOT	逻辑非
OR	逻辑或
AND	逻辑与

12.1.5　条件

PL/SQL 程序可通过条件结构来控制命令执行的流程。PL/SQL 提供了丰富的流程控制语句，有三种控制结构：①顺序控制，②条件控制，③循环控制。本节重点介绍条件控制语句。

PL/SQL 中关于条件控制的关键字有 IF-THEN、IF-THEN-ELSE、IF-THEN-ELSIF 和多分枝条件 CASE。

1) IF-THEN

该结构先判断一个条件是否为 TRUE，条件成立则执行对应的语句块，具体语法如下：

```
IF 条件 THEN
    条件语句体
END IF;
```

注意：

用 IF 关键字开始，END IF 关键字结束，END IF 后面有一个分号。

条件部分可以不使用括号，但是必须以关键字 THEN 来标识条件结束，如果条件成立，则执行 THEN 后到对应 END IF 之间的语句块内容。如果条件不成立，则不执行条件语句块的内容。

> PL/SQL 中关键字 THEN 到 END IF 之间的内容是条件结构体内容。
> 条件可以使用关系运算符和逻辑运算符。

【示例 12-7】查询 Den 的工资，如果大于 5000 元，则发奖金 800 元。

```
DECLARE
    newSal EMPLOYEES.SALARY%TYPE;
BEGIN
    SELECT SALARY INTO newSal FROM EMPLOYEES
    WHERE FIRST_NAME='Den';
    IF newSal>5000 THEN
        UPDATE EMPLOYEES
        SET SALARY=SALARY+800
        WHERE FIRST_NAME='Den';
    END IF;
    COMMIT ;
END;
/
```

在 PL/SQL 块中可以使用事务控制语句，该 COMMIT 同时也能把 PL/SQL 块外没有提交的数据一并提交，使用时需要注意。

2) IF-THEN-ELSE

把 ELSE 与 IF-THEN 连在一起使用，如果 IF 条件不成立则执行就会执行 ELSE 部分的语句，具体语法如下：

```
IF 条件 THEN
    --条件成立结构体
ELSE
    --条件不成立结构体
END IF;
```

3) IF-THEN-ELSIF

PL/SQL 中的再次条件判断中使用关键字 ELSIF，具体语法如下：

```
IF 条件 1 THEN
    --条件 1 成立结构体
ELSIF 条件 2 THEN
    --条件 2 成立结构体
ELSE
    --以上条件都不成立结构体
END IF;
```

【示例 12-8】查询 Den 的工资，如果大于 8000 元，则发奖金 1000 元，如果大于 5000 元，则发奖金 800 元，否则发奖金 400 元。

```
DECLARE
```

```
        newSal EMPLOYEES.SALARY%TYPE;
BEGIN
    SELECT SALARY INTO newSal FROM EMPLOYEES
    WHERE FIRST_NAME='Den';
    IF newSal>8000 THEN
        UPDATE EMPLOYEES
        SET SALARY=SALARY+1000
        WHERE FIRST_NAME='Den';
ELSIF newSal>5000 THEN
        UPDATE EMPLOYEES
        SET SALARY=SALARY+800
        WHERE FIRST_NAME='Den';
ELSE
        UPDATE EMPLOYEES SET SALARY=SALARY+400
        WHERE FIRST_NAME='Den';
    END IF;
    COMMIT;
END;
/
```

4) CASE

CASE 是一种选择结构的控制语句，可以根据条件从多个执行分支中选择相应的执行动作。也可以作为表达式使用，返回一个值。具体语法如下：

```
CASE [selector]
WHEN 表达式1 THEN 语句序列1;
WHEN 表达式2 THEN 语句序列2;
WHEN 表达式3 THEN 语句序列3;
...
[ELSE 语句序列N];
END CASE;
```

【示例 12-9】输入一个字母 A、B、C 分别输出对应的级别信息。

```
SET SERVEROUTPUT ON
DECLARE
    v_grade CHAR(1):=UPPER('&grade');
BEGIN
    CASE v_grade
        WHEN 'A' THEN
            dbms_output.put_line('Excellent');
        WHEN 'B' THEN
```

第12章 PL/SQL 语言

```
                dbms_output.put_line('Very Good');
            WHEN 'C' THEN
                dbms_output.put_line('Good');
            ELSE
                dbms_output.put_line('No such grade');
        END CASE;
END;
/
输入 grade 的值：A
原值   2:      v_grade CHAR(1):=UPPER('&grade');
新值   2:      v_grade CHAR(1):=UPPER('A');
Excellent
```

> 注意：&grade 是替换变量，表示在运行时由键盘输入字符串到 grade 变量中。v_grade 分别于 WHEN 后面的值匹配，如果成功就执行 WHEN 后的程序序列。

12.1.6 循环

PL/SQL 提供了丰富的循环结构来重复执行一些列语句。Oracle 提供的循环类型有：
- 无条件循环 LOOP … END LOOP 语句
- WHILE 循环语句
- FOR 循环语句

在上面的三类循环中 EXIT 用来强制结束循环。

1）LOOP 循环

LOOP 循环是最简单的循环，也称为无限循环，LOOP 和 END LOOP 是关键字。具体语法如下：

```
LOOP
    --循环体
END LOOP;
```

> 注意：循环体在 LOOP 和 END LOOP 之间，在每个 LOOP 循环体中，首先执行循环体中的语句序列，执行完后再重新开始执行。
>
> 在 LOOP 循环中可以使用 EXIT 或者[EXIT WHEN 条件]的形式终止循环。否则该循环就是死循环。

【示例 12-10】用 LOOP 循环结构计算 1+2+3+…+100 的值。

```
DECLARE
    counter number(3):=0;
    sumResult number:=0;
BEGIN
    LOOP
```

```
        counter: = counter+1;
        sumResult: = sumResult+counter;
        IF counter>=100 THEN
            EXIT;
        END IF;
        --EXIT WHEN counter>=100;
    END LOOP;
    dbms_output.put_line('result is: '||to_char(sumResult));
END;
/
```

> **注意**：LOOP 循环中可以使用 IF 结构嵌套 EXIT 关键字退出循环。
>
> **注释行**-- *EXIT WHEN counter>=100;*，该行可以代替下面的循环结构，WHEN 后面的条件成立时跳出循环。
>
> IF counter>=100 THEN
>   EXIT;
> END IF;

2) WHILE 循环

先判断条件，条件成立再执行循环体，具体语法如下：

```
WHILE 条件 LOOP
    --循环体
END LOOP;
```

【示例 12-11】用 WHILE 循环结构计算 1+2+3+…+100 的值。

```
DECLARE
    counter number(3): =0;
    sumResult number: =0;
BEGIN
    WHILE counter<100 LOOP
        counter: = counter+1;
        sumResult: = sumResult+counter;
    END LOOP;
    dbms_output.put_line('result is: '||sumResult);
END;
/
```

3) FOR 循环

FOR 循环需要预先确定的循环次数，可通过给循环变量指定下限和上限来确定循环运行的次数，然后循环变量在每次循环中递增(或者递减)。具体语法如下：

```
FOR 循环变量 IN [REVERSE] 循环下限..循环上限 LOOP
    --循环体
```

END LOOP;

> 注意:循环变量的值每次循环根据上下限的 REVERSE 关键字进行加 1 或者减 1。REVERSE 指明循环从上限向下限依次循环。

【示例 12-12】 用 FOR 循环结构计算 1+2+3+...+100 的值。

```
DECLARE
   counter number(3): =0;
   sumResult number: =0;
BEGIN
   FOR counter IN 1..100 LOOP
      sumResult: = sumResult+counter;
   END LOOP;
   dbms_output.put_line('result is: '||sumResult);
END;
/
```

12.2 异常处理

在程序运行时出现的错误,称为异常。发生异常后,语句将停止执行,PL/SQL 引擎立即将控制权转到 PL/SQL 块的异常处理部分。异常处理机制简化了代码中的错误检测,每一个异常都对应一个异常码和异常信息。

12.2.1 预定义异常

为了 Oracle 开发和维护的方便,在 Oracle 异常中,为常见的异常码定义了对应的异常名称,称为预定义异常,常见的预定义异常如表 12-3 所示。

表 12-3 PL/SQL 中预定义异常

异常名称	异常码	描述
DUP_VAL_ON_INDEX	ORA-00001	试图向唯一索引列插入重复值
INVALID_CURSOR	ORA-01001	试图进行非法游标操作
INVALID_NUMBER	ORA-01722	试图将字符串转换为数
NO_DATA_FOUND	ORA-01403	SELECT INTO 语句中没有返回任何记录
TOO_MANY_ROWS	ORA-01422	SELECT INTO 语句中返回多于 1 条记录
ZERO_DIVIDE	ORA-01476	试图用 0 作为除数
CURSOR_ALREADY_OPEN	ORA-06511	试图打开一个已经打开的游标

PL/SQL 中用 EXCEPTION 关键字开始异常处理。具体语法是：

```
BEGIN
  --可执行部分
    EXCEPTION    -- 异常处理开始
    WHEN 异常名1 THEN
        --对应异常处理
    WHEN 异常名2 THEN
        --对应异常处理
    ...
    WHEN OTHERS THEN
        --其他异常处理
END;
```

异常发生时，进入异常处理部分，具体的异常与若干个 WHEN 子句中指明的异常名匹配，匹配成功就进入对应的异常处理部分，如果对应不成功，则进入 OTHERS 进行处理。

【示例 12-13】异常处理

```
DECLARE
    newSal employees.salary % TYPE;
BEGIN
    SELECT salary INTO newSal FROM employees;
    EXCEPTION
    WHEN TOO_MANY_ROWS THEN
        dbms_output.put_line('返回的记录太多了');
    WHEN OTHERS THEN
        dbms_output.put_line('未知异常');
END;
/
```

12.2.2 自定义异常

除了预定义异常外，用户还可以在开发中自定义异常，自定义异常可以让用户采用与 PL/SQL 引擎处理错误相同的方式进行处理，用户自定义异常的两个关键点：

（1）异常定义：在 PL/SQL 块的声明部分采用 EXCEPTION 关键字声明异常，定义方法与定义变量相同。比如声明一个 MYEXP 异常方法是：

```
MYEXP EXCEPTION;
```

（2）异常引发：在程序可执行区域，使用 RAISE 关键字进行引发。比如引发 MYEXP 方法是：

```
RAISE MYEXP;
```

【示例 12-14】自定义异常

```
SET SERVEROUTPUT ON
DECLARE
   sal employees.salary %TYPE;
   MYEXP EXCEPTION;
BEGIN
   SELECT salary INTO sal FROM employees WHERE first_name='Valli';
   IF sal<5000 THEN
      RAISE MYEXP;
   END IF;
   EXCEPTION
   WHEN NO_DATA_FOUND THEN
      dbms_output.put_line ('NO RECORDSET FIND!');
   WHEN MYEXP THEN
      dbms_output.put_line ('SAL IS TO LESS!');
END;
/
SAL IS TO LESS!
```

12.2.3 引发应用程序异常

在 Oracle 开发中,遇到的系统异常都有对应的异常码,在应用系统开发中,用户自定义的异常也可以指定一个异常码和异常信息,Oracle 系统为用户预留了自定义异常码,其范围是-20000 到-20999 之间的负整数。引发应用程序异常的语法是:

```
RAISE_APPLICATION_ERROR(异常码,异常信息)
```

【示例 12-15】引发系统异常

```
DECLARE
   sal employees.salary %TYPE;
   myexp EXCEPTION;
BEGIN
   SELECT salary INTO sal FROM employees WHERE first_name='Valli';
   IF sal<5000 THEN
      RAISE myexp;
   END IF;
   EXCEPTION
   WHEN NO_DATA_FOUND THEN
      dbms_output.put_line ('NO RECORDSET FIND!');
   WHEN MYEXP THEN
```

```
        RAISE_APPLICATION_ERROR(-20001, 'SAL IS TO LESS!');
END;
/
ORA-20001: SAL IS TO LESS!
ORA-06512：在 line 13
```

如果要处理未命名的内部异常，必须使用 OTHERS 异常处理器。也可以利用 PRAGMA EXCEPTION_INIT 把一个异常码与异常名绑定。

PRAGMA 由编译器控制，PRAGMA 在编译时处理，而不是在运行时处理。EXCEPTION_INIT 告诉编译器将异常名与 Oracle 错误码绑定起来，这样可以通过异常名引用任意的内部异常，并且可以通过异常名为异常编写适当的异常处理器。PRAGMA EXCEPTION_INIT 的语法是：

```
PRAGMA EXCEPTION_INIT(异常名，异常码)
```

这里的异常码可以是用户自定义的异常码，也可以是 Oracle 系统的异常码。

【示例 12-16】 PRAGMA EXCEPTION_INIT 异常

本示例是 SCOTT 用户登录，查询表 emp，在测试的时候，通过输入替换变量&empno 不同的值，观察不同的异常处理。7369 员工的 comm 为空，抛出了异常 null_salary，7499 员工的 comm 不为空，因此正常输出。

```
SQL>
<<outterseg>>
  DECLARE
  null_salary EXCEPTION;
  PRAGMA EXCEPTION_INIT(null_salary, -20101);
BEGIN
  <<innerStart>>
    DECLARE
    curr_comm NUMBER;
  BEGIN
    SELECT comm INTO curr_comm FROM emp WHERE empno = &empno;
    IF curr_comm IS NULL THEN
       RAISE_APPLICATION_ERROR(-20101, 'Salary is missing');
    ELSE
       dbms_output.put_line('有津贴');
    END IF;
  END;
  EXCEPTION
  WHEN NO_DATA_FOUND THEN
    dbms_output.put_line('没有发现行');
  WHEN null_salary THEN
```

```
        dbms_output.put_line('津贴未知');
    WHEN OTHERS THEN
        dbms_output.put_line('未知异常');
END;
SQL>/
Enter value for empno: 7369
old  10:  SELECT comm INTO curr_comm FROM emp WHERE empno = &empno;
new  10:  SELECT comm INTO curr_comm FROM emp WHERE empno = 7369;
津贴未知
SQL> //
Enter value for empno: 7499
old  10:  SELECT comm INTO curr_comm FROM emp WHERE empno = &empno;
new  10:  SELECT comm INTO curr_comm FROM emp WHERE empno = 7499;
有津贴
```

> 注意：把异常名称 null_salary 与异常码-20101 关联，该语句由于是预编译语句，必须放在声明部分。也就是说-20101 的异常名称就是 null_salary。
>
> 在内部 PL/SQL 语句块中引发应用系统异常-20101。
>
> 在外部 PL/SQL 语句块中就可以用异常名 null_salary 进行捕获。

12.3 游 标

游标(Cursor)是一种能从包括多条数据记录的结果集中每次提取一条记录的机制。游标充当指针的作用。尽管游标能遍历结果中的所有行，但它一次只指向一行。SQL 的游标是一种临时的数据库对象，即可以用来存放在数据库表中的数据行副本，也可以指向存储在数据库中的数据行的指针。游标提供了在逐行的基础上操作表中数据的方法。

Oracle 中游标的类型分为静态游标和引用游标(也叫动态游标)两类。静态游标又分为显示游标和隐式游标。

(1)隐式游标：所有 DML 语句(增加、删除、修改、查询单条记录)为隐式游标，使用时不需要声明隐式游标，它由系统定义。

(2)显示游标：用户显示声明的游标，即指定结果集。当查询返回结果超过一行时，就需要一个显式游标。

12.3.1 游标的基本操作

显示游标的基本操作过程是：声明游标、打开游标、提取数据和关闭游标。如图 12-4 所示。

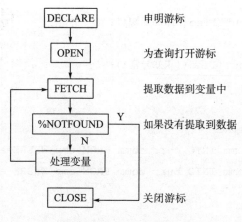

图 12-4　游标的基本操作过程

1）声明游标

声明游标的语法如下：

```
CURSOR 游标名 [(参数1 数据类型[,参数2 数据类型...])]
IS SELECT 子句;
```

参数是可选部分，所定义的参数可以出现在 SELECT 语句的 WHERE 子句中。如果定义了参数，则必须在打开游标时传递相应的实际参数。

SELECT 语句是对表或视图的查询语句，甚至也可以是联合查询。可以带 WHERE 条件、ORDER BY 或 GROUP BY 等子句，但不能使用 INTO 子句。在 SELECT 语句中可以使用在定义游标之前定义的变量。

2）打开游标

声明游标后还不能使用它，必须使用 OPEN 语句打开游标。打开游标的语法如下：

```
OPEN 游标名[(实际参数1[,实际参数2...])];
```

打开游标时，SELECT 语句的查询结果就被传送到了游标工作区。

注意，OPEN 打开游标的方式是只读的，是为查询而打开的。如果为了修改而打开游标，需要加上 FOR UPDATE 子句。

3）读取游标中的数据

打开游标之后，就可以读取游标中的数据了，FETCH 语句用来从游标中提取数据，该语句每次返回一行数据，并自动将记录指针前移到下一行，即不能后退。用 FETCH 提取字段数据时，必须保证提取游标中所有的字段。读取游标中的数据语法格式如下：

```
FETCH 游标名 INTO 变量名1[,变量名2...];
```

或

```
FETCH 游标名 INTO 记录变量;
```

游标打开后有一个指针指向数据区，FETCH 语句一次返回指针所指的一行数据，要返回多行需重复执行，可以使用循环语句来实现。控制循环可以通过判断游标的属性来进行。

下面对这两种格式进行说明：

第一种格式中的变量名是用来从游标中接收数据的变量,需要事先定义。变量的个数和类型应与 SELECT 语句中的字段变量的个数和类型一致。

第二种格式一次将一行数据取到记录变量中,需要使用%ROWTYPE 事先定义记录变量,这种形式使用起来比较方便,不必分别定义和使用多个变量。

定义记录变量的方法如下:

变量名 表名|游标名%ROWTYPE;

其中的表必须存在,游标名也必须先定义。

4)关闭游标

显式游标打开后,必须显式地关闭。游标一旦关闭,游标占用的资源就被释放,游标变成无效,必须重新打开才能使用。关闭游标语法格式如下:

CLOSE 游标名;

下面通过一个简单的案例来学习使用游标的完整过程,其中包含了声明游标、打开游标、提取游标和关闭游标的全过程。

【示例 12-17】用游标提取 EMPLOYEES 表中一个职员信息。

```
SET SERVEROUTPUT ON
DECLARE
  v_fname VARCHAR2(10);
  v_lname VARCHAR2(10);
  v_salary employees.salary%TYPE;
  CURSOR emp_cursor  --声明一个名称为 emp_cursor 的游标
    IS SELECT first_name, last_name, salary FROM employees;
  BEGIN
  OPEN emp_cursor;   --打开游标
  --利用 FETCH 语句从结果集中提取指针指向的当前行记录。
  FETCH emp_cursor INTO v_fname, v_lname, v_salary;
      DBMS_OUTPUT.PUT_LINE(v_fname||', '||v_lname||', '|| v_salary);
  CLOSE emp_cursor;    --关闭游标
END;
/
Steven, King, 24000
PL/SQL 过程已成功完成。
```

本例中游标的 SELECT 语句没有 WHERE 子句,查询所有的职工,但最后只提取了一个职工的数据,原因是因为没有循环提取数据。如果用户想使用显式游标提取多条记录,就需要使用 LOOP 语句遍历结果集,在循环的过程中需要使用游标属性,判断游标在循环过程的状态。显示游标有以下 4 个属性,可按照游标名%属性的形式取得游标的属性值:

%ROWCOUNT:获得 FETCH 语句返回的数据行数。

%FOUND:最近的 FETCH 语句返回一行数据则为真,否则为假。

%NOTFOUND:与%FOUND 属性返回值相反。

%ISOPEN：游标已经打开时值为真，否则为假。

【示例 12-18】提取 EMPLOYEES 表中所有职员信息。

本示例使用 LOOP 语句循环提取每一行数据，直到所有数据取完，判断游标数据取完的语句是"EXIT WHEN emp_cursor%NOTFOUND；"，这条语句使用了%NOTFOUND 游标属性，表示如果没有找到数据，就退出循环。

```
SET SERVEROUTPUT ON
DECLARE
  v_fname VARCHAR2(100);
  v_lname VARCHAR2(100);
  v_salary employees.salary%TYPE;
  CURSOR emp_cursor  --声明一个名称为 emp_cursor 的游标
     IS SELECT first_name, last_name, salary FROM employees;
BEGIN
  OPEN emp_cursor;   --打开游标
  LOOP
    --利用 FETCH 语句从结果集中提取指针指向的当前行记录。
    FETCH emp_cursor INTO v_fname, v_lname, v_salary;
    EXIT WHEN emp_cursor%NOTFOUND;
    DBMS_OUTPUT.PUT_LINE(v_fname||', '||v_lname||', '|| v_salary);
  END LOOP;
  CLOSE emp_cursor; --关闭游标
END;
/
Steven, King, 24000
Neena, Kochhar, 17000
Lex, De Haan, 17000
...
```

在使用游标循环取数据的时候一定要设置正确的退出循环条件，不要出现无限循环。另外，如果需要的话，还可以使用游标的其他属性。

12.3.2 游标 FOR 循环

游标的 FOR IN 方式是游标使用的快捷方式，在实现同样的功能情况下，语句更简单。这种方式游标会自动打开，自动取所有数据，自动关闭。

【示例 12-19】使用 FOR 循环提取 EMPLOYEES 表中所有职员信息。

```
SET SERVEROUTPUT ON
BEGIN
  FOR v in (SELECT first_name, last_name, salary FROM employees)
```

```
   LOOP
     DBMS_OUTPUT.PUT_LINE(v.first_name||', '
       ||v.last_name||', '|| v.salary);
   END LOOP;
END;
/
Steven, King, 24000
Neena, Kochhar, 17000
Lex, De Haan, 17000
...
```

本示例是提取 EMPLOYEES 表中所有职员信息的最简单的方式,甚至都没有定义变量,唯一的变量 v 不需要在 DECLARE 中定义,而是在 FOR 中定义后立即使用。从这里可以看出,Oracle 的游标功能既强大又方便。

12.3.3 引用游标

引用游标(REF 游标)是一种动态游标,它不依赖于指定的 SQL 语句,比普通游标更灵活。一个引用游标在运行时可以与任何 SQL 相关联,可以定义为游标变量。在程序开发中,常用引用游标变量将结果集返回给应用程序,显示到用户界面中。在第 14 章"小型商品销售系统"的案例中,使用了引用游标以分页的形式返回员工的信息。

引用游标的类型是 REF CURSOR 或者 SYS_REFCURSOR,可以用 SYS_REFCURSOR 定义一个引用类型的游标变量。

【示例 12-20】 使用引用游标分页提取 EMPLOYEES 表的记录。

本示例使用引用游标提取 EMPLOYEES 表中第二页的记录。其中 rc 变量是引用游标变量,该变量与 v_sql 语句动态绑定,之所以称为"动态",是因为 v_sql 语句中有绑定变量 v_offset,v_pagesize,这两个变量的值是可以动态改变的,如果将这个示例应用于实际的应用程序中,可以让用户方便地分页查询表的记录。

```
SET SERVEROUTPUT ON
DECLARE
  rc SYS_REFCURSOR;
  v_pagesize number;
  v_pageidx number;
  v_offset number;
  v_sql varchar2(2000);
  lrow employees%rowtype;
BEGIN
  v_pagesize: =3; --每页 3 条记录
  v_pageidx: =2; --取第二页数据
```

```
      v_offset: =v_pagesize*(v_pageidx-1);
      v_sql: ='select * from employees offset: v_offset rows
         fetch next: v_pagesize rows only';
      OPEN rc FOR v_sql using v_offset, v_pagesize;
      LOOP
         FETCH rc INTO lrow;
         EXIT WHEN rc%NOTFOUND;
         DBMS_OUTPUT.PUT_LINE(lrow.employee_id ||
            ' '|| lrow.first_name);
      END LOOP;
      CLOSE rc;       --关闭游标
   END;
   /
   103    Alexander
   104    Bruce
   105    David
   PL/SQL 过程已成功完成。
```

12.3.4 修改或删除游标结果集

游标除了可以用于逐行查询数据之外，还可以用来逐行修改或者删除行记录。

1) 修改游标结果集中的行

定义游标时使用"FOR UPDATE"子句指定更新的列，在修改数据时使用"UPDATE 表名 SET 子句 WHERE CURRENT OF 游标名；"方式修改一行记录，而不需要在 WHERE 之后写其他条件，SET 子句中只能出现"FOR UPDATE"中定义过的字段。

【示例 12-21】使用游标修改 MYEMP 表的记录。

本示例使用游标逐行修改 MYEMP 表每行的 salary，在原值的基础上加 100。MYEMP 表是 employees 表的一个副本，可以通过命令"CREATE TABLE myemp AS SELECT * FROM employees；"创建。

```
SET SERVEROUTPUT ON
DECLARE
   CURSOR cur  --声明游标 cur
      IS SELECT employee_id, salary FROM myemp FOR UPDATE OF salary;
BEGIN
   FOR rec1 IN cur
   LOOP
      UPDATE myemp SET salary=salary+100 WHERE CURRENT OF cur;
   END LOOP;
```

```
END;
/
```

其实，可以用一条 SQL 语句"UPDATE myemp SET salary=salary+100；"实现本例的功能，为什么非要用游标来完成呢？原因是游标可以逐行更新，对于一些复杂的更新逻辑，如果没有办法用一条 SQL 语句来完成，就只要依靠游标了。

2）删除游标结果集中的行

删除游标结果集中的一行的命令是：

```
DELETE FROM 表名 WHERE CURRENT OF 游标名
```

【示例 12-22】删除 MYEMP 的奇数行。

本示例删除 MYEMP 表的奇数行。用变量 R 表示行号。由于这个逻辑比较复杂，这个任务用一条 DELETE 语句无法完成。

```
SET SERVEROUTPUT ON
DECLARE
   R NUMBER;
   CURSOR cur --声明游标 cur
     IS SELECT employee_id, salary FROM myemp FOR UPDATE OF salary;
BEGIN
   R：=1;
   for rec1 in cur
   loop
     IF R MOD 2 =1 THEN
        DELETE myemp WHERE CURRENT OF cur;
     END IF;
     R：=R+1;
   end loop;
END;
/
```

> 注意：在使用游标逐行删除记录的时候，定义游标时也必须有"FOR UPDATE OF"关键字。

12.4 存储过程

存储过程与函数（另外还有包与触发器）是命名的 PL/SQL 块（也是用户的方案对象），被编译后存储在数据库中，以备执行。因此，其他 PL/SQL 块可以按名称来使用他们。所以，可以将商业逻辑、企业规则写成函数或过程保存到数据库中，以便共享。

存储过程和函数统称为 PL/SQL 子程序，他们是被命名的 PL/SQL 块，均存储在数据库中，并通过输入、输出参数或输入/输出参数与其调用者交换信息。过程和函数的唯一区别是函数必须向调用者返回数据，而过程则不返回数据。

12.4.1 创建存储过程

在 Oracle 上建立存储过程，可以被多个应用程序调用，可以向存储过程传递参数，也可以向存储过程传回参数。

创建存储过程语法：

```
CREATE [OR REPLACE] PROCEDURE procedure_name
([arg1 [ IN | OUT | IN OUT ]] type1 [DEFAULT value1],
 [arg2 [ IN | OUT | IN OUT ]] type2 [DEFAULT value1]],
 ...
 [argn [ IN | OUT | IN OUT ]] typen [DEFAULT valuen])
    [ AUTHID DEFINER | CURRENT_USER ]
{ IS | AS }
    <声明部分>
BEGIN
    <执行部分>
EXCEPTION
    <可选的异常错误处理程序>
END procedure_name;
```

其中，IN，OUT，IN OUT 是形参的模式。若省略，则为 IN 模式，表示输入类型参数。OUT 是输出类型参数，返回时应被赋值。IN OUT 模式既是输入也是输出。调用时，对于 IN 模式的实参可以是常量或变量，但对于 OUT 和 IN OUT 模式的实参必须是变量。

"OR REPLACE"选项表示替换原过程，如果原过程存在，必须有这个选项才能覆盖原过程。

【示例 12-23】删除指定员工记录。

```
CREATE OR REPLACE PROCEDURE DelEmp
  (v_empno IN employees.employee_id%TYPE)
AS
   No_result EXCEPTION;
BEGIN
   DELETE FROM employees WHERE employee_id = v_empno;
   IF SQL%NOTFOUND THEN
      RAISE no_result;
   END IF;
   DBMS_OUTPUT.PUT_LINE('编码为'||v_empno||'的员工已被删除！');
EXCEPTION
   WHEN no_result THEN
```

```
        DBMS_OUTPUT.PUT_LINE('温馨提示：你需要的数据不存在！');
    WHEN OTHERS THEN
        DBMS_OUTPUT.PUT_LINE(SQLCODE||'---'||SQLERRM);
END DelEmp;
```

【示例 12-24】 插入员工记录：

```
CREATE OR REPLACE PROCEDURE InsertEmp(
    v_empno     in employees.employee_id%TYPE,
    v_firstname in employees.first_name%TYPE,
    v_lastname  in employees.last_name%TYPE,
    v_deptno    in employees.department_id%TYPE
    )
AS
    empno_remaining EXCEPTION;
    PRAGMA EXCEPTION_INIT(empno_remaining, -1);
    /* -1 是违反唯一约束条件的错误代码 */
BEGIN
    INSERT INTO EMPLOYEES(EMPLOYEE_ID, FIRST_NAME, LAST_NAME,
        HIRE_DATE, DEPARTMENT_ID)
    VALUES(v_empno, v_firstname, v_lastname, sysdate, v_deptno);
    DBMS_OUTPUT.PUT_LINE('温馨提示：插入数据记录成功！');
EXCEPTION
    WHEN empno_remaining THEN
        DBMS_OUTPUT.PUT_LINE('温馨提示：违反数据完整性约束！');
    WHEN OTHERS THEN
        DBMS_OUTPUT.PUT_LINE(SQLCODE||'---'||SQLERRM);
END InsertEmp;
```

【示例 12-25】 使用存储过程向 DEPARTMENTS 表中插入数据。

```
CREATE OR REPLACE PROCEDURE insert_dept
    (v_dept_id IN departments.department_id%TYPE,
    v_dept_name IN departments.department_name%TYPE,
    v_mgr_id IN departments.manager_id%TYPE,
    v_loc_id IN departments.location_id%TYPE)
IS
    ept_null_error EXCEPTION;
    PRAGMA EXCEPTION_INIT(ept_null_error, -1400);
    ept_no_loc_id EXCEPTION;
    PRAGMA EXCEPTION_INIT(ept_no_loc_id, -2291);
BEGIN
```

```
      INSERT INTO departments
        (department_id, department_name, manager_id, location_id)
      VALUES
        (v_dept_id, v_dept_name, v_mgr_id, v_loc_id);
      DBMS_OUTPUT.PUT_LINE('插入部门'||v_dept_id||'成功');
    EXCEPTION
      WHEN DUP_VAL_ON_INDEX THEN
        RAISE_APPLICATION_ERROR(-20000,'部门编码不能重复');
      WHEN ept_null_error THEN
        RAISE_APPLICATION_ERROR(-20001,'部门编码、部门名称不能为空');
      WHEN ept_no_loc_id THEN
        RAISE_APPLICATION_ERROR(-20002,'没有该地点');
    END insert_dept;
```

12.4.2 调用存储过程

存储过程建立完成后，只要通过授权，用户就可以调用运行。对于参数的传递也有三种：按位置传递、按名称传递和组合传递，传递方法与函数的一样。Oracle 使用 EXECUTE 语句来实现对存储过程的调用：

```
    EXEC[UTE] procedure_name( parameter1, parameter2…);
```

【示例12-26】查询指定员工记录的存储过程的创建与调用

本过程中有 3 个参数，v_empno 是输入类型，其他两个是输出类型，用于接收存储查询的两个结果值。

```
    CREATE OR REPLACE PROCEDURE QueryEmp
       (v_empno IN  employees.employee_id%TYPE,
      v_ename OUT employees.first_name%TYPE,
      v_sal OUT employees.salary%TYPE)
    AS
    BEGIN
      SELECT last_name || last_name, salary INTO v_ename, v_sal
      FROM employees WHERE employee_id = v_empno;
      DBMS_OUTPUT.PUT_LINE('温馨提示：编码为'||v_empno||'的员工已经查到!');
    EXCEPTION
      WHEN NO_DATA_FOUND THEN
        DBMS_OUTPUT.PUT_LINE('温馨提示：你需要的数据不存在!');
        WHEN OTHERS THEN
        DBMS_OUTPUT.PUT_LINE(SQLCODE||'---'||SQLERRM);
    END QueryEmp;
```

```
/
--调用
DECLARE
    v1 employees.first_name%TYPE;
    v2 employees.salary%TYPE;
BEGIN
    QueryEmp(100, v1, v2);
    DBMS_OUTPUT.PUT_LINE('姓名:'||v1);
    DBMS_OUTPUT.PUT_LINE('工资:'||v2);
END;
/
温馨提示：编码为100的员工已经查到！
姓名：KingKing
工资：24000
```

> 注意：本示例之所以使用存储过程，而没有使用函数，是因为返回的参数有两个，函数无法返回两个值。

12.5 自定义函数

12.5.1 函数的创建与调用

函数是只返回一个值的命名程序块，比存储过程的调用更方便，除了自身的函数之外，Oracle允许自定义函数，自定义函数创建的语法如下：

```
CREATE [OR REPLACE] FUNCTION function_name
 (arg1 [ { IN | OUT | IN OUT }] type1 [DEFAULT value1],
 [arg2 [ { IN | OUT | IN OUT }] type2 [DEFAULT value1]],
 ...
 [argn [ { IN | OUT | IN OUT }] typen [DEFAULT valuen]])
 [ AUTHID DEFINER | CURRENT_USER ]
RETURN return_type
 IS | AS
    <类型.变量的声明部分>
BEGIN
    执行部分
    RETURN expression
EXCEPTION
    异常处理部分
```

```
END function_name;
```

函数的参数类型与过程相似，只是必须有 RETURN 函数返回类型的申明。另外，由于函数可以在 SQL 语句中直接使用，所以应该尽量避免使用 OUT，IN OUT 类型的参数。

【示例 12-27】使用函数获取某部门的工资总和

本示例创建 get_salary() 函数获取部门 Dept_No 的工资总和，通过 RETURN 返回 Number 类型的值。最后，通过 SQL 语句查询 10 号部门的工资总和。

```
CREATE OR REPLACE FUNCTION get_salary(Dept_no NUMBER)
RETURN NUMBER
IS
  V_sum NUMBER;
BEGIN
  SELECT SUM(SALARY)INTO V_sum
    FROM EMPLOYEES WHERE DEPARTMENT_ID=dept_no;
  RETURN v_sum;
EXCEPTION
  WHEN NO_DATA_FOUND THEN
    DBMS_OUTPUT.PUT_LINE('你需要的数据不存在!');
  WHEN OTHERS THEN
    DBMS_OUTPUT.PUT_LINE(SQLCODE||'---'||SQLERRM);
END get_salary;
/
SQL> SELECT get_salary(10)FROM dual;
GET_SALARY(10)
--------------
  4400
```

12.5.2 函数参数的调用形式

函数声明时所定义的参数称为形式参数，应用程序调用时为函数传递的参数称为实际参数。应用程序在调用函数时，可以使用位置表示法或者名称表示法向函数传递参数：

1）位置表示法

位置表示法即在调用时按形参的排列顺序，依次写出实参的名称，而将形参与实参关联起来进行传递。用这种方法进行调用，形参与实参的名称是相互独立、没有关系的，强调次序才是重要的。

格式为：

```
argument_value1[, argument_value2 …]
```

2）名称表示法

名称表示法即在调用时按形参的名称与实参的名称，写出实参对应的形参，而将形参与实参关联起来进行传递。这种方法,形参与实参的名称是相互独立、没有关系的,名称的对应关系才是最重要的,次序并不重要。

格式为:

```
argument1=>value1 [, …]
```

其中:argument1 为形式参数,它必须与函数定义时所声明的形式参数名称相同,value1 为实际参数。在这种格式中,形势参数与实际参数成对出现,相互间关系唯一确定,所以参数的顺序可以任意排列。显然,这种方式更直观。

【示例 12-28】get_salary()函数的两种参数传递方式

本示例在 PL/SQL 中使用函数,并使用了两种参数传传递方式。

```
DECLARE
  v NUMBER;
BEGIN
  v: =get_salary(10);
  v: =get_salary(dept_no=>10);
END;
```

位置表示法和名称表示法还可以组合使用,但不提倡,容易引起混淆。

12.6 删除过程和函数

删除过程可以使用 DROP PROCEDURE,语法如下:

```
DROP PROCEDURE [user.]Procudure_name;
```

删除函数可以使用 DROP FUNCTION,语法如下:

```
DROP FUNCTION [user.]Function_name;
```

12.7 块内存储过程和函数

在 PL/SQL 程序中还可以在块内建立本地函数和过程,这些函数和过程不存储在数据库中,但可以在创建它们的 PL/SQL 程序中被调用。本地函数和过程在 PL/SQL 块的声明部分定义,语法格式相同,但不能使用"CREATE OR REPLACE"关键字。

【示例 12-29】定义块内的存储过程

建立本地过程,用于计算指定部门的工资总和,并统计其中的职工数量。

```
DECLARE
  V_num NUMBER;
  V_sum NUMBER(8, 2);
  PROCEDURE proc_demo(
    Dept_no NUMBER DEFAULT 10,
    Sal_sum OUT NUMBER,
```

```
      Emp_count OUT NUMBER
   )
   IS
   BEGIN
      SELECT SUM(salary),COUNT(*)INTO sal_sum,emp_count
      FROM employees WHERE department_id=dept_no;
   EXCEPTION
      WHEN NO_DATA_FOUND THEN
         DBMS_OUTPUT.PUT_LINE('你需要的数据不存在!');
      WHEN OTHERS THEN
         DBMS_OUTPUT.PUT_LINE(SQLCODE||'---'||SQLERRM);
   END proc_demo;
BEGIN
   --调用块内过程Proc_demo
   Proc_demo(30,v_sum,v_num);
   DBMS_OUTPUT.PUT_LINE('30号部门工资总和:'||v_sum||',人数:'
      ||v_num);
   Proc_demo(sal_sum => v_sum, emp_count => v_num);
   DBMS_OUTPUT.PUT_LINE('10号部门工资总和:'||v_sum||',人数:'
      ||v_num);
END;
/
30号部门工资总和:29700,人数:6
10号部门工资总和:4400,人数:1
PL/SQL 过程已成功完成。
```

12.8 过程与函数的比较

1)过程与函数的优点

(1)共同使用的代码可以只需要被编写和测试一次,可被需要该代码的任何应用程序(如:.NET、C++、JAVA、VB程序,也可以是DLL库)调用。

(2)集中编写、集中维护更新、大家共享(或重用)的方法,简化了应用程序的开发和维护,提高了效率与性能。

(3)这种模块化的方法,使得可以将一个复杂的问题、大的程序逐步简化成几个简单的、小的程序部分,进行分别编写、调试。使程序的结构变得清晰、简单,也容易实现。

(4)可以保证在开发者之间提供处理数据、控制流程和提示信息等方面的一致性。

(5)节省内存空间。它们以一种压缩的形式被存储在外存中,当被调用时才被放入内存进行处理。并且，如果多个用户要执行相同的过程或函数时，就只需要在内存中加载一个该过程或函数。

(6) 提高数据的安全性与完整性。通过把一些对数据的操作放到过程或函数中，就可以通过是否授予用户有执行该过程或的权限，来限制某些用户对数据进行这些操作。

2) 过程与函数的相同功能

(1) 都使用 IN 模式的参数传入数据、OUT 模式的参数返回数据。
(2) 输入参数都可以接受默认值，都可以传值或传引导。
(3) 调用时的实际参数都可以使用位置表示法、名称表示法或组合方法。
(4) 都有声明部分、执行部分和异常处理部分。
(5) 其管理过程都有创建、编译、授权、删除、显示依赖关系等。

3) 过程与函数的主要区别

(1) 标识符不同。函数的标识符为 FUNCTION，过程为 PROCEDURE。
(2) 函数中一般不用变量形参，用函数名直接返回函数值；而过程如有返回值，则必须用变量形参返回。
(3) 过程无类型，不能给过程名赋值；函数有类型，最终要将函数值传送给函数名。
(4) 函数在定义时一定要进行函数的类型说明，过程则不进行过程的类型说明。
(5) 调用方式不同。函数的调用出现在表达式中，过程调用，由独立的过程调用语句来完成。
(6) 过程一般会被设计成求若干个运算结果，完成一系列的数据处理，或与计算无关的各种操作；而函数往往只为了求得一个函数值。

> 注意：
> 函数可以在表达式中使用 x:= func()；过程不能。
> 函数可以作为列计算表达式 select func() from dual；过程不能。

如表 12-4 说明了它们之间的区别。

表 12-4 过程和函数区别对比

过程	函数
用于在数据库中完成特定的操作或者任务（如插入、更新、删除等）	只用于特定的数据（如选择等）
头部声明用 procedure	头部声明用 function
头部申明时不需要描述返回类型	头部申明时需要描述返回类型，而且 PL/SQL 至少要包含一个有效的 return 语句
可以使用 int/out/in out 三种模式的参数	可以使用 int/out/in out 三种模式的参数
可作为一个独立的 PL/SQL 语句来执行	不能独立执行，必须作为表达式的一部分调用
可以通过 out/int out 返回零个或多个值	通过 return 语句返回一个值，而且该值要与声明部分一致，也可以是通过 OUT 类型的参数带出的变量
SQL 语句（DML 或 SELECT）中不可以调用任何存储过程	而函数中的 SQL（DML 或 SELECT）语句中可以调用函数

12.9 包

包是一组相关过程、函数、变量、常量和游标等 PL/SQL 程序设计元素的组合,作为一个完整的单元存储在数据库中,用名称来标识包。将不同功能的程序,数据类型分别存储在不同的包中,有利于项目开发和团队协作。

一个包由两个分开的部分组成:

(1)包声明(Package):声明包内数据类型、变量、常量、游标、子程序和异常错误处理等元素,这些元素为包的公有元素。

(2)包主体(Package Body):是包定义部分的具体实现,它实现了包定义部分所声明的游标和子程序,在包主体中还可以声明包的私有元素。

12.9.1 创建包

1)创建包的声明

包声明可以使用 CREATE PACKAGE 语句来定义,语法格式如下:

```
CREATE OR REPLACE PACKAGE package_name
IS|AS
   package_specification
END package_name;
```

其中:package_name 是包名,package_specification 列出了包可以使用的仅有存储过程、函数、类型和游标等元素。定义包头应当遵循以下原则:

(1)包元素位置可以任意安排,在声明部分,对象必须在引用前进行声明。

(2)包头可以不对任何类型的元素进行说明。例如:包头可以只带过程和函数说明语句,而不声明任何异常和类型。

(3)对过程和函数的任何声明都必须只对子程序和其参数进行描述,不能有任何代码的说明,代码的实现只能在包体中出现。它不同于块声明,在块声明中,过程和函数的代码可同时出现在声明部分。

【示例 12-30】 定义 test_pkg 包的声明部分

```
CREATE OR REPLACE PACKAGE test_pkg
IS
   PROCEDURE update_sal(e_name VARCHAR2, newsal NUMBER);
   FUNCTION ann_income(e_name VARCHAR2)RETURN NUMBER;
END test_pkg;
```

在 test_pkg 包定义了 update_sal()存储过程和 ann_income()函数。

2)创建包主体

```
CREATE [OR REPLACE] PACKAGE BODY package_name
IS|AS
```

```
    [Public type and item declarations]
    [Subprogram bodies]
BEGIN
    [Initialization_statements]
END [package_name];
```

其中，package_name 是包的名称。Public type and item declarations 是声明类型、常量、变量、异常和游标等。Subprogram bodies 是定义公共和私有 PL/SQL 子程序。

【示例 12-31】 实现 test_pkg 包体部分

```
CREATE OR REPLACE PACKAGE BODY test_pkg
IS
    PROCEDURE update_sal(e_name VARCHAR2, newsal NUMBER)
    IS
    BEGIN
        UPDATE employees SET salary=newsal WHERE first_name=e_name;
    END;
    FUNCTION ann_income(e_name VARCHAR2)
        RETURN NUMBER IS
        annsal NUMBER;
    BEGIN
        SELECT salary+2000 INTO annsal FROM employees
        WHERE first_name=e_name;
        RETURN annsal;
    END;
END test_pkg;
```

> 注意：可以通过数据字典 user_source、all_source 和 dba_source 分别查询包声明与包主体的详细信息。

12.9.2 调用包

对包内共有元素的调用格式为：包名.元素名称。

```
SET SERVEROUTPUT ON
DECLARE
    v_annsal NUMBER(7, 2);
    BEGIN
    v_annsal:=test_pkg.ann_income('Lex');
    dbms_output.put_line('工资为：'||v_annsal);
END;
/
```

```
工资为：19000
```

我们可以用"DROP PACKAGE"命令对不需要的包进行删除，语法如下：
```
DROP PACKAGE package_name;
```

12.10 触 发 器

触发器(Trigger)是一种特殊类型的 PL/SQL 程序块。触发器类似于函数和过程，也具有声明部分、执行部分和异常处理部分。触发器作为 Oracle 对象存储在数据库中。触发器在事件发生时被隐式触发，而且触发器不能接受参数，不像过程一样显式调用并传递参数。

触发器类型主要有 DML 触发器、替代触发器、DDL 触发器和系统触发器四种类型。

(1) DML 触发器，是由 DML 语句触发的触发器，例如 INSERT，UPDATE 和 DELETE 语句。针对 DML 所包含的触发事件，DML 触发器可以为这些触发事件创建 BEFORE 触发器(发生前)和 AFTER 触发器(发生后)，分别表示在 DML 事件发生之前与之后执行。DML 触发器可以在语句级或行级操作上被触发，语句级触发器对于每一个 SQL 语句只触发一次；行级触发器对 SQL 语句受影响的表中的每一行都触发一次。

(2) INSTEAD OF 触发器，又称替代触发器。用于执行一个替代操作来触发事件的操作。例如，针对 INSERT 事件的 INSTEAD OF 触发器，它由 INSERT 语句触发，当出现 INSERT 语句时，该语句不会执行，而是执行 INSTEAD OF 触发器定义的语句。

(3) DDL 触发器是由 DDL 语句(CREATE、ALTER 或 DROP 等)触发的触发器。可以在这些 DDL 语句之前(或之后)定义 DDL 触发器。

(4) 系统触发器，分为数据库级(Database)和模式级(Schema)两种。数据库级触发器的触发事件对于所有用户均有效，模式级触发器仅被指定模式的用户触发。系统触发器支持的触发事件有：LOGON、LOGOFF、SERVERERROR、STARTUP 和 SHUTDOWN 等。

12.10.1 创建触发器

创建触发器的语法如下：
```
CREATE [OR REPLACE] TRIGGER 模式.]触发器名
{BEFORE|AFTER|INSTEAD OF}
{DML 触发事件
|DDL 触发事件 [OR DDL 触发事件]...
|DATABASE 事件 [OR DATABASE 事件]...}
ON
{ [模式.]表| [模式.]视图|DATABASE}
 [FOR EACH ROW [WHEN (触发条件)]]
触发体
```

其中，TRIGGER 表示触发器对象，BEFORE 或 AFTER 表示在事件发生之前触发还是事件发生之后触发。DML 触发事件可以是 INSERT，UPDATE 或 DELETE，DDL 触发事件可以是 CREATE，ALTER 或 DROP，DATABASE 事件可以是 SERVERERROR，LOGON，LOGOFF，STARTUP 和 SHUTDOWN。"FOR EACH ROW"选项可选，表示触发器是行级触发器，如果没有此选项，则默认是语句级触发器。触发体类似于程序体，由 PL/SQL 语句组成。

1）创建 DML 触发器

DML 触发器，由 DML 语句触发的触发器，例如 INSERT，UPDATE 和 DELETE 语句。针对 DML 所包含的触发事件，DML 触发器可以为这些触发事件创建 BEFORE 触发器（发生前）和 AFTER 触发器（发生后），分别表示在 DML 事件发生之前与之后执行。DML 触发器可以在语句级或行级操作上被触发，语句级触发器对于每一个 SQL 语句只触发一次；行级触发器对 SQL 语句受影响的表中的每一行都触发一次。

【示例 12-32】用触发器限制对表的修改

本示例创建表级触发器 tr_dept_time，只允许对 DEPARTMENTS 表在工作时间内进行 DML 操作（包括 INSERT，DELETE，UPDATE），如果在非工作时间进行操作，将会产生异常。

```
CREATE OR REPLACE TRIGGER tr_dept_time
    BEFORE INSERT OR DELETE OR UPDATE
    ON departments
BEGIN
    IF (TO_CHAR(sysdate,'DAY')IN ('星期六','星期日'))OR
        (TO_CHAR(sysdate,'HH24:MI')NOT BETWEEN '08:30' AND '18:00')
    THEN
        RAISE_APPLICATION_ERROR(-20001,'不是上班时间,
        不能修改 departments 表');
    END IF;
END;
/
```

下面测试一下触发器的效果，在星期六运行：

```
SQL> !date
2017 年 06 月 24 日 星期六 07:14:57 CST
SQL> UPDATE departments SET manager_id=manager_id;
UPDATE departments SET manager_id=manager_id
       *
ERROR at line 1:
ORA-20001: 不是上班时间,
不能修改 departments 表
ORA-06512: 在 "HR.TR_DEPT_TIME", line 5
ORA-04088: 触发器 'HR.TR_DEPT_TIME' 执行过程中出错
```

2) 创建 INSTEAD OF 触发器

INSTEAD OF 触发器,又称替代触发器。用于执行一个替代操作来触发事件的操作。例如,针对 INSERT 事件的 INSTEAD OF 触发器,它由 INSERT 语句触发,当出现 INSERT 语句时,该语句不会执行,而是执行 INSTEAD OF 触发器定义的语句。在 Oracle 系统中,如果视图由多个表连接而成,则该视图不允许 INSERT、DELETE 和 UPDATE 操作。而 Oracle 提供的替代触发器就是用于对视图进行 INSERT、DELETE 和 UPDATE 操作的触发器。通过编写替代触发器对视图进行 DML 操作,从而实现对基表数据的修改。替代触发器只能定义在视图上,而 DML 触发器只能定义在表上。

【示例 12-33】实现在视图上完成 DELETE 操作

本示例创建视图 emp_view,通过删除视图记录,自动删除对应的 EMPLOYEES 表中的记录。

```
CREATE OR REPLACE VIEW emp_view AS
  SELECT department_id , count(*)total_employeer, sum(SALARY)
  total_salary
  FROM employees GROUP BY department_id;
```

在此视图中直接删除是非法的:

```
SQL>DELETE FROM emp_view WHERE department_id =20;
DELETE FROM emp_view WHERE department_id =20
            *
ERROR at line 1:
ORA-01732:此视图的数据操纵操作非法
```

但是我们可以创建 INSTEAD_OF 触发器来为 DELETE 操作执行所需的处理,即删除 employees 表中所有基准行:

```
CREATE OR REPLACE TRIGGER emp_view_delete
  INSTEAD OF DELETE ON emp_view FOR EACH ROW
BEGIN
  DELETE FROM employees WHERE department_id =: old.department_id;
END emp_view_delete;
```

测试删除 emp_view 视图的行记录:

```
SQL>DELETE FROM emp_view WHERE department_id=20;
1 row deleted.
```

【示例 12-34】向复杂视图插入数据。

本示例创建视图 V_REG_COU 和触发器 TR_I_O_REG_COU,向视图 V_REG_COU 插入一行数据,实际结果是如果 region_id 在 REGIONS 表中不存在,会自动向 REGIONS 表插入一条数据;如果 country_id 在 COUNTRIES 表中不存在,会自动向 COUNTRIES 表插入一条数据;

```
CREATE OR REPLACE FORCE VIEW "HR"."V_REG_COU"
   ("R_ID", "R_NAME", "C_ID", "C_NAME")
```

```
    AS
        SELECT r.region_id, r.region_name, c.country_id, c.country_name
            FROM regions r, countries c
            WHERE r.region_id = c.region_id;
```
创建触发器：
```
CREATE OR REPLACE TRIGGER "HR"."TR_I_O_REG_COU" INSTEAD OF
    INSERT ON v_reg_cou FOR EACH ROW DECLARE v_count NUMBER;
BEGIN
    SELECT COUNT(*)INTO v_count FROM regions
        WHERE region_id =: new.r_id;
    IF v_count = 0
    THEN
        INSERT INTO regions(region_id, region_name)
            VALUES (: new.r_id, : new.r_name);
    END IF;
    SELECT COUNT(*)INTO v_count FROM countries
        WHERE country_id =: new.c_id;
    IF v_count = 0 THEN
        INSERT INTO countries (country_id, country_name, region_id)
            VALUES(: new.c_id, : new.c_name, : new.r_id);
    END IF;
END;
```
由此可以看出，INSTEAD OF 触发器可以实现复杂的逻辑，在实际的项目工程中特别有用。

> 注意：
> 创建 INSTEAD OF 触发器需要注意以下几点：
> (1) 只能被创建在视图上，并且该视图没有指定 WITH CHECK OPTION 选项。
> (2) 不能指定 BEFORE 或 AFTER 选项。
> (3) INSTEAD OF 触发器只能在行级上触发，FOR EACH ROW 是否指定都一样。
> (4) 没有必要在针对一个表的视图上创建 INSTEAD OF 触发器，只要创建 DML 触发器就可以了。

3）创建 DDL 触发器

DDL 触发器在 DDL 语句（CREATE、ALTER 或 DROP 等）之前或之后触发。

【示例 12-35】创建的 DDL 触发器，阻止对 EMPLOYEES 表的删除。
```
CREATE OR REPLACE TRIGGER NODROP_EMP BEFORE DROP ON SCHEMA
BEGIN
    IF SYS.DICTIONARY_OBJ_NAME='employees'
    THEN
```

```
        RAISE_APPLICATION_ERROR(-20005,'不允许能删除 EMP 表!');
    END IF;
END;
/
```
下面的语句通过删除 EMPLOYEES 表验证触发器:

```
SQL>DROP TABLE employees;
    DROP TABLE employees
    *
    第 1 行出现错误:
    ORA-00604:递归 SQL 级别 1 出现错误 ORA-20005:不允许能删除 employees 表!
    ORA-06512:在 line3
```

12.10.2 触发器的管理

1)查看触发器

创建成功的触发器存放在数据库中,与触发器有关的数据字典有:USER_TRIGGERS、ALL_TRIGGERS 和 DBA_TRIGGERS 等。其中,USER_TRIGGERS 存放当前用户的所有触发器,ALL_TRIGGERS 存放当前用户可以访问的所有触发器,DBA_TRIGGERS 存放数据库中的所有触发器。

【示例 12-36】 查看 JOB_COUNT 触发器的类型、触发事件及所在表名称。

```
SQL>SELECT TRIGGER_TYPE, TRIGGERING_EVENT, TABLE_NAME
  2 FROM USER_TRIGGERS WHERE TRIGGER_NAME='JOB_COUNT';
```

下面的代码块查看 JOB_COUNT 触发器的触发体

```
SQL>SELECT TRIGGER_BODY FROM USER_TRIGGERS
  2 WHERE TRIGGER_NAME = 'JOB_COUNT';
```

2)禁用和启用触发器

触发器有 Enabled(有效)和 Disabled(无效)两种状态。新建的触发器默认是 Enabled 状态。无效的触发器暂时没有触发的功能,有效之后才能触发。使触发器有效或无效的语句分别如下:

```
ALTER TRIGGER 触发器名称 ENABLE;
ALTER TRIGGER 触发器名称 DISABLE;
```

如果要使一个表上的所有触发器都有效或无效,可以使用下面的语句.

```
ALTER TABLE 表名 ENABLE ALL TRIGGERS;
ALTER TABLE 表名 DISABLE ALL TRIGGERS;
```

3)删除触发器

删除触发器的语法如下:

```
DROP TRIGGER 触发器名;
```

注意:删除表或者视图的时候也将删除表或者视图对应的所有触发器。

12.10.3 行级触发器

行级触发器是 Oracle 的特色，对表的每一行操作都可以进行触发。行级触发器的关键词是"FOR EACH ROW"，在行级触发器中有两个行变量":NEW"和":OLD"，":NEW"表示修改之后的行，":OLD"表示修改之前的行。在行级触发器中，可以使用 UPDATING、DELETING 和 INSERTING 判断正在进行的操作类型是修改、删除和插入中的哪一种。

":OLD"关键字只对 UPDATE 和 DELETE 操作有效，对 INSERT 操作无效；":NEW"关键字只对 UPDATE 和 INSERT 操作有效，对 DELETE 操作无效。

【示例 12-37】行级触发器示例 1

本示例限定只对部门号为 80 的记录进行行级触发器操作，当对表的部门号为 80 的记录进行修改（UPDATE）或者删除（DELETE）的时候触发。

```
CREATE OR REPLACE TRIGGER tr_emp_sal_comm
   BEFORE UPDATE OF salary, commission_pct OR DELETE
   ON employees
   FOR EACH ROW
   WHEN (OLD.department_id = 80)
BEGIN
   CASE
   WHEN UPDATING ('salary')THEN
      IF: NEW.salary <: OLD.salary THEN
         RAISE_APPLICATION_ERROR(-20001,'部门80的人员的工资不能降');
      END IF;
   WHEN UPDATING('commission_pct')THEN
      IF: NEW.commission_pct <: OLD.commission_pct THEN
         RAISE_APPLICATION_ERROR(-20002,'部门80的人员的奖金不能降');
      END IF;
   WHEN DELETING THEN
      RAISE_APPLICATION_ERROR(-20003,'不能删除部门80的人员记录');
   END CASE;
END;
/
```

下面测试触发器的效果，可以正常修改 112 号员工，但不能修改 177 号员工，这是因为 177 号员工的部门号为 80：

```
SQL>UPDATE employees SET salary = 8000 WHERE employee_id = 112;
1 row updated.
SQL> UPDATE employees SET salary = 8000 WHERE employee_id = 177;
UPDATE employees SET salary = 8000 WHERE employee_id = 177
```

```
             *
ERROR at line 1:
ORA-20001: 部门 80 的人员的工资不能降 ORA-06512:
在 "HR.TR_EMP_SAL_COMM", line 5
ORA-04088: 触发器 'HR.TR_EMP_SAL_COMM' 执行过程中出错

SQL>DELETE FROM employees WHERE employee_id in (177, 170);
DELETE FROM employees WHERE employee_id in (177, 170)
             *
ERROR at line 1:
ORA-20003: 不能删除部门 80 的人员记录 ORA-06512:
在 "HR.TR_EMP_SAL_COMM", line 12
ORA-04088: 触发器 'HR.TR_EMP_SAL_COMM' 执行过程中出错
```

【示例 12-38】行级触发器示例 2

本示例利用行触发器实现级联更新。在修改了主表 REGIONS 中的 region_id 之后（AFTER），自动更新子表 COUNTRIES 表中原来在该地区的国家的 region_id。

```
CREATE OR REPLACE TRIGGER tr_reg_cou
   AFTER update OF region_id
   ON regions
   FOR EACH ROW
BEGIN
   DBMS_OUTPUT.PUT_LINE('旧的 region_id 值是'||:old.region_id
      ||'、新的 region_id 值是'||:new.region_id);
   UPDATE countries SET region_id =:NEW.region_id
      WHERE region_id =:OLD.region_id;
END;
/
```

> 注意: :OLD 和 :NEW 关键字只能用于行级触发器(FOR EACH ROW)，不能用在语句级触发器，因为在语句级触发器中一次触发涉及多行数据，无法指定是哪一个新旧值。

12.10.4 系统级触发器

系统触发器分为数据库级(Database)和模式级(Schema)两种。前者定义在整个数据库上，触发事件是数据库事件，如数据库的启动、关闭，对数据库的登录或退出。后者定义在模式上，触发事件包括用户的登录或退出，或对数据库对象的创建和修改(DDL 事件)。系统触发器的触发事件的种类和级别如表 12-5 所示。

表 12-5 系统触发器的触发事件的种类和级别

种类	关键字	说明
模式级	CREATE	在创建新对象时触发
	ALTER	修改数据库或数据库对象时触发
	DROP	删除对象时触发
数据库级	STARTUP	数据库打开时触发
	SHUTDOWN	在使用 NORMAL 或 IMMEDIATE 选项关闭数据库时触发
	SERVERERROR	发生服务器错误时触发
数据库级与模式级	LOGON	当用户连接到数据库，建立会话时触发
	LOGOFF	当会话从数据库断开时触发

【示例 12-39】创建系统级触发器，记录用户登录情况

本示例创建系统级触发器，记录 PDBORCL 数据库中的用户的登录信息。登录信息存储在新建表 USERLOG 中。首先以 SYSTEM 用户登录 PDBORCL 数据库，创建记录登录事件表 USERLOG。

```
SQL>CREATE TABLE USERLOG(USERNAME VARCHAR2(20), LOGON_TIME DATE);
表已创建。
```

接下来再以 SYSTEM 身份创建了两个数据库级事件触发器 INIT_LOGON 和 DATABASE_LOGON，其中，INIT_LOGON 在数据库启动时触发，清除 USERLOG 表中记录的数据；DATABASE_LOGON 在用户登录时触发，向表 USERLOG 中增加一条记录，记录登录用户名和登录时间。

```
CREATE OR REPLACE TRIGGER INIT_LOGON
    AFTER STARTUP ON DATABASE
BEGIN
    DELETE FROM USERLOG;
END;
/
触发器已创建
CREATE OR REPLACE TRIGGER DATABASE_LOGON
    AFTER LOGON ON DATABASE
BEGIN
    INSERT INTO userlog VALUES(SYS.LOGIN_USER, SYSDATE);
END;
/
触发器已创建
```

下面验证 DATABASE_LOGON 触发器，分别以 HR 和 SCOTT 用户登录后，再以 SYSTEM 身份执行查询：

```
SQL>SELECT * FROM userlog;
USERNAME              LOGON_TIME
--------------------  --------------------
HR                    2017-06-24 08:38:03
SCOTT                 2017-06-24 08:38:12
```

注意：创建触发器中的"ON DATABASE"仅针对用户登录的当前数据库，可能是 CDB 也可能是一个 PDB，不是针对所有 PDB，本示例的触发器只针对数据库 PDBORCL 这一个数据库有效，对 CDB 和其他插接式数据库是无效的。

练习

一、判断题

1. PL/SQL 允许不同包中的子程序具有同一名称。（   ）
2. 可以在表上创建 INSTEAD OF 触发器。（   ）
3. 程序包是一种数据库对象，它是对相关 PL/SQL 类型、子程序、游标、异常、变量和常量的封装。（   ）
4. 在函数内可以修改表数据。（   ）
5. 只要在存储过程中有增删改语句，一定要加自治事务。（   ）

二、单项选择题

1. 在 Oracle 中，关于 PL/SQL 下列描述正确的是（   ）
   A．PL/SQL 代表 Power Language/SQL
   B．PL/SQL 不支持面向对象编程
   C．PL/SQL 块包括声明部分、可执行部分和异常处理部分
   D．PL/SQL 的四种内置数据类型是 character，integer，float，boolean
2. 关于触发器，下列说法正确的是（   ）
   A.可以在表上创建 INSTEAD OF 触发器
   B.语句级触发器不能使用":old"和":new"
   C.行级触发器不能用于审计功能
   D.触发器可以显式调用
3. Oracle 内置程序包由（   ）用户所有
   A.sys   B.system   C.Scott   D.Public
4. 在 Oracle 中，当 FETCH 语句从游标获得数据时，下面叙述正确的是（   ）
   A.游标打开   B.游标关闭
   C.当前记录的数据加载到变量中   D.创建变量保存当前记录的数据
5. 下列哪个语句可以在 SQL*Plus 中直接调用一个过程（   ）
   A.RETURN   B.CALL   C.SET   D.EXEC

三、填空题

1. 调用存储过程可以使用_____命令或 EXECUTE 命令。

2. 创建触发器要使用 CREATE_____语句。

3. 通过数据字典 user_source、_____和 dba_source 分别查询包声明与包主体的详细信息。

4. 在 Oracle 中游标的操作，包括声明游标、打开游标、_____和关闭游标。

5. 在 PL/SQL 代码段的异常处理块中，捕获所有其他异常的关键词是_____。

四、简答题

1. Oracle 中 function 和 procedure 的区别？
2. 简述一个包的创建和使用过程？
3. 举例说明 DML 触发器 BEFORE 和 AFTER 的区别？
4. 简述 PL/SQL 提供了哪些类型的集合？

训练任务

1. 编写程序块

培养能力	工程能力、问题分析、使用现代工具、工程与社会、职业规范、项目管理、项目管理、终身学习		
掌握程度	★★★★★	难度	中
结束条件	能够创建和执行程序块		
训练内容： 编写一个程序块，处理 employees 表中职工号 200 的职工，如果工资小于 3000 那么把工资更改为 3000			

2. 创建游标

培养能力	工程能力、使用现代工具、工程与社会、职业规范、项目管理、沟通、项目管理、终身学习		
掌握程度	★★★★★	难度	中
结束条件	能够创建和执行游标		

训练内容：
1. 创建 fruit 表并插入数据
● 创建 fruit 表，输入语句如下：
   CREATE TABLE fruit(
   f_id varchar2(10)NOT NULL,
   f_name varchar2(255)NOT NULL,
   f_price number (8, 2)NOT NULL);
● 插入如下数据：
   INSERT INTO fruit VALUES ('a1', 'apple', 5.2);
   INSERT INTO fruit VALUES ('b1', 'blackberry', 10.2);
   INSERT INTO fruit VALUES ('bs1', 'orange', 11.2);
   INSERT INTO fruit VALUES ('bs2', 'melon', 8.2);
   INSERT INTO fruit VALUES ('t1', 'banana', 10.3);
   INSERT INTO fruit VALUES ('t2', 'grape', 5.3);
   INSERT INTO fruit VALUES ('o2', 'coconut', 9.2);
2. 创建表 fruitage
表 fruitage 和表 fruit 的字段一致，利用以下语句创建：
   CREATE TABLE fruitage AS SELECT * FROM fruit WHERE 2=3;
如果没有 WHERE，或者 WHERE 后面的条件为真，则复制表时把数据也一起复制。
3. 创建 PL/SQL 语句块，使用游标，将 fruit 中 f_price>10 的记录复制到 fruitage 表中			

3. 创建函数和过程

培养能力	工程知识、问题分析、设计/开发解决方案研究、使用现代工具、工程与社会、环境和可持续发展、职业规范、个人和团队、项目管理、终身学习		
掌握程度	★★★★★	难度	高
结束条件	能够创建和调用函数和过程		

训练内容：
1. 创建函数
创建基于 hr.jobs 表的函数 get_emp_name()，输入职工编号，输出姓名。
2. 创建过程
- 创建一个 sch 表，并且向 sch 表中插入表格中的数据，代码如下：

```
CREATE TABLE sch(
id NUMBER(10), name VARCHAR2(50), glass VARCHAR2(50));
INSERT INTO sch VALUES(1, 'xiaoming', 'glass 1');
INSERT INTO sch VALUES(2, 'xiaojun', 'glass 2');
```

- 创建存储过程 count_sch 统计表 sch 中的记录数和 id 的和

4. 创建包和触发器

培养能力	工程知识、问题分析、设计/开发解决方案研究、使用现代工具、工程与社会、环境和可持续发展、职业规范、个人和团队、项目管理、终身学习		
掌握程度	★★★★★	难度	高
结束条件	能够创建和调用包和触发器		

训练内容：
1. 包的创建和使用
定义一个名为 pack_op 的包，共有三个元素，pro_print_ename()存储过程通过职工号查询职工姓名，pro_print_sal()存储过程通过职工号查询职工工资，fun_re_date()函数通过接受的职工号返回聘用日期。完成三个部分的代码：包声明，包主体，使用包。

2. 触发器的创建和使用
- 创建一个业务统计表 persons，代码如下：

```
CREATE TABLE persons(name VARCHAR2(40), num NUMBER(11));
```

- 创建一个销售额表 sales，代码如下：

```
CREATE TABLE sales (name VARCHAR2(40), sum NUMBER(11));
```

- 创建一个触发器 NUM_SUM，要求在插入过 persons 表的 num 字段后，更新 sales 表的 sum 字段，sum 与 num 的关系为：sum=num*7 请自行完成代码编写。
- 测试语句：
向 persons 表中插入记录，观察触发器的效果，代码如下：

```
INSERT INTO persons VALUES ('xiaoxiao', 20);
INSERT INTO persons VALUES ('xiaohua', 69);
```

第13章 备份与恢复

本章目标

知识点	理解	掌握	应用
1.归档模式	✓	✓	
2.冷备份和热备份	✓	✓	
3.完全恢复和不完全恢复	✓	✓	✓
4.备份失效与备份过期	✓		
5.rman	✓	✓	✓
6.flashback	✓	✓	✓
7.导出与导入	✓	✓	✓

训练任务

- 能够完成整个数据库的冷备份或者热备份。
- 能够完成整个数据库系统的恢复或者某个数据库的恢复。
- 能够完成逻辑备份，即导出与导入。

知识能力点

能力点 \ 知识点	知识点1	知识点2	知识点3	知识点4	知识点5	知识点6	知识点7
工程知识		✓	✓		✓	✓	✓
问题分析	✓	✓	✓	✓			
设计/开发解决方案		✓	✓		✓	✓	✓
研究	✓			✓			
使用现代工具					✓	✓	✓
工程与社会					✓	✓	✓
环境和可持续发展		✓	✓				
职业规范					✓	✓	✓
个人和团队					✓	✓	✓
沟通					✓	✓	✓
项目管理					✓	✓	✓
终身学习		✓	✓				

13.1 备份与恢复概述

备份与恢复是数据库管理中最重要的方面之一。如果数据库崩溃却没有办法恢复它,那么对企业造成的毁灭性结果可能会是数据丢失、收入减少、客户不满等。在任何情况下,无论数据库服务器有多稳定,可靠,我们都需要制订一个备份与恢复方案来备份重要数据并使自身免于灾难。

备份就是把数据库复制到永久转储设备的过程,是备份时间点的数据库的副本。恢复就是备份的逆向过程,是把数据库的备份数据重新转储到数据库中的过程。Oracle 的备份方式分为物理备份和逻辑备份两大类,其中物理备份又分为脱机备份(也叫冷备份)和联机备份(也叫热备份),见图 13-1。

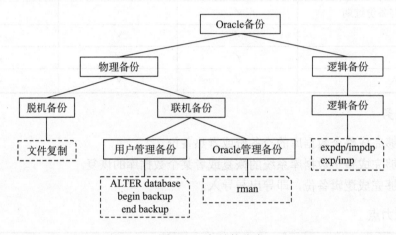

图 13-1 备份的分类

1)物理备份

物理备份是指将数据库的各种操作系统中的文件(如数据文件、控制文件、归档日志文件和系统参数文件等)从一处复制到另一处的过程,一旦数据库发生故障,可以利用这些文件进行还原。进行物理备份的工具可以是操作系统的复制工具,也可以是 Oracle 提供的 rman 等。

2)逻辑备份

逻辑备份就是对数据库对象(如用户、表、存储过程等对象)利用 expdp 等工具进行导出工作,可以利用 impdp 等工具把逻辑备份文件导入到数据库。

物理备份的优点是备份速度快,可以进行完全恢复,缺点就是平台移植性差。逻辑备份的优点是备份集的可移植性比较强,可以把数据库的逻辑备份恢复到不同版本不同平台的数据库上,但缺点是备份时间长,另外备份之后的数据库更改不能反映到逻辑备份中,只能恢复到备份的时间点。

数据库的恢复方式有完全恢复和不完全恢复两种。

1) 完全恢复

完全恢复是指将数据库恢复到数据库失败时的状态，数据没有损失，这种恢复是通过装载(Restore)备份文件与恢复(Recover)全部的日志文件实现的。

2) 不完全恢复

不完全恢复是指将数据库恢复到数据库失败前的某一时间点的状态，这种方法只应用了部分重做日志文件或者不使用重做日志文件。不完全恢复会丢失恢复时间点之后的修改数据。进行不完全恢复之后，必须用 resetlogs 选项重新设置重做日志。

备份之后的完全恢复与不完全恢复的总体流程，如图 13-2 所示。如果只进行数据文件的复制或者只运行 restore 命令装载数据文件，只是进行了不完全恢复。如果再运行 recover 命令，则可以进行完全恢复。

数据库的恢复方式与备份方式密切相关，通常，采用什么样的备份方式，就应该采用对应的恢复方式，比如用 empdp 方式导出数据，就应该用 impdp 方式导入数据；用 rman 的 backup 命令备份数据，就应该用 rman 的 restore 和 recover 恢复数据。

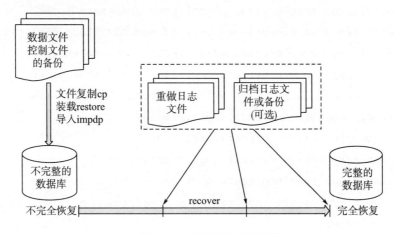

图 13-2　数据库的恢复流程

13.2　脱机备份与恢复

脱机备份是物理备份的一种，由于脱机备份时必须关闭数据库，所以也叫冷备份。冷备份操作的过程是先一致性停机(Shutdown 或者 Shutdown Immediate)，然后将所有数据文件、控制文件、联机重做日志文件和归档日志文件(可选)拷贝到另一目录保存。

这里需要注意的是必须将数据库里所有的数据文件、控制文件和联机重做日志文件拷贝到另一目录保存，如果遗漏任何一个文件，恢复的时候都会遇到错误，且是致命的错误，它将会使数据库无法打开，无法继续任何操作。为了确保所有文件被复制，首先可以在停机前查询这些文件及存储位置，主要通过三个视图：v$datafile、v$logfile 和 v$controlfile。

【示例 13-1】查询所有的数据文件、控制文件和联机重做日志文件

```
$ sqlplus / as sysdba
SQL> SELECT NAME FROM v$datafile
```

```
  2  UNION ALL
  3  SELECT MEMBER as NAME FROM v$logfile
  4  UNION ALL
  5  SELECT NAME FROM v$controlfile;
NAME
--------------------------------------------------------------------
/home/oracle/app/oracle/oradata/orcl/system01.dbf
/home/oracle/app/oracle/oradata/orcl/sysaux01.dbf
/home/oracle/app/oracle/oradata/orcl/undotbs01.dbf
/home/oracle/app/oracle/oradata/orcl/pdbseed/system01.dbf
/home/oracle/app/oracle/oradata/orcl/users01.dbf
/home/oracle/app/oracle/oradata/orcl/pdbseed/sysaux01.dbf
/home/oracle/app/oracle/oradata/orcl/pdborcl/system01.dbf
/home/oracle/app/oracle/oradata/orcl/pdborcl/sysaux01.dbf
/home/oracle/app/oracle/oradata/orcl/pdborcl/SAMPLE_SCHEMA_users01.dbf
/home/oracle/app/oracle/oradata/orcl/pdborcl/example01.dbf
/home/oracle/app/oracle/oradata/orcl/pdborcl/pdbtest_users02_1.dbf
/home/oracle/app/oracle/oradata/orcl/pdborcl/pdbtest_users02_2.dbf
/home/oracle/app/oracle/oradata/orcl/redo03.log
/home/oracle/app/oracle/oradata/orcl/redo02.log
/home/oracle/app/oracle/oradata/orcl/redo01.log
/home/oracle/app/oracle/oradata/orcl/control01.ctl
/home/oracle/app/oracle/fast_recovery_area/orcl/control02.ctl
17 rows selected.
```

以上查询出的所有文件都是必须复制的，另外，最好也备份初始化文件 $ORACLE_HOME/dbs 目录中的 init.ora 和 initorcl.ora。脱机备份之后，如果遇到数据损坏或者需要恢复到备份时的数据，可以进行脱机恢复。过程是首先停机，然后将所有脱机备份的所有文件复制到原来的目录中，最后再重启数据库（startup）即可。

> 注意：脱机备份后只能做不完全恢复，只能将数据库恢复到脱机备份的时刻，备份之后的数据修改是无法恢复的。另外，脱机备份和恢复都需要停机，对于需要 24 小时不停机的数据库系统，这样做是不能接受的。

13.3 用户管理备份与恢复

用户管理备份是一种联机备份，不需要停机，备份期间数据库可以正常读写，在恢复的时候可以做完全恢复，不丢失备份之后的数据。同脱机备份一样，用户管理备份也需要复制文件，但过程却完全不同。用户管理备份的流程见图 13-3 所示：

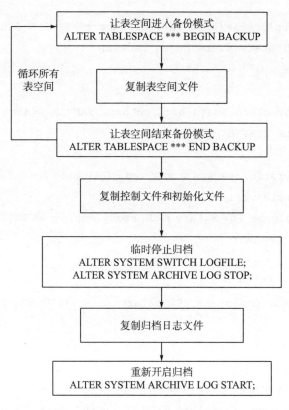

图 13-3　用户管理备份流程

用户管理备份之所以可以做到完全恢复，不丢失数据，是因为 Begin Backup 会锁住表空间对应的数据文件头的 scn，不管后来怎么改变，恢复都会从这个 scn 开始。同时，备份过程中和备份之后，任何数据的改变都会被写入重做日志文件和归档日志文件中，在恢复的时候只要这些文件不损失，数据库就可以自动从日志文件中恢复所有修改后的数据。所以联机备份的前提是数据库必须工作在归档日志(Archivelog)模式下，备份的时候数据库必须是打开的。可以通过命令"archive log list；"查看数据库是否工作在归档日志模式下。

【示例 13-2】用户管理备份：备份 USERS 表空间

本例用 SYSTEM 用户联机备份 USERS 表空间，通过命令"SELECT file_name FROM dba_data_files WHERE tablespace_name='USERS'；"查询出表空间的数据文件，表空间 users 进入备份模式之后，通过调用 cp 命令复制数据文件 SAMPLE_SCHEMA_users01.dbf 到目录/home/oracle，然后修改数据，这是模拟联机备份，备份结束之后，退出 Sqlplus，转到备份目录，通过 ls 查看备份文件。

```
$ sqlplus system/***@pdborcl
SQL> SELECT file_name FROM dba_data_files WHERE  tablespace_name='USERS';
FILE_NAME
--------------------------------------------------------------------------------
```

```
/home/oracle/app/oracle/oradata/orcl/pdborcl/SAMPLE_SCHEMA_users01.dbf
SQL> ALTER tablespace users begin backup;
Tablespace altered.
SQL> !cp
/home/oracle/app/oracle/oradata/orcl/pdborcl/SAMPLE_SCHEMA_users01.dbf
/home/oracle/SAMPLE_SCHEMA_users01.dbf
SQL> !cp
/home/oracle/app/oracle/oradata/orcl/control01.ctl
/home/oracle/control01.ctl
SQL> SELECT employee_id, salary FROM study.employees WHERE  employee_id=1;
EMPLOYEE_ID        SALARY
-----------     ----------
          1         50000
SQL> UPDATE study.employees SET salary=salary+100 WHERE  employee_id=1;
1 row updated.
SQL> commit;
Commit complete.
SQL> ALTER tablespace users end backup;
SQL> SELECT employee_id, salary FROM study.employees WHERE  employee_id=1;
EMPLOYEE_ID        SALARY
-----------     ----------
          1         50100
SQL> exit
$ cd /home/oracle/
$ ls -l SAMPLE_SCHEMA_users01.dbf control01.ctl
-rw-r-----. 1 oracle oracle 17973248 4月  16 07:06 control01.ctl
-rw-r-----. 1 oracle oracle 32776192 4月  16 07:05 SAMPLE_SCHEMA_users01.dbf
```

1号员工新的 salary 由 50000 修改为 50100，由于是在复制文件之后修改的，所以备份文件/home/oracle/SAMPLE_SCHEMA_users01.dbf 中肯定没有新数据，新数据在日志文件(重做日志文件和归档日志文件)中。如果这时原数据文件损坏，而日志文件没有损坏，就可以通过 Recover Database 完全恢复。

> 注意：在 backup 模式下，可能导致 redo log file 中的信息量大增(有用户写数据等)，影响性能，所以备份完以后，应该立即 END BACKUP，也不推荐使用 ALTER DATABASE BEGIN BACKUP 命令，以免等待备份的表空间过多。

【示例13-3】用户管理备份：完全恢复

首先关闭数据库，通过删除数据文件模拟数据库损坏。再通过普通的 startup 命令启动数据库，再打开 PDBORCL 插接式数据库，就会出现 ORA-01157 错误，提示找不到数据文件。

```
$ sqlplus / as sysdba
SQL> shutdown immediate
SQL> !rm /home/oracle/app/oracle/oradata/orcl/pdborcl/SAMPLE_SCHEMA_users01.dbf
SQL> startup
SQL> ALTER pluggable database pdborcl open;
ALTER pluggable database pdborcl open;
ERROR at line 1:
ORA-01157：无法标识/锁定数据文件 10 - 请参阅 DBWR 跟踪文件 ORA-01110：
数据文件 10：
'/home/oracle/app/oracle/oradata/orcl/pdborcl/SAMPLE_SCHEMA_users01.dbf'
```

接下来，将原数据库文件的备份文件复制回原来的目录，然后进行数据恢复，这之前要重新关闭数据库，并以 startup mount 方式启动 CDB，该方式只是读取控制文件，不打开数据文件，然后再运行 recover database 命令从日志中恢复最新数据，在运行 recover database 命令后，Oracle 建议从归档日志文件(*.arc)中恢复数据，用户可以选择 4 种恢复方式：{<RET>=suggested | filename | AUTO | CANCEL}，即直接按"回车"键使用建议的 arc 文件恢复，或者指定文件名，或者输入"AUTO"，或者输入"CANCEL"取消这个更改的恢复。下面全部直接按"回车键"进行建议的恢复：

```
SQL> !cp /home/oracle/SAMPLE_SCHEMA_users01.dbf /home/oracle/app/oracle/oradata/orcl/pdborcl/SAMPLE_SCHEMA_users01.dbf
SQL> shutdown immediate
SQL> startup mount
Database mounted.
SQL> recover database;
ORA-00279：更改 8376635 (在 04/14/2017 08：54：08 生成)对于线程 1 是必需的
ORA-00289：
建议：
/home/oracle/app/oracle/fast_recovery_area/ORCL/archivelog/2017_04_14/o1_mf_1_216_dh0ots4p_.arc
ORA-00280：更改 8376635 (用于线程 1)在序列 #216 中
Specify log: {<RET>=suggested | filename | AUTO | CANCEL}
ORA-00279：更改 8386363 (在 04/14/2017 13：00：09 生成)对于线程 1 是必需的
ORA-00289：
建议：
/home/oracle/app/oracle/fast_recovery_area/ORCL/archivelog/2017_04_14/o1_mf_1_217_dh1knx83_.arc
ORA-00280：更改 8386363 (用于线程 1)在序列 #217 中
Specify log: {<RET>=suggested | filename | AUTO | CANCEL}
```

```
    ORA-00279：更改 8496297（在 04/14/2017 20：54：53 生成）对于线程 1 是必需的
ORA-00289：
    建议：
/home/oracle/app/oracle/fast_recovery_area/ORCL/archivelog/2017_04_14/o1_mf
_1_218_dh1om5q8_.arc
    ORA-00280：更改 8496297（用于线程 1)在序列 #218 中
    Specify log: {<RET>=suggested | filename | AUTO | CANCEL}

    ORA-00279：更改 8506305（在 04/14/2017 22：02：13 生成）对于线程 1 是必需的
ORA-00289：
    建议：
/home/oracle/app/oracle/fast_recovery_area/ORCL/archivelog/2017_04_15/o1_mf
_1_219_dh2blmxh_.arc
    ORA-00280：更改 8506305（用于线程 1)在序列 #219 中
    Specify log: {<RET>=suggested | filename | AUTO | CANCEL}

    ORA-00279：更改 8517918（在 04/15/2017 04：00：19 生成）对于线程 1 是必需的
ORA-00289：
    建议：
/home/oracle/app/oracle/fast_recovery_area/ORCL/archivelog/2017_04_15/o1_mf
_1_220_dh2kq950_.arc
    ORA-00280：更改 8517918（用于线程 1)在序列 #220 中

    Specify log: {<RET>=suggested | filename | AUTO | CANCEL}
    Log applied.
    Media recovery complete.
```

当看到"Log applied."和"Media recovery complete."就表示日志文件应用成功，介质恢复成功。这样就可以打开数据库了。打开后，通过 show pdbs 命令可以看出，PDBORCL 插接式数据库的打开模式为正常值"READ WRITE"。

```
    SQL> ALTER database open;
    Database altered.
    SQL> show pdbs;
        CON_ID CON_NAME                       OPEN MODE  RESTRICTED
    ---------- ------------------------------ ---------- ----------
             2 PDB$SEED                       READ ONLY  NO
             3 PDBORCL                        READ WRITE NO
    SQL>
```

最后，观察完全恢复的效果：1 号员工的工资确实是最新值 50100，而不是备份之前的 50000。

```
$ sqlplus system/***@pdborcl
SQL> SELECT employee_id, salary FROM study.employees WHERE  employee_id=1;
EMPLOYEE_ID SALARY
----------- ----------
          1      50100
```

本例只是备份了 PDBORCL 插接式数据库的 USERS 表空间，并没有走完。

图 13-3 的所有流程，既没有备份 CDB 根数据库以及其他表空间，也没有备份控制文件，初始化文件与归档日志文件。因此，当没有备份的表空间或者其他文件损坏以后，就无法恢复相应的表空间或者数据库了。另外，既然备份了文件，就必须保证备份后的所有文件不损失，否则就跟没有备份一样。

在联机备份方式中，对于日志文件，我们一般只会备份归档日志文件，不会备份重做日志文件，但这意味重做日志文件不重要，相反，重做日志文件更重要，在进行完全恢复时，系统总是先恢复重做日志，再恢复归档日志。之所以不手工备份重做日志文件，完全是因为重做日志文件有"动态备份"的机制："重做日志文件组"，重做日志文件被分为多个组，每一组中包含多个重做日志文件。同一组中的文件是互为备份的，即文件内容和大小完全相同，因此不需要手工备份重做日志文件，只要这些文件不同时损坏即可。

13.4 RMAN 工具

RMAN 是 Recovery Manager 的缩写，是 Oracle 的主要备份和恢复工具。可以备份和恢复数据文件、日志文件、控制文件及参数文件。可以进行完全恢复和不完全恢复。RMAN 是 Oracle 管理备份工具，比起用户管理备份方式，RMAN 有方便、高效、可选择方式多的特点，在逻辑上 RMAN 可以备份/恢复整个数据库，也可以只备份/恢复 CDB；可以只备份/恢复部分 PDB、部分表空间、部分数据文件、部分归档日志文件，也可以备份/恢复控制文件和参数文件 spfile，因此 RMAN 是日常备份和恢复数据库的主要手段。

13.4.1 备份集与镜像复制

RMAN 有两种不同类型的备份方式：创建镜像复制和创建备份集。

1) 备份集(Backup Sets)

备份集为 RMAN 默认备份选项。备份集是 RMAN 创建的具有特定格式的逻辑备份对象，备份集在逻辑上由一个或多个备份片段(Backup Piece)组成，每个备份片段在物理上对应一个操作系统文件，一个备份片段中可能包含多个数据文件、控制文件或归档文件。通过 RMAN 创建备份集的优势在于，备份时只读取数据库中已经使用的数据块，因此不管是从备份效率，或是从节省存储空间的角度，创建备份集的方式都更有优势。

2) 镜像复制(Image Copies)

镜像复制实际上就是创建数据文件、控制文件或归档文件的备份文件，与用户管理备份方式的过程相似，只不过 RMAN 是利用目标数据库中的服务进程来完成文件复制，而用户管理备份方式则是用操作系统命令。这种方式本质仍是复制数据库中的物理文件，包括数据文件、控制文件等，复制出的文件与原始文件一模一样，所以镜像复制的方式体现不出 RMAN 的优势。

通过 RMAN 中的命令"show device type"可以查看当前的默认的备份方式，Oracle 默认的备份方式是备份集。通过"show all"可以查看所有 RMAN 配置参数。

13.4.2　启动 RMAN 并连接到数据库

Oracle 提供了命令 RMAN 用以启动 RMAN 并连接到数据库。Oracle 提供了两种方式登录和管理备份，一种是恢复目录方式(Catalog)，另一种是非恢复目录方式(Nocatalog)，默认是 Nocatalog 方式。RMAN 需要有一个地方存储备份和恢复需要的元数据信息。在 Catalog 方式，元数据信息存储在单独的数据库中，在 Nocatalog 方式，元数据信息存储在控制文件中。这里只介绍以 Nocatalog 方式启动 RMAN。

连接到本地数据库的命令是"rman target /"，连接到其他数据库的方式是"rman target sys/password@连接字符串"。

【示例 13-4】　连接本地数据库，查看所有配置。

```
$ rman target /
Recovery Manager: Release 12.1.0.2.0 - Production on 星期三 4月 26 23:28:39 2017
Copyright (c)1982, 2014, Oracle and/or its affiliates.  All rights reserved.
connected to target database: ORCL (DBID=1392946895)
using target database control file instead of recovery catalog

RMAN> show all;
RMAN configuration parameters for database with db_unique_name ORCL are:
CONFIGURE RETENTION POLICY TO REDUNDANCY 1; # default
CONFIGURE BACKUP OPTIMIZATION OFF; # default
CONFIGURE DEFAULT DEVICE TYPE TO DISK; # default
CONFIGURE CONTROLFILE AUTOBACKUP ON; # default
CONFIGURE CONTROLFILE AUTOBACKUP FORMAT FOR DEVICE TYPE DISK TO '%F';#default
CONFIGURE DEVICE TYPE DISK PARALLELISM 1 BACKUP TYPE TO BACKUPSET; # default
CONFIGURE DATAFILE BACKUP COPIES FOR DEVICE TYPE DISK TO 1; # default
CONFIGURE ARCHIVELOG BACKUP COPIES FOR DEVICE TYPE DISK TO 1; # default
CONFIGURE MAXSETSIZE TO UNLIMITED; # default
CONFIGURE ENCRYPTION FOR DATABASE OFF; # default
CONFIGURE ENCRYPTION ALGORITHM 'AES128'; # default
```

```
    CONFIGURE COMPRESSION ALGORITHM 'BASIC' AS OF RELEASE 'DEFAULT' OPTIMIZE
FOR LOAD TRUE ; # default
    CONFIGURE RMAN OUTPUT TO KEEP FOR 7 DAYS; # default
    CONFIGURE ARCHIVELOG DELETION POLICY TO NONE; # default
    CONFIGURE SNAPSHOT CONTROLFILE NAME TO
'/home/oracle/app/oracle/product/12.1.0/dbhome_1/dbs/snapcf_orcl.f';
# default
```

在本例中，命令"rman target /"等价于"rman target / nocatalog"，表示以 Nocatalog 模式启动 RMAN。因为 Nocatalog 是默认的模式，所以这个选项可以省略。查看所有配置的命令是"show all"，注意，RMAN 要求命令以分号结束。

13.4.3 备份失效（Expired）

Oracle 的备份可能失效（Expired）。失效是指备份文件丢失了，但元数据信息还在。在 RMAN 之外，通过操作系统命令删除备份文件，就会产生备份失效，因为删除文件之后，在 RMAN 中还保留了这些文件的元数据信息，换句话说，RMAN 认为这些被删除的文件还存在。

可以通过 crosscheck 发现并标记这些失效的文件，通过"delete expired backup"删除失效的备份元数据。常用的操作是：

```
RMAN> CROSSCHECK BACKUP;
RMAN> DELETE EXPIRED BACKUP;
```

所以，最好养成在 RMAN 中删除备份文件的习惯，比如使用命令"delete backup"，这样就不会产生失效的备份。

13.4.4 备份过期（Obsolete）

Oracle 的备份可能过期（Obsolete）。过期是指备份文件没有被删除，是以前备份的，但现在不再需要它了。比如，在前后两个时间点完整备份了数据库，很明显，后面备份集中的数据才是最新的，而前面备份的数据就过期了。

Oracle 通过备份保留策略"CONFIGURE RETENTION POLICY"来设置备份过期的规则。可以使用配置命令"CONFIGURE RETENTION POLICY"创建一个永久的和自动的备份保留策略。当设置一个备份保留策略有效以后，RMAN 根据配置命令指定的规则，将数据文件、归档日志文件或者控制文件的备份认为是过期的备份（即不再需要恢复的备份）。

可以使用命令"REPORT OBSOLETE"查看过期的文件和命令"DELETE OBSOLETE"删除它们，Oracle 不会自动删除过期的备份。

Oracle 实现备份保留策略有两个互斥选项：冗余度（Redundancy）和恢复窗口期（Recovery Window），两个选项只能选择其一。

1）冗余度（Redundancy）

Oracle 默认的备份保留策略是冗余度为 1（redundancy=1）的方式，即每个备份文件的副本数量为 1，只保留最近一次的备份，以前的副本将被视为过期(Obsolete)。RMAN 配置冗余度为 1 的命令是：

```
RMAN> CONFIGURE RETENTION POLICY TO REDUNDANCY 1;
```

在保留冗余度为 1 的情况下，当做了一次完整备份或者 0 级备份之后，以前的完整备份、0 级备份以及归档日志文件就过期了，可以删除。

如图 13-4 所示，当前时间是 1 月 30 日，之前在 3 个时间点都进行了完整备份，最近一次的备份是 t3 时间点(1 月 28 日)。那么 t3 时间点之前的备份和归档日志文件将过期。如果要进行数据库的恢复，只能从 t3 点开始恢复到当前日期。

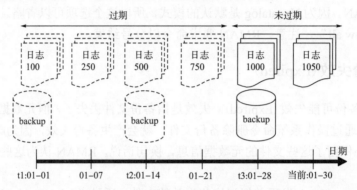

图 13-4　冗余度为 1 方式下的备份过期

2) 恢复窗口期(Recovery Window)

恢复窗口是指从当前时间开始倒退到以前的一个时间长度范围。恢复窗口的目的是为了让备份晚一点过期，保持一段时间再过期，以便可以恢复到更早的时间点。如图所示，一旦 t3 时间点备份完成后，之前时间的备份和归档日志就过期了，不能恢复了。如果设置为恢复窗口方式，比如设置为 7 天，命令如下：

```
RMAN> CONFIGURE RETENTION POLICY TO RECOVERY WINDOW OF 7 DAYS;
```

如果当前时间是 01-30，那么 t3 之前 5 天的范围以及 t3 之后的所有时间都是可以恢复的，如图 13-5 所示。由于窗口期也在 t2 到 t3 时间范围之内，所以 t2 时间点的备份，以及 t2 到 t3 之间的所有归档日志文件也是有效的，不会过期。

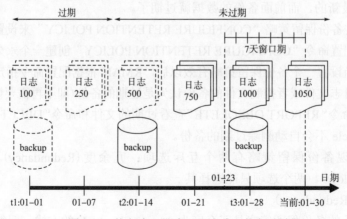

图 13-5　恢复窗口期为 7 天方式下的备份过期

在删除过期备份之前，可以自由配置不同的冗余度(Redundancy)，也可以自由地在冗余度和恢复窗口期(Recovery Window)之间切换，更改配置之后，可以通过命令"report obsolete"随时查看过期备份的变化情况。当然，配置改变之后，命令"delete obsolete"删除过期备份文件的范围也会相应改变。通常，应该先执行 report，看清楚之后再执行 delete。

```
RMAN> REPORT OBSOLETE;
RMAN> DELETE OBSOLETE;
```

RMAN 在执行命令"delete obsolete;"时，要提示手工输入 yes 或者 no，如果使用"delete noprompt obsolete;"就不会提示，直接删除。了解备份过期的原理，对于进行数据库的完全恢复和不完全恢复都非常重要。

13.4.5　RMAN 备份和恢复命令

RMAN 工具中的备份命令是 backup，恢复命令是 restore 和 recover，查看备份的命令是 list backup，删除备份的命令是 delete backup。这些命令的选项非常多，常见的命令如表 13-1 所示。

表 13-1　RMAN 部分常用命令一览表

RMAN 命令	说明
show all	查看所有参数
configure	配置参数
report schema	报告整个数据库(CDB+PDBs)的所有表空间和所有数据文件
report need backup	报告哪些数据库文件需要备份
report obsolete	报告过期的备份
list backup	显示所有备份集，含数据文件，控制文件，参数文件，归档日志文件的备份集
list backup of archivelog all	只显示归档日志文件的备份集
list copy	显示所有镜像复制文件
list archivelog all	查看所有归档日志文件
backup database	备份整个数据库(CDB+PDBs)
backup as compressed backupset database	压缩方式备份整个数据库(CDB+PDBs)
backup database root	只备份 CDB 根数据库
backup pluggable database pdb1，pdb2，…	只备份部分 PDB 数据库 pdb1，pdb2，…
backup incremental level 0 database	0 级备份整个数据库(CDB+PDBs)
backup incremental level 1 database	1 级差异增量备份备份整个数据库(CDB+PDBs)
backup incremental level 1 cumulative database	1 级累积增量备份备份整个数据库(CDB+PDBs)
backup as copy database	直接以镜像复制的方式备份整个数据库(CDB+PDBs)

RMAN 命令	说明
backup as copy pluggable database pdb1，pdb2，…	直接以镜像复制的方式备份部分 PDB 数据库 pdb1，pdb2，…
restore database	装载整个数据库（CDB+PDBs）
restore database root	只装载 CDB 根数据库
restore pluggable database pdb1，pdb2，…	只装载部分 PDB 数据库 pdb1，pdb2，…
recover database	恢复整个数据库（CDB+PDBs）
recover database root	只恢复 CDB 根数据库
recover pluggable database pdb1，pdb2，…	只恢复部分 PDB 数据库 pdb1，pdb2，…
set until time …	restore 或者 recover 到指定的时间点
crosscheck backup	交叉检查备份集
crosscheck copy	交叉检查镜像复制
crosscheck archivelog all	交叉检查归档日志文件
backup validate check logical database archivelog all	检查所有数据库文件和归档重做日志文件是否存在物理和逻辑损坏
delete backup	删除所有备份集
delete expired backup	删除失效的备份集
delete copy	删除镜像复制文件
delete expired copy	删除失效的镜像复制文件
delete archivelog all	删除所有归档日志文件
delete expired archivelog all	删除失效的归档日志文件
delete obsolete	删除过期的备份

在 configure 命令中经常使用格式串，在 backup，resotre，allocate channel 等其他 RMAN 命令中也会经常看到格式串。RMAN 提供了与格式串关联的一些语法元素。这些元素可以看作是占位符，RMAN 将使用相应的定义值来替换他们。如下所示（注意大小写有区别）：

%a：Oracle 数据库的 activation ID 即 RESETLOG_ID。

%c：备份片段的复制数（从 1 开始编号，最大不超过 256）。

%d：Oracle 数据库名称。

%D：当前时间中的日，格式为 DD。

%e：归档序号。

%f：绝对文件编号。

%F：基于"DBID+时间"确定的唯一名称，格式的形式为 c-IIIIIIIIII-YYYYMMDD-QQ，其中 IIIIIIIIII 为该数据库的 DBID，YYYYMMDD 为日期，QQ 是一个 1～256 的序列。

%h：归档日志线程号。

%I：Oracle 数据库的 DBID。

%M：当前时间中的月，格式为 MM。

%N：表空间名称。

%n：数据库名称，并且会在右侧用 x 字符进行填充，使其保持长度为 8。比如数据库名 JSSBOOK，则生成的名称则是 JSSBOOKx。

%p：备份集中备份片段的编号，从 1 开始。

%s：备份集号。

%t：备份集时间戳。

%T：当前时间的年月日格式（YYYYMMDD）。

%u：是一个由备份集编号和建立时间压缩后组成的 8 字符名称。利用%u 可以为每个备份集生成一个唯一的名称。

%U：默认是%u_%p_%c 的简写形式，利用它可以为每一个备份片段（即磁盘文件）生成一个唯一名称，这是最常用的命名方式，执行不同备份操作时，生成的规则也不同，如下所示：

生成备份片段时，%U=%u_%p_%c。

生成数据文件镜像复制时，%U=data-D-%d_id-%I_TS-%N_FNO-%f_%u。

生成归档文件镜像复制时，%U=arch-D_%d-id-%I_S-%e_T-%h_A-%a_%u。

生成控制文件镜像复制时，%U=cf-D_%d-id-%I_%u。

%Y：当前时间中的年，格式为 YYYY。

> 注意：如果在 BACKUP 命令中没有指定 FORMAT 选项，则 RMAN 默认使用%U 为备份片段命名。

【示例 13-5】典型的完整备份和完全恢复案例

本例演示一个典型的备份和恢复过程：首先完整备份数据库（CDB+所有 PDB），备份之后，删除一些数据文件（模拟数据丢失），最后完全恢复整个数据库，恢复之后，打开数据库，数据不会有任何损失。

本例执行的前提是：数据库在归档方式下正常运行，RMAN 配置为自动备份控制文件，即"CONFIGURE CONTROLFILE AUTOBACKUP ON；"。

首先 RMAN 连接数据库，然后完全备份，备份后查看备份结果：

```
$ rman target /
RMAN> SHOW ALL;
CONFIGURE RETENTION POLICY TO REDUNDANCY 1;
CONFIGURE CONTROLFILE AUTOBACKUP ON;
RMAN> BACKUP DATABASE;
RMAN> LIST BACKUP;
```

通过 list backup 可以看到备份文件所在的目录和名称，缺省的文件名称是按参数%U 格式生成的。还可以看到备份文件的类型以及备份文件中包含的备份信息。

现在，删除一些或者全部数据文件，模拟数据丢失。这些文件名会在 list backup 命令的输出中显示出来。通过 host 运行操作系统的删除文件命令：

```
RMAN> host "rm /home/oracle/app/oracle/oradata/orcl/*.dbf";
```

```
host command complete
RMAN> host "ls /home/oracle/app/oracle/oradata/orcl/*.dbf";
ls: 无法访问/home/oracle/app/oracle/oradata/orcl/*.dbf: 没有那个文件或目录
host command complete
```

可以看见，删除文件之后，文件就没有了。注意，删除文件的工作是在数据库联机的情况下完成的。这时候，数据库仍然是联机的，但与这些被删除文件相关的读写操作都会失败。需要进行数据恢复了，数据恢复的步骤是，SHUTDOWN→STARTUP MOUNT→RESTORE→RECOVER→OPEN DATABASE，实际的命令是：

```
RMAN> SHUTDOWN IMMEDIATE;
RMAN-03002: failure of shutdown command at 04/28/2017 01: 20: 35
ORA-01116：打开数据库文件 1 时出错
ORA-01110：数据文件 1: '/home/oracle/app/oracle/oradata/orcl/system01.dbf'
ORA-27041：无法打开文件
Linux-x86_64 Error: 2: No such file or directory
Additional information: 3
RMAN> SHUTDOWN ABORT;
Oracle instance shut down
RMAN> STARTUP MOUNT;
RMAN> RESTORE DATABASE;
Finished restore at 28-4月 -17
RMAN> recover database;
Starting recover at 28-4月 -17
using channel ORA_DISK_1
starting media recovery
media recovery complete, elapsed time: 00: 00: 01
Finished recover at 28-4月 -17
RMAN> ALTER DATABASE OPEN;
Statement processed
RMAN> host "ls /home/oracle/app/oracle/oradata/orcl/*.dbf";
/home/oracle/app/oracle/oradata/orcl/sysaux01.dbf
/home/oracle/app/oracle/oradata/orcl/system01.dbf
/home/oracle/app/oracle/oradata/orcl/temp01.dbf
/home/oracle/app/oracle/oradata/orcl/undotbs01.dbf
/home/oracle/app/oracle/oradata/orcl/users01.dbf
host command complete
```

上面的实际操作中"SHUTDOWN IMMEDIATE；"失败了，无法正常关闭，原因是正好删除了 CDB 的数据文件，所以不得不使用"SHUTDOWN ABORT；"强制关闭数据库。

"STARTUP MOUNT；"方式启动数据库(只需要读取控制文件，不读数据文件)。

这一步必须要求控制文件是正常的。当然如果控制文件丢失，也不要紧，就必须先从备份集中恢复控制文件，因为在备份的时候自动备份了控制文件。

"RESTORE DATABASE；"是指装载数据库，将备份集还原为数据文件。所以，备份集文件是千万不能丢失的，否则这个命令会失败，那么也无法进行下面的操作。

"RECOVER DATABASE；"是指恢复数据库，将备份之后的数据库修改（存储在重做日志文件或者归档文件中）应用到数据文件中。所以，从这里可以看出，即使做了数据文件的完全备份，重做日志文件和归档日志文件也是不能丢失的，否则下面的数据库打开也无法完成。

"ALTER DATABASE OPEN；"这个命令运行成功，没有出现错误，就说明了数据已经完全恢复，没有任何损失。与完全恢复相反，如果这个命令带了参数 RESETLOGS（重置日志），变成"ALTER DATABASE OPEN RESETLOGS；"，那么就是不完全恢复，会有数据损失。

数据库打开之后，通过命令"host ls"可以看出，被删除的数据文件又出现了。

本例的优点是操作简单易用，只需要在 RMAN 一个工具中就可以完成。另外，本例安全可靠。在备份之后的任何时期，只要做到：不丢失备份集文件、控制文件、重做日志文件和归档日志文件，并且归档日志空间没有溢出，都能完全恢复，不损失数据。

但本例也有两个比较明显的缺点：

缺点 1：备份空间大，容易生产归档日志溢出。完整备份所需要空间较大，也没有压缩备份。另外由于没有备份归档日志文件，随着时间推移，会造成归档日志文件越来越多，最后可能超出 Oracle 的允许空间范围（db_recovery_file_dest_size）。

缺点 2：数据库整体恢复。本例是恢复 CDB 和所有的 PDB，在恢复之前需要将整个数据库停机。实际情况可能是只需要恢复部分丢失数据的 PDB，在这样的情况下，更好的方法应该是只需要将需要恢复的 PDB 停机，然后只恢复这些 PDB，不影响未出故障的 PDB 的正常运行，数据库不会整体停机。

13.4.6　实用案例：完全恢复一个 PDB

本节将提供一个比较接近实用场景的备份和恢复案例。

在备份的策略上，采用完整的压缩备份集方式备份，并且备份完成之后，删除过期的备份文件和归档日志文件，从而节约出 db_recovery_file_dest_size 允许的空间，这样数据库运行期间就不会产生归档日志满的错误。

在恢复的时候，只模拟一个 PDB 数据库损坏，只完全恢复一个 PDB 数据库。这样的好处是在恢复的时候只需要使用这个 PDB 停机，而不会将整个数据库停机，保证其他数据库一直保持正常使用。

首先 RMAN 连接数据库，然后完整压缩备份整个数据库，删除过期的备份和归档日志，查看备份结果，主要查看 PDBORCL 的备份集，然后删除 PDBORCL 插接式数据库的所有文件，模拟 PDBORCL 出故障。

```
$ rman target /
RMAN> SHOW ALL;
CONFIGURE RETENTION POLICY TO REDUNDANCY 1;
CONFIGURE CONTROLFILE AUTOBACKUP ON;
RMAN> BACKUP AS COMPRESSED BACKUPSET DATABASE;
RMAN> LIST BACKUP;
RMAN> REPORT OBSOLETE;
RMAN> DELETE OBSOLETE;
RMAN> host "rm /home/oracle/app/oracle/oradata/orcl/pdborcl/*.dbf";
host command complete
RMAN> host "ls /home/oracle/app/oracle/oradata/orcl/pdborcl/*.dbf";
ls: 无法访问/home/oracle/app/oracle/oradata/orcl/pdborcl/*.dbf: 没有那个文件或目录
host command complete
```

这时，PDBORCL 的数据文件被全部删除了，PDBORCL 已经不能正常工作了，需要进行仅针对 PDBORCL 的完全恢复。在恢复之前，必须先关闭 PDBORCL，仅关闭 PDB 的操作只能在 Sqlplus 工具中完成，打开一个新的操作系统命令窗口，运行 Sqlplus：

```
$ sqlplus / as sysdba
SQL> show pdbs;
    CON_ID    CON_NAME                       OPEN MODE  RESTRICTED
    ----------  ------------------------------  ----------  ----------
         2    PDB$SEED                       READ ONLY  NO
         3    PDBORCL                        READ WRITE NO
SQL> ALTER SESSION SET CONTAINER =pdborcl;
Session altered.
SQL> UPDATE hr.jobs SET min_salary=min_salary+10;
UPDATE hr.jobs SET min_salary=min_salary+10
       *
ERROR at line 1:
ORA-00604: 递归 SQL 级别 1 出现错误
ORA-01116: 打开数据库文件 8 时出错 ORA-01110:
数据文件 8: '/home/oracle/app/oracle/oradata/orcl/pdborcl/system01.dbf'
ORA-27041: 无法打开文件
SQL> shutdown immediate;
Pluggable Database closed.
SQL> exit
```

通过"show pdbs;"可以看出，虽然数据文件都被删除了，但是 PDBORCL 的状态仍然是正常的读写状态(READ WRITE)，但当 session 切换到 PDBORCL 里，执行 UPDATE

语句的时候,就会报"无法打开文件"的错误了。最后执行"shutdown immediate;"看见"Pluggable Database closed."的时候,就表示 PDBORCL 被关闭了。这时候,就可以回到 RMAN 中单独恢复 PDBORCL 了:

```
RMAN> restore pluggable database pdborcl;
RMAN> recover pluggable database pdborcl;
starting media recovery
media recovery complete, elapsed time: 00: 00: 00
Finished recover at 28-4月-17
RMAN> ALTER pluggable database pdborcl open;
Statement processed
```

最后"ALTER pluggable database pdborcl open;"命令打开 PDBORCL 没有报错,也没有加 resetlogs 选项,这就说明了 PDBORCL 完全恢复成功,没有损失数据。

除了精准恢复 PDB 之外,Oracle 还支持精准恢复一个表空间,在这样的方式下,连 PDB 都不用停机,只需要将要恢复的表空间离线,再恢复表空间,恢复成功后将表空间在线即可。

13.4.7 实用案例:不完全恢复一个 PDB

所谓不完全恢复就是指将数据库恢复到备份后的一个指定的时间点。这种情况的应用场景之一是:数据文件并没有损坏,数据库本身没有故障,但由于用户在修改数据的时候进行了错误的操作,比如错误删除了一些记录,甚至删除了数据表,错误修改了表中的数据等。这时,用户希望将数据库恢复到误操作之前的时间点,这就是不完全恢复。

关于不完全恢复的时间点是有限制的,不是指可以恢复到任何的时间点。这个时间点只能是未过期的备份时间点开始之后的所有时间点,见图 13-4 和图 13-5。如果试图恢复备份时间点之前的时间点,恢复过程将无法进行,并提示错误。

下面以误修改 HR.JOBS 表的数据为例,演示恢复到时间点的操作。jobs_id='JOB_REP'的工作的最低工资 min_salary 是 4500 元,我们在时间点 t1 增加 100 元,变成 4600 元,在随后的时间点 t2 增加 1000 元,变成 5600 元。我们将 t2 的修改看作是误操作,希望恢复到 t1 时刻的值 4600 元,见图 13-6。

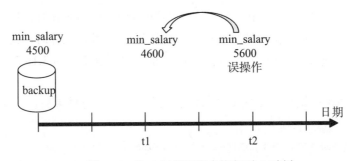

图 13-6 从 t2 时刻不完全恢复到 t1 时刻

在做这个实验之前，必须先备份数据库，否则无法恢复到时间点，备份之后，删除过期备份(可选)。

```
RMAN> backup as compressed backupset database;
RMAN> list backup;
RMAN> delete obsolete;
RMAN> report obsolete;
```

备份之后，可以修改数据了，新开一个 Sqlplus 窗口修改数据，修改数据的时候一定要注意当前时间，通过"set time on"打开当前时间提示：

```
$ sqlplus hr/***@pdborcl;
SQL> set time on
03: 15: 04 SQL> SELECT job_id, min_salary FROM jobs WHERE job_id='PR_REP';
JOB_ID       MIN_SALARY
----------   ----------
PR_REP       4500
03 : 15 : 13 SQL> UPDATE jobs SET min_salary=min_salary+100 WHERE job_id='PR_REP';
1 rows updated.
03: 16: 51 SQL> commit;
Commit complete.
03 : 17 : 00 SQL> UPDATE jobs SET min_salary=min_salary+1000 WHERE job_id='PR_REP';
1 rows updated.
03: 17: 04 SQL> commit;
Commit complete.
03: 17: 08 SQL> SELECT job_id, min_salary FROM jobs WHERE job_id='PR_REP';
JOB_ID       MIN_SALARY
----------   ----------
PR_REP       5600
03: 19: 00 SQL> exit
```

从上面的修改操作可以看出，第一次修改的时间点 t1=03:16:51，第二次修改的时间是 t2=03:17:04。为了确保恢复时间点正确，往往延后 1 秒以上计时，比如我要希望恢复到 t1，t1 的时间最好在 03:16:51 之后，在 t2 之前。这里就取 t1=03:16:55，在实际的命令中，还要在 t1 中加上年月日。

在回到 RMAN 中做不完全恢复之前，必须在以 sys 身份登录后，关闭 PDBORCL：

```
$ sqlplus / as sysdba
SQL> ALTER SESSION SET CONTAINER =pdborcl;
Session altered.
SQL> shutdown immediate;
```

第 13 章 备份与恢复

```
Pluggable Database closed.
SQL> exit
```

关闭 PDBORCL 之后，就可以回到 RMAN 中做数据库的不完全恢复了：

```
RMAN> restore pluggable database pdborcl until time "to_date('2017-04-28
03: 16: 55', 'yyyy-mm-dd hh24: mi: ss')";
Finished restore at 28-4月-17
RMAN> recover pluggable database pdborcl until time "to_date('2017-04-28
03: 16: 55', 'yyyy-mm-dd hh24: mi: ss')";
Finished recover at 28-4月-17
RMAN> ALTER pluggable database pdborcl open resetlogs;
Statement processed
```

注意，"ALTER pluggable database pdborcl open resetlogs；"中的 resetlogs 表示重置日志，这是不完全恢复后必须的选项，否则无法打开数据库。打开数据库成功之后，看看数据是否恢复到了 t1 时刻：

```
$ sqlplus hr/***@pdborcl;
SQL> SELECT job_id, min_salary FROM jobs WHERE  job_id='PR_REP';
JOB_ID       MIN_SALARY
----------   ----------
PR_REP       4600
```

从这个结果可以看出，实验成功，确实恢复到了 t1 时间的 4600 元。

这里要特别强调的是，不完全恢复是不可逆的，由于使用了 resetlogs 打开数据库，重置了日志文件，恢复到 t1 时间点之后，最新的数据就是 4600 了，不再有恢复之前 t2 时刻的 5600 的痕迹，数据无法恢复到原来 t2 时刻的数据，所以做不完全恢复的时候要特别小心，不要把 until time 的时间点搞错了。

另外，不完全恢复还可以应用在丢失了部分归档日志文件、重做日志文件、控制文件、甚至丢失了备份集文件的情况下，这种情况的恢复就要困难得多，而且不保证一定成功，所以要保护好数据库中各种文件的安全。

13.4.8　RMAN 批处理

RMAN 执行备份和恢复的时候，可以使用 run 定义一组需要执行的语句，以批处理的方式运行。这种方式的好处是所有批处理中的命令被视为一个作业，如果作业中任何一条命令执行失败，则整个命令停止住下执行。run 命令一个最大的用处是手工分配连接通道（Channel），如果希望增加备份的效率，可以分配多个通道。

【示例 13-6】使用 run 批处理备份

本例设置恢复窗口期为 8 天，自动备份控制文件，分配通道 ch1，备份到磁盘文件，然后开始以压缩方式 0 级备份整个数据库，最后释放通道 ch1。

```
$ rman target /
RMAN> run
2> {
3>   configure retention policy to recovery window of 8 days;
4>   configure controlfile autobackup on;
5>   allocate channel ch1 device type disk;
6>   backup as compressed backupset incremental level 0 database ;
7>   release channel ch1;
8> }
```

13.5 闪回技术 Flashback

为了简化对备份和恢复的日常操作，Oracle 提供了闪回技术（Flashback），该技术自动管理快速恢复区，实现自动备份与恢复，大大减小了管理开销。当 Oracle 发生人为误操作的时候，不需要事先备份数据库，就可以快速而方便地进行恢复。

Oracle 有 3 个参数与 flashback 有关，它们是：

db_recovery_file_dest：快速恢复区的位置。

db_recovery_file_dest_size：快速恢复区的大小。

db_flashback_retention_target：闪回数据保留时间，单位是分钟。

Oracle 安装的时候默认没有打开 Flashback 功能，如果需要这个功能，必须手工打开，打开 Flashback 功能的命令是："ALTER database flashback on；"。

【示例 13-7】打开 Flashback 功能，并查询相关参数

```
$ sqlplus / as sysdba
SQL> ALTER database flashback on;
Database altered.
SQL> show parameter db_recovery

NAME                                 TYPE         VALUE
------------------------------------ ------------ ------------------
db_recovery_file_dest                string       /home/oracle/app/oracle/fast_recovery_area
db_recovery_file_dest_size           big integer  4560M
SQL> show parameter db_flashback
NAME                                 TYPE         VALUE
------------------------------------ ------------ ------------------
db_flashback_retention_target        integer      1440
SQL>
```

13.5.1　Flashback Database

Flashback Database（闪回数据库）可以方便地让整个数据库前滚到当前的前一个时间点或者 SCN，而不需要做时间点的恢复。闪回数据库可以迅速将数据库回到误操作或人为错误的前一个时间点，不利用备份就快速地实现基于时间点的恢复。Oracle 通过创建新的 Flashback Logs（闪回日志），记录数据库的闪回操作。

【示例 13-8】 闪回数据库

本示例在打开了 Flashback 的情况下，通过 Flashback Database 命令将恢复到修改之前的时间点。

首先用 SYSTEM 登录到 PDBORCL，修改一些数据，并记录下修改之前的时间。

```
$ sqlplus system/***@pdborcl
SQL> set time on
08:56:57 SQL> SELECT job_id, min_salary FROM hr.jobs WHERE job_id='PR_REP';

JOB_ID       MIN_SALARY
----------   ----------
PR_REP       4500


08:57:01 SQL> UPDATE hr.jobs SET min_salary=min_salary+100;
19 rows updated.
08:57:06 SQL> commit;
Commit complete.
08:57:10 SQL> SELECT job_id, min_salary FROM hr.jobs WHERE job_id='PR_REP';
JOB_ID       MIN_SALARY
----------   ----------
PR_REP       4600
08:57:14 SQL>exit
```

可以看出，在时间点"08：57：06"将 MIN_SALARY 的值从 4500 修改成了 4600，现在我们将通过 Flashback Database 命令把这个值恢复回 4500，恢复的时间点是"08：56：57"。首先以 sys 用户登录，恢复的过程必须按下面的顺序依次完成：

```
$ sqlplus / as sysdba
SQL> shutdown immediate
SQL> startup mount
Database mounted.
SQL> flashback database to timestamp to_timestamp(
  2  '2017-04-28 08:56:57', 'yyyy-mm-dd hh24:mi:ss');
Flashback complete.
```

```
SQL> ALTER database open resetlogs;
Database altered.
SQL> exit;
```
恢复完成后,查询一下 MIN_SALARY 的值是否恢复到了 4500:
```
$ sqlplus system/***@pdborcl
SQL> SELECT job_id, min_salary FROM hr.jobs WHERE job_id='PR_REP';
JOB_ID      MIN_SALARY
----------  ----------
PR_REP            4500
```
可以看出,恢复成功。

Flashback Database 的优点是操作简便(相对 RMAN),可以在 Sqlplus 中操作,不必通过 rman 操作。但也有一些局限性:

(1)打开这个功能会增加数据库的闪回写入负担,降低数据库性能。

(2)恢复的时候必须使整个数据库停机。

(3)只能对整个数据库(CDB+所有 PDB)一起恢复,在恢复误操作的时候,很可能将其他用户的正常数据修改恢复到修改之前,这就是所谓的"误恢复"。

13.5.2 Flashback Table

Flashback Table(闪回表)是指只恢复一个表,而不是整个数据库。闪回数据库是将整个数据库恢复到指定的时间点,但有时仅仅是对一个表做了误操作,就没有必要恢复整个数据库,而只需要恢复这个表。必须设置某个表为行移动(Row Movement)方式,这个表才可以进行闪回操作。任何一个表的缺省方式都没有这个属性,必须手工设置,设置行移动的命令是"ALTER table 表名 enable row movement;",取消行移动的命令是"ALTER table 表名 disable row movement;",例如:
```
SQL> ALTER table hr.jobs enable row movement;
Table altered.
```
【示例 13-9】闪回表

本例以 HR 身份登录,修改表 JOBS,然后闪回到修改之前的状态。
```
$ sqlplus hr/***@pdborcl
SQL> ALTER table jobs enable row movement;
Table altered.
SQL> set time on
15: 26: 34 SQL> UPDATE hr.jobs SET min_salary=min_salary+10;
19 rows updated.
15: 26: 45 SQL> commit;
Commit complete.
```

```
15:26:47 SQL> SELECT * FROM hr.jobs WHERE rownum<3;
JOB_ID     JOB_TITLE                        MIN_SALARY MAX_SALARY
---------- -------------------------------- ---------- ----------
AD_PRES    President                             20090      40000
AD_VP      Administration Vice President         15010      30000
SQL> flashback table jobs to timestamp
2 to_timestamp('2017-04-28 15:26:33', 'yyyy-mm-dd hh24:mi:ss');
Flashback complete.
SQL> SELECT * FROM hr.jobs WHERE rownum<3;
JOB_ID     JOB_TITLE                        MIN_SALARY MAX_SALARY
---------- -------------------------------- ---------- ----------
AD_PRES    President                             20080      40000
AD_VP      Administration Vice President         15000      30000
SQL>exit
```

本例在时间点"15:26:34"修改了数据,Flashback Table 的时间点是前一秒,即"15:26:33",可以看出当闪回成功后,数据又恢复到了修改前的值。为了说明的简洁,本例在查询的时候只查询前两行(where rownum<3)。

13.5.3 回收站

Flashback Table 除了可以恢复表到某个时间点之外,还可以恢复表的删除(To Before Drop),这样,删除表以后也能快速恢复。所以,当启用了数据库闪回之后,Oracle 会将删除的表保存到回收站中,回收站中的表可以被恢复出来。

【示例 13-10】删除表并恢复。

本例先创建一个表 MYJOBS,然后删除这个表。

```
$ sqlplus hr/***@pdborcl
SQL> CREATE TABLE myjobs as SELECT * FROM jobs;
Table created.
SQL> SELECT count(*)FROM myjobs;
  COUNT(*)
----------
        19
SQL> drop table myjobs;
Table dropped.
```

删除表 MYJOBS 之后,MYJOBS 就放到了回收站中,通过查询 user_recyclebin 视图可以查询该用户回收站中的表,运行 flashback table 恢复 MYJOBS 表:

```
SQL> SELECT object_name, original_name FROM user_recyclebin;
OBJECT_NAME                                      ORIGINAL_NAME
```

```
---------------------------------------------  ---------------
BIN$TjZNNLvxLjDgUwEAAH/dbA==$0                 MYJOBS
SQL> flashback table myjobs to before drop;
Flashback complete.
SQL> SELECT count(*)FROM myjobs;
  COUNT(*)
----------
        19
```

可以看到，恢复后可以查询出数据了，表示恢复成功。除了通过查询 user_recyclebin 查询回收站之外，还可以通过 SQL Developer 工具在 GUI 界面中操作回收站，如图 13-7 所示：

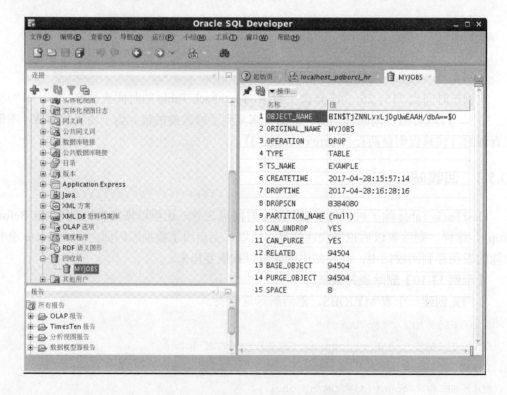

图 13-7 在 SQL Developer 中查看回收站

如果在删除一个表的时候加上 PURGE 选项，该表就直接删除，不会进入回收站，当然也就不能通过回收站的方式恢复：

```
SQL> DROP TABLE myjobs PURGE;
```

13.6 数据导出与导入

Oracle 的数据导出与导入功能是通过数据泵(DataPump)技术实现的，导出的命令是 expdp，导入的命令是 impdp。数据库泵技术提供快速数据迁移功能，其主要特性有：

(1) 支持并行处理导入、导出任务。
(2) 支持暂停和重启动导入、导出任务。
(3) 支持通过 Database Link 的方式导出或导入远端数据库中的对象。
(4) 支持在导入时通过 Remap_schema、Remap_datafile、Remap_tablespace 几个参数实现导入过程中自动修改对象属主、数据文件或数据所在表空间。
(5) 导入/导出时提供了非常细粒度的对象控制。通过 Include、Exclude 两个参数，甚至可以详细制定是否包含或不包含某个对象。

13.6.1　Oracle 目录对象

数据库的导出与导入的原理是：expdp 命令将用户的对象从数据库导出到文件中保存，impdp 命令将这个文件导入回数据库中。导出文件存储的目录必须首先在操作系统中创建好，然后再到 Oracle 中创建一个"链接"，这个"链接"在 Oracle 中仍然叫目录(Directory)。

因此，Oracle 的目录(Directory)对象是一个操作系统目录在 Oracle 中的一个代名词，一个目录对象只是指向操作系统的一个路径。每个 Oracle Directory 都包含 Read，Write 两个权限，可以通过 Grant 命令授权给指定的用户或角色。拥有读写权限的用户才可以读写该 Directory 对象指定的操作系统路径下的文件。

Oracle 用户要具有"CREATE ANY DIRECTORY"的系统权限才创建目录对象,普通用户不具有创建目录的权限。一般来说，创建 Oracle Directory 的流程是以下三步：首先在操作系统中创建一个目录，再由 system 用户创建指向操作系统目录的 Oracle 目录对象，最后将目录对象授权给其他用户读写(Read，Write)。

【示例 13-11】创建 Oracle 目录 expdir

本示例创建一个操作系统目录"/home/oracle/expdir"，由 SYSTEM 用户创建一个对应的 Oracle 目录 expdir，再将它授权给 HR 和 STUDY 两个用户，权限是读加写。

```
$ mkdir expdir
$ cd expdir
$ pwd
/home/oracle/expdir
$ sqlplus system/***@pdborcl
SQL> create or replace directory expdir as '/home/oracle/expdir';
Directory created.
SQL> GRANT read, write on directory expdir to hr, study;
Grant succeeded.
```

注意，Oracle 的目录对象是数据库级别的对象，表对象是用户级别的对象。比如，同一个数据库中有两个用户 A 和 B，都允许 A，B 用户各有一张表，表名都叫 T，表 T 虽然名称相同，但其实在物理上是两个不相同的表。但对于目录对象，在一个数据库中，就不允许出现同名的目录对象。

这里说的同一个数据库是指 CDB 或者同一个 PDB。也就是说，在同一个 PDB 中不允许同名的目录对象，但在不同的 PDB 中，还是允许同名的。

要删除一个目录对象，可以使用命令"drop directory 目录对象名；"，删除目录对象不会删除操作系统中的目录和文件。

13.6.2 数据导出

expdp 命令导出数据，可以一次性导出整个数据库，或者导出一个数据库，或者导出几个用户，也可以导出指定的表。默认的情况是导出用户自己的所有对象。

【示例 13-12】导出用户所有对象

本示例导出 study 用户的所有对象，导出文件名是 study.dmp，目录是 expdir。

```
$ expdp study/***@pdborcl directory=expdir dumpfile=study.dmp
Export: Release 12.1.0.2.0 - Production on 星期六 4月 29 10：57：48 2017
Copyright (c)1982，2014，Oracle and/or its affiliates. All rights reserved.
Connected to: Oracle Database 12c Enterprise Edition Release 12.1.0.2.0 - 64bit Production
With the Partitioning, OLAP, Advanced Analytics and Real Application Testing options
启动 "STUDY"."SYS_EXPORT_SCHEMA_01": study/********@pdborcl directory=expdir dumpfile=study.dmp
正在使用 BLOCKS 方法进行估计...
处理对象类型 SCHEMA_EXPORT/TABLE/TABLE_DATA
使用 BLOCKS 方法的总估计：32.31 MB
处理对象类型 SCHEMA_EXPORT/PRE_SCHEMA/PROCACT_SCHEMA
处理对象类型 SCHEMA_EXPORT/SEQUENCE/SEQUENCE
处理对象类型 SCHEMA_EXPORT/TABLE/TABLE
处理对象类型 SCHEMA_EXPORT/TABLE/COMMENT
处理对象类型 SCHEMA_EXPORT/PACKAGE/PACKAGE_SPEC
处理对象类型 SCHEMA_EXPORT/PACKAGE/COMPILE_PACKAGE/PACKAGE_SPEC/ALTER_PACKAGE_SPEC
处理对象类型 SCHEMA_EXPORT/VIEW/VIEW
处理对象类型 SCHEMA_EXPORT/PACKAGE/PACKAGE_BODY
处理对象类型 SCHEMA_EXPORT/TABLE/INDEX/INDEX
处理对象类型 SCHEMA_EXPORT/TABLE/CONSTRAINT/CONSTRAINT
处理对象类型 SCHEMA_EXPORT/TABLE/INDEX/STATISTICS/INDEX_STATISTICS
处理对象类型 SCHEMA_EXPORT/TABLE/CONSTRAINT/REF_CONSTRAINT
处理对象类型 SCHEMA_EXPORT/TABLE/TRIGGER
处理对象类型 SCHEMA_EXPORT/TABLE/STATISTICS/TABLE_STATISTICS
处理对象类型 SCHEMA_EXPORT/STATISTICS/MARKER
```

处理对象类型 SCHEMA_EXPORT/POST_SCHEMA/PROCOBJ
. . 导出了 "STUDY"."ORDERS": "PARTITION_BEFORE_2016"   268.4 KB   5000 行
. . 导出了 "STUDY"."ORDERS": "PARTITION_BEFORE_2017"   268.5 KB   5000 行
. . 导出了 "STUDY"."EMPLOYEES"   8.859 KB   7 行
. . 导出了 "STUDY"."DEPARTMENTS"   5.593 KB   3 行
. . 导出了 "STUDY"."PRODUCTS"   6.156 KB   9 行
. . 导出了 "STUDY"."ORDERS": "PARTITION_BEFORE_2018"   0 KB   0 行
. . 导出了 "STUDY"."ORDER_DETAILS": "PARTITION_BEFORE_2016"   425.7 KB   15000 行
. . 导出了 "STUDY"."ORDER_DETAILS": "PARTITION_BEFORE_2017"   426.1 KB   15000 行
. . 导出了 "STUDY"."ORDER_DETAILS": "PARTITION_BEFORE_2018" 0 KB   0 行
已成功加载/卸载了主表 "STUDY"."SYS_EXPORT_SCHEMA_01"
****
STUDY.SYS_EXPORT_SCHEMA_01 的转储文件集为:
  /home/oracle/expdir/study.dmp
作业 "STUDY"."SYS_EXPORT_SCHEMA_01" 已于 星期六 4 月 29 10∶58∶38 2017 elapsed 0 00∶00∶48 成功完成

13.6.3 数据导入

使用 expdp 导出后,可以使用 impdp 导入。导入的操作就是根据导出文件中的内容在数据库的用户中创建对象和插入数据。因此,在导入之前,要保证用户中没有这些对象和数据,否则就要报错。

如果用户导入之前已经存在这些对象,就要删除,如果对象很多,这种删除对象的操作是很麻烦的。常见的办法是先把整个用户连同对象一起删除,然后再重建用户,重新授权。删除用户及其所有对象的命令是"DROP USER 用户名 CASCADE;"。

【示例 13-13】导入自己的文件

本示例是用户 STUDY 导入由自己导出生成的文件。导入之前先由 SYSTEM 用户删除和重建用户 STUDY 用户,不然导入的时候因为对象的存在而引发错误:

```
$ sqlplus system/***@pdborcl
SQL> DROP USER study CASCADE;
SQL> CREATE USER study IDENTIFIED BY 123 default
2 tablespace "USERS" temporary tablespace "TEMP";
SQL> ALTER USER study quota unlimited on USERS;
SQL> ALTER USER study quota unlimited on USERS02;
SQL> GRANT "CONNECT" to study with admin option;
SQL> GRANT "RESOURCE" to study with admin option;
```

```
SQL> ALTER USER study default role "CONNECT", "RESOURCE";
SQL> GRANT CREATE JOB TO study with admin option;
SQL> GRANT CREATE VIEW TO study with admin option;
SQL> GRANT READ, WRITE ON DIRECTORY expdir TO study;
SQL> exit;
```

在给用户授权及分配表空间的时候,应该根据导出文件中的对象类型选择不同的权限和存储限额,如果权限没有给够,导入的时候要报错。接下来,STUDY 用户可以导入整个方案了。

```
$ impdp study/***@pdborcl directory=expdir dumpfile=study.dmp
...
已成功加载/卸载了主表 "STUDY"."SYS_IMPORT_FULL_01"
启动 "STUDY"."SYS_IMPORT_FULL_01": study/********@pdborcl directory=expdir dumpfile=study.dmp
处理对象类型 SCHEMA_EXPORT/PRE_SCHEMA/PROCACT_SCHEMA
处理对象类型 SCHEMA_EXPORT/SEQUENCE/SEQUENCE
处理对象类型 SCHEMA_EXPORT/TABLE/TABLE
处理对象类型 SCHEMA_EXPORT/TABLE/TABLE_DATA
. . 导入了 "STUDY"."ORDERS":"PARTITION_BEFORE_2016"    268.4 KB    5000 行
. . 导入了 "STUDY"."ORDERS":"PARTITION_BEFORE_2017"    268.5 KB    5000 行
. . 导入了 "STUDY"."EMPLOYEES"                          8.859 KB      7 行
. . 导入了 "STUDY"."DEPARTMENTS"                        5.593 KB      3 行
. . 导入了 "STUDY"."PRODUCTS"                           6.156 KB      9 行
. . 导入了 "STUDY"."ORDERS":"PARTITION_BEFORE_2018"      0 KB        0 行
. . 导入了 "STUDY"."ORDER_DETAILS":"PARTITION_BEFORE_2016"  425.7 KB  15000 行
. . 导入了 "STUDY"."ORDER_DETAILS":"PARTITION_BEFORE_2017"  426.1 KB  15000 行
. . 导入了 "STUDY"."ORDER_DETAILS":"PARTITION_BEFORE_2018" 0 KB 0 行
处理对象类型 SCHEMA_EXPORT/TABLE/COMMENT
处理对象类型 SCHEMA_EXPORT/PACKAGE/PACKAGE_SPEC
处理对象类型 SCHEMA_EXPORT/PACKAGE/COMPILE_PACKAGE/PACKAGE_SPEC/ALTER_PACKAGE_SPEC
处理对象类型 SCHEMA_EXPORT/VIEW/VIEW
处理对象类型 SCHEMA_EXPORT/PACKAGE/PACKAGE_BODY
处理对象类型 SCHEMA_EXPORT/TABLE/INDEX/INDEX
处理对象类型 SCHEMA_EXPORT/TABLE/CONSTRAINT/CONSTRAINT
处理对象类型 SCHEMA_EXPORT/TABLE/INDEX/STATISTICS/INDEX_STATISTICS
处理对象类型 SCHEMA_EXPORT/TABLE/CONSTRAINT/REF_CONSTRAINT
```

处理对象类型 SCHEMA_EXPORT/TABLE/TRIGGER
处理对象类型 SCHEMA_EXPORT/TABLE/STATISTICS/TABLE_STATISTICS
处理对象类型 SCHEMA_EXPORT/STATISTICS/MARKER
处理对象类型 SCHEMA_EXPORT/POST_SCHEMA/PROCOBJ
作业 "STUDY"."SYS_IMPORT_FULL_01" 已于 星期六 4月 29 12：56：29 2017 elapsed 0 00：00：03 成功完成

【示例 13-14】导入到其他用户

expdp 和 impdp 的一个重要用途之一就是做数据迁移，比如将数据迁移到其他服务器或者用户之中。STUDY.DMP 中的对象是由 STUDY 用户创建的，本示例就是将 STUDY.DMP 导入到 STUDY1 新用户中，导入成功之后，所有对象的用户会由 STUDY 变为 STUDY1。下面首先创建用户 STUDY1：

```
$ sqlplus system/***@pdborcl
SQL> CREATE USER study1 IDENTIFIED BY 123 default
2 tablespace "USERS" temporary SQL> tablespace "TEMP";
SQL> ALTER USER study1 quota unlimited on USERS;
SQL> ALTER USER study1 quota unlimited on USERS02;
SQL> GRANT "CONNECT" to study1 with admin option;
SQL> GRANT "RESOURCE" to study1 with admin option;
SQL> ALTER USER study1 default role "CONNECT","RESOURCE";
SQL> GRANT CREATE JOB TO study1 with admin option;
SQL> GRANT CREATE VIEW TO study1 with admin option;
SQL> GRANT READ,WRITE ON DIRECTORY expdir TO study1;
```

用户 STUDY1 创建成功后，可以由 SYSTEM 用户来做导入操作（STUDY 用户的权限不够），导入命令中使用参数"remap_schema=study：study1"将 STUDY 用户对象改变为 STUDY1 的对象，使用参数"transform=disable_archive_logging：y"在导入时禁用归档日志，提高导入的速度。

```
$ impdp system/***@pdborcl directory=expdir dumpfile=study.dmp remap_schema=study: study1 transform=disable_archive_logging: y
...
. . 导入了 "STUDY1"."ORDERS": "PARTITION_BEFORE_2016"   268.4 KB    5000 行
. . 导入了 "STUDY1"."ORDERS": "PARTITION_BEFORE_2017"   268.5 KB    5000 行
. . 导入了 "STUDY1"."EMPLOYEES"                          8.859 KB     7 行
. . 导入了 "STUDY1"."DEPARTMENTS"                        5.593 KB     3 行
. . 导入了 "STUDY1"."PRODUCTS"                           6.156 KB     9 行
. . 导入了 "STUDY1"."ORDERS": "PARTITION_BEFORE_2018"   0 KB        0 行
. . 导入了 "STUDY1"."ORDER_DETAILS": "PARTITION_BEFORE_2016"   425.7 KB  15000 行
. . 导入了 "STUDY1"."ORDER_DETAILS": "PARTITION_BEFORE_2017"   426.1 KB
```

```
15000 行
. . 导入了 "STUDY1"."ORDER_DETAILS":"PARTITION_BEFORE_2018" 0 KB 0 行
处理对象类型 SCHEMA_EXPORT/TABLE/COMMENT
处理对象类型 SCHEMA_EXPORT/PACKAGE/PACKAGE_SPEC
处理对象类型
SCHEMA_EXPORT/PACKAGE/COMPILE_PACKAGE/PACKAGE_SPEC/ALTER_PACKAGE_SPEC
处理对象类型 SCHEMA_EXPORT/VIEW/VIEW
处理对象类型 SCHEMA_EXPORT/PACKAGE/PACKAGE_BODY
处理对象类型 SCHEMA_EXPORT/TABLE/INDEX/INDEX
处理对象类型 SCHEMA_EXPORT/TABLE/CONSTRAINT/CONSTRAINT
处理对象类型 SCHEMA_EXPORT/TABLE/INDEX/STATISTICS/INDEX_STATISTICS
处理对象类型 SCHEMA_EXPORT/TABLE/CONSTRAINT/REF_CONSTRAINT
处理对象类型 SCHEMA_EXPORT/TABLE/TRIGGER
处理对象类型 SCHEMA_EXPORT/TABLE/STATISTICS/TABLE_STATISTICS
处理对象类型 SCHEMA_EXPORT/STATISTICS/MARKER
处理对象类型 SCHEMA_EXPORT/POST_SCHEMA/PROCOBJ
作业 "SYSTEM"."SYS_IMPORT_FULL_01" 已于 星期六 4 月 29 12：08：17 2017 elapsed 0 00：00：03 成功完成
```

注意：expdp 和 impdp 有很多选择项，可以通过 expdp -help 和 impdp -help 获取更多的帮助。

练习

一、判断题

1. Oracle 的备份方式分为物理备份和逻辑备份两大类。（ ）
2. Oracle 的备份过期是指备份文件丢失了，但元数据信息还在。（ ）
3. 数据库的完全恢复是恢复到数据库失败时的状态，数据没有损失。（ ）
4. 脱机备份可以作为完全恢复的依据。（ ）
5. 数据导出与导入可以作为完全恢复的依据。（ ）

二、单项选择题

1. 闪回恢复数据库的语句是（   ）

   A. flashback database   B. recovery database
   C. restore database   C. rman

2. 表示 Oracle 快速恢复区位置的系统参数是（   ）

   A. db_recovery_file_dest_size   B. db_recovery_file_dest
   C. db_flashback_retention_target   C. log_archive_dest

3. 只恢复 CDB 根数据库的命令是（   ）

   A.recover database;   B.recover pluggable database root;

C.recover database root； D.restore database root；

4. 删除失效的镜像复制文件（　　）

A.delete backup； B.delete expired backup；

C.delete copy； D.delete expired copy；

5. 在 restore database 命令执行之前，必须要执行命令（　　）启动数据库，只读取控制文件，不读取数据文件。

A.startup mount； B.startup nomount；

C.startup force； D.startup；

三、填空题

1. RMAN 执行批处理命令的开始关键词是_____。
2. 联机备份的前题是数据库必须工作在_____日志模式下，备份的时候数据库必须是打开的。
3. Oracle 的备份方式分为_____和_____两大类。
4. 数据库的物理备份是指将数据库的各种操作系统中的文件（如_____、_____、_____和系统参数文件等）从一处复制到另一处的过程。
5. 脱机备份时必须_____数据库，所以也叫冷备份。

四、简答题

1. Oracle 的物理备份与逻辑备份有什么区别？
2. restore database 和 recover database 有何区别？
3. rman 的备份恢复和 flashback 的备份恢复有什么区别？
4. 什么是完全恢复，什么是不完全恢复，各有何应用场景？

训练任务

1. 脱机备份和恢复

培养能力	工程能力、设计/开发解决方案		
掌握程度	★★★	难度	中
结束条件	编写脱机备份和恢复脚本		
训练内容：脱机备份时必须关闭数据库，所以也叫冷备份。训练要求：编写文件复制的 Linux 脚本，具有两个功能。功能 1 是备份功能，即通过脚本可以将所要备份的文件复制到备份目录；功能 2 是恢复功能，即将备份目录中的文件复制回到原始目录。需要复制的文件有：整个数据库的所有数据文件、控制文件、联机重做日志文件、归档日志文件（可选）。注意，在备份与恢复之前，必须关闭数据库。			

2. 自动联机备份

培养能力	工程能力、设计/开发解决方案		
掌握程度	★★★★★	难度	高
结束条件	看到备份后生成的自动备份文件表示实验成功		

训练内容:
数据库的备份是非常重要而繁琐的工作,做一个自动定时备份能节省管理员的日常备份工作量。本实验要求以每周 7 天为一个循环,每天 23 点钟都作一个备份:每周日 23 点钟作一次针对整个数据库的 0 级完整备份,每周的其余 6 天(周一到周六)每天都作一次 1 级增量备份,每次备份之后都要清除过期和失效的备份和归档日志

3. 数据迁移

培养能力	工程能力、设计/开发解决方案		
掌握程度	★★★★★	难度	高
结束条件	将一个 PDB 数据库迁移到其他服务器中		

训练内容:
数据迁移是指将数据从一个服务器迁移到另一个服务器的过程。总体的过程是在源服务器端运行 expdp 命令导出数据,在目的服务器端运行 impdp 命令导入数据

第 14 章 小型商品销售系统

本章目标

知识点	理解	掌握	应用
1.熟悉 E-R 图到关系模型的转换方法	✓	✓	
2.掌握系统所需的表空间和用户的创建	✓	✓	
3.掌握指定表名、列名、数据类型约束的方法	✓	✓	✓
4.掌握分区表的创建方法	✓		
5.掌握视图、普通存储过程和带参数存储过程	✓	✓	✓
6.掌握触发器的创建和使用	✓	✓	✓
7.掌握基于 Oracle+JSP 开发一套 Web 管理系统的过程和方法	✓	✓	✓

训练任务

- 本章模拟一个小型商品销售系统,通过需求分析,然后绘制出 E-R 模型图和数据表关系图。在 Oracle 12c 中创建表空间和用户、创建普通表、分区表和视图、编写程序包和存储过程,通过触发器计算订单的应收货款金额。数据库设计完成后,还设计出了一个基于 JSP 调用 Oracle 的 Web 管理系统。这样就从头到尾实现了基于 Oracle+JSP 完成一个小型项目的全过程。

知识能力点

能力点 \ 知识点	知识点 1	知识点 2	知识点 3	知识点 4	知识点 5	知识点 6	知识点 7
工程知识		✓	✓	✓	✓	✓	✓
问题分析	✓	✓	✓	✓	✓	✓	✓
设计/开发解决方案	✓						✓
研究	✓						✓
使用现代工具		✓	✓	✓	✓	✓	
工程与社会	✓						
环境和可持续发展		✓	✓	✓	✓	✓	
职业规范							✓
个人和团队	✓						✓
沟通	✓	✓	✓	✓	✓	✓	✓
项目管理	✓				✓	✓	
终身学习	✓						✓

14.1 小型商品销售系统 E-R 模型

本系统的应用场景是模拟一个企业销售其商品的过程。该企业的销售活动通过订单进行，企业员工要销售产品给客户，必须要新开一张订单，在订单中记录销售过程的关键信息。比如哪个客户购买了哪些产品，订单的经手员工是谁，何时销售的，订单的应收款是多少等等。

14.1.1 实体模型

根据应用场景分析，共有 3 个原始的实体（Entity），它们是部门、员工和产品。

部门（DEPARTMENTS）：部门包括部门 ID（DEPARTMENT_ID）和部门名称（DEPARTMENT_NAME），如图 14-1。

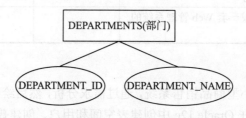

图 14-1　DEPARTMENTS 实体

员工（EMPLOYEES）：员工包括员工 ID（EMPLOYEE_ID），姓名（NAME），照片（PHOTO），工资（SALARY）等。员工的属性中还应包括员工所属的部门 ID（DEPARTMENT_ID），部门 ID 不能为空，表示员工必须属于某一个部门。员工的属性中还有员工的上司（MANAGER_ID），该属性可以为空，表示没有上司。员工的实体如图 14-2。

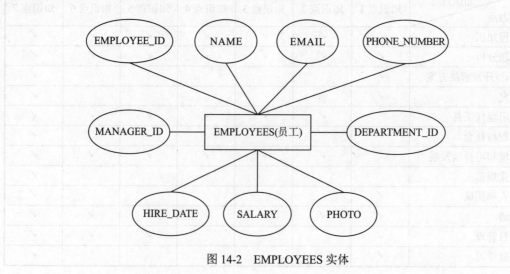

图 14-2　EMPLOYEES 实体

产品(PRODUCTS)：产品包括 3 个属性：产品 ID(PRODUCT_ID)，产品名称(PRODUCT_NAME)和产品类型(PRODUCT_TYPE)，见图 14-3。

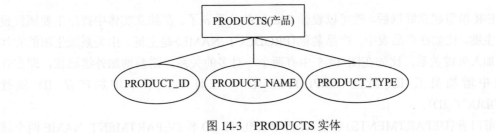

图 14-3　PRODUCTS 实体

14.1.2　实体联系模型

企业员工的工作是销售产品，因此员工和产品之间就有一个"销售"的联系，如图 14-4 所示，员工与产品之间的关系是多对多的关系，如图 14-4。

图 14-4　员工与产品销售关系简图

考虑到销售活动中有一些重要属性，比如折扣，客户信息，销售时间，产品的销售数量和销售价格，我们把销售关系细分为订单和订单详单两个实体，订单中存储：订单 ID(ORDER_ID)、折扣(DISCOUNT)、客户信息(CUSTOMER_NAME，CUSTOMER_TEL)、订单时间(ORDER_DATE)以及应收货款总额(TRADE_RECEIVABLE)，订单详单中存储订单详单的 ID，以及订单中的全部产品信息，包括：销售数量(PRODUCT_NUM)和销售价格(PRODUCT_PRICE)，见图 14-5。

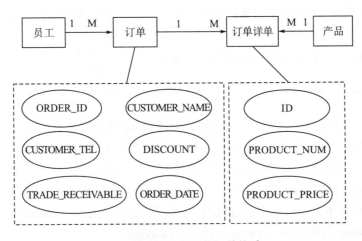

图 14-5　员工的订单关系

14.2 数据表的设计

E-R 模型建立好以后，就可以设计 Oracle 的关系表了。在独立实体中找出主要属性设置为主键，比如在产品表中，产品名称(PRODUCT_NAME)是主键。由关系派生出的实体中要加入外键关系，比如在图 14-5 中有两个一对多的关系，需要增加外键属性，即在订单表中增加员工 ID 属性(EMPLOYEE_ID)，在订单详单中增加产品 ID 属性(PRODUCT_ID)。

部门表(DEPARTMENTS)包括 DEPARTMENT_ID 和 DEPARTMENT_NAME 两个属性，见表 14-1。

表 14-1 部门表 DEPARTMENTS

字段名	数据类型	可以为空	注释
DEPARTMENT_ID	NUMBER(6, 0)	NO	部门 ID，主键
DEPARTMENT_NAME	VARCHAR2(40 BYTE)	NO	部门名称，非空

产品表 PRODUCTS 包括产品 ID(PRODUCT_ID)、产品名称(PRODUCT_NAME)和产品类型(PRODUCT_TYPE)，见表 14-2。注意产品表中的产品类型只能取值：耗材，手机，电脑。

表 14-2 产品表 PRODUCTS

字段名	数据类型	可以为空	注释
PRODUCT_ID	VARCHAR2(40 BYTE)	NO	产品 ID 号，产品表的主键
PRODUCT_NAME	VARCHAR2(40 BYTE)	NO	产品名称
PRODUCT_TYPE	VARCHAR2(40 BYTE)	NO	产品类型，只能取值：耗材，手机，电脑

员工表 EMPLOYEES 包括员工的属性，要注意 MANAGER_ID 是员工的上司，是员工表 EMPOLYEE_ID 的外键，MANAGER_ID 不能等于 EMPLOYEE_ID，即员工的领导不能是自己。主键删除时 MANAGER_ID 设置为空值。见表 14-3。

订单表 ORDERS 是订货信息，见表 14-4。表中 TRADE_RECEIVABLE 是订单的应收货款，是从订单详单表中自动汇总过来的。计算公式是：Trade_Receivable= sum(订单详单表.Product_Num*订单详单表.Product_Price)-Discount。

表 14-3 员工表 EMPLOYEES

字段名	数据类型	可以为空	注释
EMPLOYEE_ID	NUMBER(6, 0)	NO	员工 ID，员工表的主键
NAME	VARCHAR2(40 BYTE)	NO	员工姓名，不能为空，创建不唯一 B 树索引
EMAIL	VARCHAR2(40 BYTE)	YES	电子信箱
PHONE_NUMBER	VARCHAR2(40 BYTE)	YES	电话
HIRE_DATE	DATE	NO	雇佣日期
SALARY	NUMBER(8, 2)	YES	月薪，必须>0
MANAGER_ID	NUMBER(6, 0)	YES	员工的上司，员工表 EMPOLYEE_ID 的外键，MANAGER_ID 不能等于 EMPLOYEE_ID
DEPARTMENT_ID	NUMBER(6, 0)	YES	员工所在部门，是部门表 DEPARTMENTS 的外键
PHOTO	BLOB	YES	员工照片

表 14-4 订单表 ORDERS

字段名	数据类型	可以为空	注释
ORDER_ID	NUMBER(10, 0)	NO	订单编号，主键，值来自于序列：SEQ_ORDER_ID
CUSTOMER_NAME	VARCHAR2(40 BYTE)	NO	客户名称，B 树索引
CUSTOMER_TEL	VARCHAR2(40 BYTE)	NO	客户电话
ORDER_DATE	DATE	NO	订货日期，应该采用分区存储方式
EMPLOYEE_ID	NUMBER(6, 0)	NO	订单经手人，员工表 EMPLOYEES 的外键
DISCOUNT	Number(8, 2)	YES	订单整体优惠金额。默认值为 0
TRADE_RECEIVABLE	Number(8, 2)	YES	订单应收货款，默认为 0

订单详单表 ORDER_DETAILS 包含订单中全部产品的信息，见表 14-5。要注意，软件系统要通过触发器将产品的金额汇总到订单表中的 TRADE_RECEIVABLE 属性。

表 14-5 订单详单表 ORDER_DETAILS

字段名	数据类型	可以为空	注释
ID	NUMBER(10, 0)	NO	本表的主键，值来自于序列：SEQ_ORDER_DETAILS_ID
ORDER_ID	NUMBER(10, 0)	NO	所属的订单号，订单表 ORDERS 的外键
PRODUCT_ID	VARCHAR2(40 BYTE)	NO	产品 ID，是产品表 PRODUCTS 的外键
PRODUCT_NUM	NUMBER(8, 2)	NO	产品销售数量，必须>0
PRODUCT_PRICE	NUMBER(8, 2)	NO	产品销售价格

为了使用触发器计算订单的应收货款,需要创建一个临时表 ORDER_ID_TEMP,这个表的目的是暂时存储 ORDER_ID。

表 14-6　订单 ID 临时表 ORDER_ID_TEMP

字段名	数据类型	可以为空	注释
ORDER_ID	NUMBER(10,0)	NO	主键

当所有表创建成功后,通过 SQL-Developer 的 Data Modeler 工具生成数据库关系图,见图 14-6。打开 SQL-Developer 的 Data Modeler 工具方法是,从菜单:"文件"→"Data Modeler"→"导入"→"数据字典",然后根据提示操作就可以了。

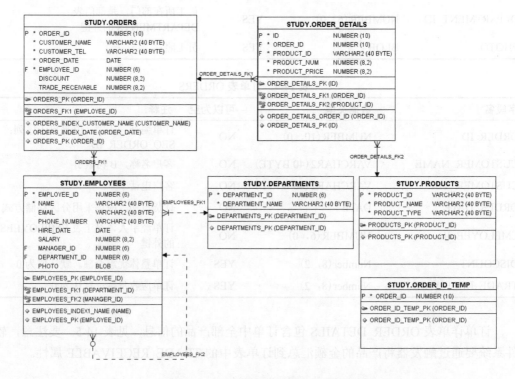

图 14-6　商品销售系统的数据库关系图

14.3　用户创建与空间分配

表的结构设计好之后,还要考虑用户和空间分配问题。我们需要为系统新建一个用户(STUDY),在磁盘空间方面,考虑将数据存储在 PDB 上,这里选择 Oracle 12c 安装时默认的 PDBORCL 数据库,另外,还需要为销售系统创建一个新表空间 USERS02 用于存储订单记录。

下面是 SYSTEM 用户创建的一个表空间 USERS02 的命令,注意给 USERS02 表空间分配了两个数据文件：pdbtest_users02_1.dbf 和 pdbtest_users02_2.dbf，这两个数据文件初始大小都是 100M，所以表空间的初始大小是 200M。

```
CREATE TABLESPACE Users02
DATAFILE
'/home/oracle/app/oracle/oradata/orcl/pdborcl/pdbtest_users02_1.dbf'
  SIZE 100M AUTOEXTEND ON NEXT 256M MAXSIZE UNLIMITED,
'/home/oracle/app/oracle/oradata/orcl/pdborcl/pdbtest_users02_2.dbf'
  SIZE 100M AUTOEXTEND ON NEXT 256M MAXSIZE UNLIMITED
EXTENT MANAGEMENT LOCAL SEGMENT SPACE MANAGEMENT AUTO;
```

表空间 USERS02 创建完成之后，就可以创建用户了。我们为本系统设计的用户名称是 STUDY。用户创建之后，给用户 STUDY 分配表空间 USERS 和 USERS02 的使用配额，再分配角色 CONNECT 和 RESOURCE，以便该用户可以连接数据库，可以创建资源(表，过程，序列等资源对象)，最后再分配一个系统权限："CREATE VIEW"，以便该用户可以创建视图。

```
CREATE USER STUDY IDENTIFIED BY 123
DEFAULT TABLESPACE "USERS"
TEMPORARY TABLESPACE "TEMP";

-- QUOTAS
ALTER USER STUDY QUOTA UNLIMITED ON USERS;
ALTER USER STUDY QUOTA UNLIMITED ON USERS02;

-- ROLES
GRANT "CONNECT" TO STUDY WITH ADMIN OPTION;
GRANT "RESOURCE" TO STUDY WITH ADMIN OPTION;
ALTER USER STUDY DEFAULT ROLE "CONNECT", "RESOURCE";

-- SYSTEM PRIVILEGES
GRANT CREATE VIEW TO STUDY WITH ADMIN OPTION;
```

到现在，我们有了两个表空间 USERS(原有的)和 USERS02(新建的)。这里我们假定企业的销售订单非常多，订单表和订单详单表的记录数量非常大，可能上千万条记录，所以我们把 ORDERS 和 ORDER_DETAILS 两张表设计为基于 ORDER_DATE 的分区表，以加快局部时间范围的查询速度。存储空间的分配规划如表 14-7 所示。

表 14-7 存储空间分配

表	表空间 USERS	表空间 USERS02
DEPARTMENTS	存储全部数据	
EMPLOYEES	存储全部数据	
PRODUCTS	存储全部数据	
ORDERS	存储 2016 年之前(不含)的数据	存储 2016 年之后(含)的数据
ORDER_DETAILS	存储 2016 年之前(不含)的数据	存储 2016 年之后(含)的数据

14.4 创建表，约束和索引

用户和空间分配完成后，可以创建表，约束和索引了。创建表的命令是 CREATE TABLE，由于命令参数和选项很复杂，命令行数非常多，所以这里就不将创建表的 DDL 语句全部写出来，读者可以下载本书中完整代码。这里仅对 ORDERS 和 ORDER_DETAILS 两个表的创建作一些说明，ORDERS 表按分区存储，分区类型选择"RANG"范围分区，创建 ORDERS 表的部分语句是：

```
CREATE TABLE ORDERS
(
  ORDER_ID NUMBER(10, 0)NOT NULL
, CUSTOMER_NAME VARCHAR2(40 BYTE)NOT NULL
, CUSTOMER_TEL VARCHAR2(40 BYTE)NOT NULL
, ORDER_DATE DATE NOT NULL
, EMPLOYEE_ID NUMBER(6, 0)NOT NULL
, DISCOUNT NUMBER(8, 2)DEFAULT 0
, TRADE_RECEIVABLE NUMBER(8，2)DEFAULT 0
)
TABLESPACE USERS
PCTFREE 10 INITRANS 1
STORAGE (   BUFFER_POOL DEFAULT )
NOCOMPRESS NOPARALLEL
PARTITION BY RANGE (ORDER_DATE)
(
  PARTITION PARTITION_BEFORE_2016 VALUES LESS THAN (
    TO_DATE(' 2016-01-01 00：00：00', 'SYYYY-MM-DD HH24：MI：SS',
    'NLS_CALENDAR=GREGORIAN'))
  NOLOGGING
  TABLESPACE USERS
```

```
  PCTFREE 10
  INITRANS 1
  STORAGE
  (
    INITIAL 8388608
    NEXT 1048576
    MINEXTENTS 1
    MAXEXTENTS UNLIMITED
    BUFFER_POOL DEFAULT
  )
  NOCOMPRESS NO INMEMORY
    'PARTITION PARTITION_BEFORE_2017 VALUES LESS THAN (
    TO_DATE(' 2017-01-01 00: 00: 00', 'SYYYY-MM-DD HH24: MI: SS',
    'NLS_CALENDAR=GREGORIAN'))
  NOLOGGING
  TABLESPACE USERS02
  ...
);
```

在这个命令中，核心的语句是"PARTITION BY RANGE（ORDER_DATE）"，表示按订单时间 ORDER_DATE 的范围进行分区。分区 PARTITION_BEFORE_2016 存储订单时间小于 2016 年的订单记录，这个分区存储在表空间 USERS 中，而分区 PARTITION_BEFORE_2017 存储的是订单时间小于 2017 年的订单记录，由于有分区 PARTITION_BEFORE_2016 的存在，PARTITION_BEFORE_2017 分区实际只存储 2016 年一年的记录。

由于 ORDER_DETAILS 是 ORDERS 的从表，ORDER_DETAILS 的记录数量比 ORDERS 还多，所以 ORDER_DETAILS 表也必须分区存储，但由于 ORDER_DETAILS 表中没有订单时间，所以不能按时间分区进行存储。好在 ORACLE 提供了引用分区的功能，引用分区功能就是将外键连接到主键，将从表按主表的分区方案与主表存储在同一分区中。创建 ORDER_DETAILS 表的部分语句如下：

```
CREATE TABLE ORDER_DETAILS
(
  ID NUMBER(10, 0)NOT NULL
, ORDER_ID NUMBER(10, 0)NOT NULL
, PRODUCT_ID VARCHAR2(40 BYTE)NOT NULL
, PRODUCT_NUM NUMBER(8, 2)NOT NULL
, PRODUCT_PRICE NUMBER(8, 2)NOT NULL
, CONSTRAINT ORDER_DETAILS_FK1 FOREIGN KEY ( ORDER_ID )
  REFERENCES ORDERS ( ORDER_ID )
```

```
    ENABLE
  )
  TABLESPACE USERS
  PCTFREE 10 INITRANS 1
  STORAGE (   BUFFER_POOL DEFAULT )
  NOCOMPRESS NOPARALLEL
  PARTITION BY REFERENCE (ORDER_DETAILS_FK1)
  (
    PARTITION PARTITION_BEFORE_2016
    NOLOGGING
    TABLESPACE USERS --必须指定表空间,否则会将分区存储在用户的默认表空间中
    ...
    )
    NOCOMPRESS NO INMEMORY,
    PARTITION PARTITION_BEFORE_2017
    NOLOGGING
    TABLESPACE USERS02
    ...
    )
    NOCOMPRESS NO INMEMORY
  );
```

创建从表 ORDER_DETAILS 的语句中的关键语句是"PARTITION BY REFERENCE (ORDER_DETAILS_FK1)",表示利用外键 ORDER_DETAILS_FK1 关联到 ORDERS 表,使用 ORDERS 表的分区方案。

使用 CREATE GLOBAL TEMPORARY TABLE 可以创建全局临时表,创建临时表 ORDER_ID_TEMP 的语句如下:

```
-- DDL for Table ORDER_ID_TEMP
CREATE GLOBAL TEMPORARY TABLE "ORDER_ID_TEMP"
   (  "ORDER_ID" NUMBER(10, 0) NOT NULL ENABLE,
   CONSTRAINT "ORDER_ID_TEMP_PK" PRIMARY KEY ("ORDER_ID") ENABLE
   ) ON COMMIT DELETE ROWS ;
COMMENT ON TABLE "ORDER_ID_TEMP"  IS '用于触发器存储临时 ORDER_ID';
```

注意:ORDER_ID_TEMP 表存储在临时表空间中,而不是存储在 USERS 和 USERS02 表空间中。

14.5 创建触发器、序列和视图

本销售系统中订单表 ORDERS 的 TRADE_RECEIVABLE 是应收货款，是自动计算字段，它的计算公式是：

Trade_Receivable=sum(ORDER_DETAILS.Product_Num*ORDER_DETAILS.Product_Price)- Discount。从这个公式可以看出，当修改订单表的折扣 Discount 或者对订单详单表 ORDER_DETAILS 作任何增加，修改，删除的操作后，都可能需要重新计算 Trade_Receivable。所以我们设计了以下 3 个触发器来实现这个功能，完整的触发器代码如下：

```
--------------------------------------------------------
-- DDL for Trigger ORDERS_TRIG_ROW_LEVEL
--------------------------------------------------------
CREATE OR REPLACE EDITIONABLE TRIGGER "ORDERS_TRIG_ROW_LEVEL"
BEFORE INSERT OR UPDATE OF DISCOUNT ON "ORDERS"
FOR EACH ROW  --行级触发器
declare
  m number(8,2);
BEGIN
  if inserting then
     : new.TRADE_RECEIVABLE: = -: new.discount;
  else
     SELECT sum(PRODUCT_NUM*PRODUCT_PRICE) INTO m FROM
        ORDER_DETAILS WHERE ORDER_ID=: old.ORDER_ID;
     if m is null then
        m: =0;
     end if;
     : new.TRADE_RECEIVABLE: = m -: new.discount;
  end if;
END;
/


--------------------------------------------------------
-- DDL for Trigger ORDER_DETAILS_ROW_TRIG
--------------------------------------------------------
CREATE OR REPLACE EDITIONABLE TRIGGER "ORDER_DETAILS_ROW_TRIG"
AFTER DELETE OR INSERT OR UPDATE  ON ORDER_DETAILS
```

```
FOR EACH ROW
BEGIN
  --DBMS_OUTPUT.PUT_LINE(: NEW.ORDER_ID);
  IF: NEW.ORDER_ID IS NOT NULL THEN
    MERGE INTO ORDER_ID_TEMP A
    USING (SELECT 1 FROM DUAL)B
    ON (A.ORDER_ID=: NEW.ORDER_ID)
    WHEN NOT MATCHED THEN
      INSERT (ORDER_ID)VALUES(: NEW.ORDER_ID);
  END IF;
  IF: OLD.ORDER_ID IS NOT NULL THEN
    MERGE INTO ORDER_ID_TEMP A
    USING (SELECT 1 FROM DUAL)B
    ON (A.ORDER_ID=: OLD.ORDER_ID)
    WHEN NOT MATCHED THEN
      INSERT (ORDER_ID)VALUES(: OLD.ORDER_ID);
  END IF;
END;
/
-----------------------------------------------------------
--  DDL for Trigger ORDER_DETAILS_SNTNS_TRIG
-----------------------------------------------------------

CREATE OR REPLACE EDITIONABLE TRIGGER "ORDER_DETAILS_SNTNS_TRIG"
AFTER DELETE OR INSERT OR UPDATE ON ORDER_DETAILS
declare
  m number(8, 2);
BEGIN
  FOR R IN (SELECT ORDER_ID FROM ORDER_ID_TEMP)
  LOOP
    --DBMS_OUTPUT.PUT_LINE(R.ORDER_ID);
    SELECT sum(PRODUCT_NUM*PRODUCT_PRICE)INTO m FROM ORDER_DETAILS
      WHERE ORDER_ID=R.ORDER_ID;
    if m is null then
      m: =0;
    end if;
    UPDATE ORDERS SET TRADE_RECEIVABLE = m - discount
      WHERE ORDER_ID=R.ORDER_ID;
```

```
    END LOOP;
    DELETE FROM ORDER_ID_TEMP;  --这句话很重要，否则可能一直不释放空间，
                                --后继插入会非常慢。
END;
/
```

在插入订单数据的时候，需要用到自增的 ID，所以需要设计两个序列。SEQ_ORDER_ID 用于给 ORDERS 表的 ORDER_ID 赋值，SEQ_ORDER_DETAILS_ID 用于给 ORDER_DETAILS 表的 ID 赋值。

```
--------------------------------------------------------
--  DDL for Sequence SEQ_ORDER_ID
--------------------------------------------------------

CREATE SEQUENCE  "SEQ_ORDER_ID"  MINVALUE 1 MAXVALUE 9999999999
  INCREMENT BY 1 START WITH 1 CACHE 2000 ORDER  NOCYCLE
  NOPARTITION ;
--------------------------------------------------------
--  DDL for Sequence SEQ_ORDER_DETAILS_ID
--------------------------------------------------------

CREATE SEQUENCE  "SEQ_ORDER_DETAILS_ID"  MINVALUE 1 MAXVALUE
  9999999999 INCREMENT BY 1 START WITH 1 CACHE 2000
  ORDER  NOCYCLE  NOPARTITION ;
```

为了方便地查询订单详单中的产品，需要设计一个视图 VIEW_ORDER_DETAILS，这个视图的代码如下：

```
--------------------------------------------------------
--  DDL for View VIEW_ORDER_DETAILS
--------------------------------------------------------

CREATE OR REPLACE VIEW VIEW_ORDER_DETAILS
AS SELECT
   d.ID,
   o.ORDER_ID,
   o.CUSTOMER_NAME, o.CUSTOMER_TEL, o.ORDER_DATE,
   d.PRODUCT_ID,
   p.PRODUCT_NAME,
   p.PRODUCT_TYPE,
   d.PRODUCT_NUM,
   d.PRODUCT_PRICE
FROM ORDERS o, ORDER_DETAILS d, PRODUCTS p WHERE d.ORDER_ID=o.ORDER_ID
   and d.PRODUCT_ID=p.PRODUCT_ID;
```

14.6 创建程序包、函数和过程

本系统设计了一些函数和过程，放在程序包 MyPack 中。Get_SaleAmount()函数的作用是计算并返回一个部门的销售总金额。Get_Employees()过程的作用是打印出某个员工的所有下属，含下属的下属。GET_EMPLOYEEBYPAGE_P()过程的作用是分页查询员工表，返回游标变量。Calc_All_TradeReceivable()过程的作用是更新订单表中每个订单的应收款。代码分为包头和包体两部分。

包头如下：

```sql
create or replace PACKAGE MyPack IS
  TYPE refcur is ref cursor;
  FUNCTION Get_SaleAmount(V_DEPARTMENT_ID NUMBER) RETURN NUMBER;
  PROCEDURE Get_Employees(V_EMPLOYEE_ID NUMBER);
  FUNCTION Get_EmployeeByPage(v_pageidx number,
    v_pagesize number) return refcur;
  PROCEDURE Get_EmployeeByPage_P(v_pageidx number,
    v_pagesize number, rc out refcur);
  PROCEDURE Calc_All_TradeReceivable;
END MyPack;
```

包体如下：

```sql
create or replace PACKAGE BODY MyPack IS
/*
取得某个部门的销售总额
*/
  FUNCTION Get_SaleAmount(V_DEPARTMENT_ID NUMBER) RETURN NUMBER
  AS
    N NUMBER(20,2);    --注意，订单ORDERS.TRADE_RECEIVABLE 的类型是
                       --NUMBER(8,2)，汇总之后，数据要大得多。
  BEGIN
    SELECT SUM(O.TRADE_RECEIVABLE) into N
      FROM ORDERS O, EMPLOYEES E
      WHERE O.EMPLOYEE_ID=E.EMPLOYEE_ID AND
      E.DEPARTMENT_ID =V_DEPARTMENT_ID;
    RETURN N;
  END;
/*
取得某个员工的所有下属(含下属的下属)
```

```
*/
PROCEDURE GET_EMPLOYEES(V_EMPLOYEE_ID NUMBER)
AS
  LEFTSPACE VARCHAR(2000);
begin
  --通过 LEVEL 判断递归的级别
  LEFTSPACE:=' ';
  --使用游标
  for v in
    (SELECT LEVEL, EMPLOYEE_ID, NAME, MANAGER_ID FROM employees
      START WITH EMPLOYEE_ID = V_EMPLOYEE_ID
      CONNECT BY PRIOR EMPLOYEE_ID = MANAGER_ID)
  LOOP
    DBMS_OUTPUT.PUT_LINE(LPAD(LEFTSPACE, (V.LEVEL-1)*4, ' ')||
      V.EMPLOYEE_ID||' '||v.NAME);
  END LOOP;
END;

/*
分页方法获取员工表的函数
*/
FUNCTION Get_EmployeeByPage(v_pageidx number, v_pagesize number)
return refcur
IS
  rc refcur;
  v_offset number;
  v_sql varchar2(2000);
BEGIN
  v_offset:=v_pagesize*(v_pageidx-1);
  v_sql:= 'select * from employees offset :v_offset rows
    fetch next :v_pagesize rows only';
  open rc for v_sql using v_offset, v_pagesize;
  return rc;
END;

/*
分页方法获取员工表的存储过程
*/
```

```
    PROCEDURE Get_EmployeeByPage_P(v_pageidx number, v_pagesize
    number, rc out refcur)
    IS
      v_offset number;
      v_sql varchar2(2000);
    BEGIN
      v_offset: =v_pagesize*(v_pageidx-1);
      v_sql: = 'select * from employees offset: v_offset rows
      fetch next: v_pagesize rows only';
      open rc for v_sql using v_offset, v_pagesize;
    END;
/*
计算所有订单的应收款
*/
    PROCEDURE Calc_All_TradeReceivable
    IS
        m number(10, 2);
    BEGIN
        FOR R IN (SELECT ORDER_ID FROM ORDERS)
        LOOP
          select sum(PRODUCT_NUM*PRODUCT_PRICE)into m from
            ORDER_DETAILS where ORDER_ID=R.ORDER_ID;
          if m is null then
            m: =0;
          end if;
          update ORDERS SET TRADE_RECEIVABLE=m-DISCOUNT
            WHERE ORDER_ID=R.ORDER_ID;
        END LOOP;
    END;
END MyPack;
```

14.7 数据库测试

所有对象的创建工作完成之后，需要插入测试数据，然后就可以进行一些必要的测试。测试的时候以 STUDY 用户身份登录。

【示例 14-1】调用程序包中的函数

调用 MyPack.Get_SaleAmount()函数，计算部门 11 的销售总金额。

```
$ sqlplus study/***@pdborcl
SQL> SELECT MyPack.Get_SaleAmount(11)AS 部门11应收金额 FROM dual;
部门11应收金额
--------------
      70327453
```

函数 Get_SaleAmount()测试方法：

```
SELECT count(*)FROM orders;
SELECT MyPack.Get_SaleAmount(11)AS 部门11应收金额,
  MyPack.Get_SaleAmount(12)AS 部门12应收金额 FROM dual;
```

【示例14-2】调用程序包中的过程

在 SQL-DEVELOPER 中调用 MYPACK.Get_Employees()过程，输出 1 号员工的所有下属及子下属。

```
set serveroutput on
DECLARE
  V_EMPLOYEE_ID NUMBER;
BEGIN
  V_EMPLOYEE_ID: = 1;
  MYPACK.Get_Employees (V_EMPLOYEE_ID => V_EMPLOYEE_ID);
  V_EMPLOYEE_ID: = 11;
  MYPACK.Get_Employees (V_EMPLOYEE_ID => V_EMPLOYEE_ID);
END;
/
输出的结果是：
1 李董事长
    11 张总
        111 吴经理
        112 白经理
    12 王总
        121 赵经理
        122 刘经理
11 张总
    111 吴经理
112 白经理
```

【示例14-3】调用过程分页查询员工表

本示例调用过程 MYPACK.Get_EmployeeByPage_P 查询员工表的第 1 页（每页 3 个员工）。

```
SQL>
set serveroutput on
```

```
declare
rc sys_refcursor;
lrow employees%rowtype;
begin
MYPACK.Get_EmployeeByPage_P(1, 3, rc);
   loop
     FETCH rc INTO lrow;
     exit when rc%notfound;
     DBMS_OUTPUT.put_line (lrow.employee_id || ' '|| lrow.name);
   end loop;
   CLOSE rc;      --关闭游标
 end;
 /
1   李董事长
11  张总
111 吴经理
```

【示例14-4】直方图统计前后的对比查询

本示例要查询 ORDERS 表，查询 TRADE_RECEIVABLE 在 10000 到 11000 之间的订单记录，由于对字段 TRADE_RECEIVABLE 没有建立索引，因此最坏的情况下，要做全表搜索，最可能的情况是做动态采样（Dynamic Statistics）。直方图统计可以更准确地找到比动态采样更准确的执行计划。

下面首先查询统计前的执行计划，为了对比直方图统计前后的结果，这里先调用 dbms_stats.delete_schema_stats 删除以前的统计信息：

```
$ sqlplus study/***@pdborcl
SQL> EXEC dbms_stats.delete_schema_stats(User);
PL/SQL procedure successfully completed.
SQL> SELECT ORDER_ID, ORDER_DATE, TRADE_RECEIVABLE
2 FROM ORDERS WHERE TRADE_RECEIVABLE BETWEEN 10000 AND 11000;
  ORDER_ID  ORDER_DATE           TRADE_RECEIVABLE
---------- -------------------- ----------------
      8040 2015-03-02 00:00:00         10993.68
      4619 2016-04-30 00:00:00         10746.37
      4723 2016-04-14 00:00:00         10851.29
      5641 2016-03-03 00:00:00         10826.78
       589 2016-04-20 00:00:00         10996.67
SQL> explain plan for SELECT ORDER_ID, ORDER_DATE, TRADE_RECEIVABLE
2 FROM ORDERS WHERE TRADE_RECEIVABLE BETWEEN 10000 AND 11000;
Explained.
```

```
SQL> SELECT * FROM table(dbms_xplan.display);
Plan hash value: 3601361376
---------------------------------------------------------------------
| Id | Operation           | Name   | Rows | Bytes | Cost (%CPU)| Time     | Pstart| Pstop |
---------------------------------------------------------------------
|  0 | SELECT STATEMENT    |        |   33 |  1155 |   547   (0)| 00:00:01 |       |       |
|  1 |  PARTITION RANGE ALL|        |   33 |  1155 |   547   (0)| 00:00:01 |     1 |     3 |
|* 2 |   TABLE ACCESS FULL | ORDERS |   33 |  1155 |   547   (0)| 00:00:01 |     1 |     3 |

Predicate Information (IDENTIFIED BY operation id):
---------------------------------------------------
   2 - filter("TRADE_RECEIVABLE">=10000 AND "TRADE_RECEIVABLE"<=11000)

Note
-----
   - dynamic statistics used: dynamic sampling (level=2)
SQL>
```

从统计前的结果可以看出，实际 ORDERS 表共有 5 行记录满足条件，执行计划的 Rows 为 33 行，字节数为 1155，Cost 为 547，使用了动态采样。下面是使用了直方图统计的结果：

```
SQL> EXEC dbms_stats.gather_schema_stats(User, method_opt=>
  2  'for all columns SIZE 250', cascade=>TRUE);
PL/SQL procedure successfully completed.
SQL> explain plan for SELECT ORDER_ID, ORDER_DATE, TRADE_RECEIVABLE
  2  FROM ORDERS WHERE TRADE_RECEIVABLE BETWEEN 10000 AND 11000;
Explained.
SQL> SELECT * FROM table(dbms_xplan.display);
| Id | Operation           | Name   | Rows | Bytes | Cost (%CPU)| Time     | Pstart| Pstop |
---------------------------------------------------------------------
|  0 | SELECT STATEMENT    |        |    9 |   162 |   547   (0)| 00:00:01 |       |       |
|  1 |  PARTITION RANGE ALL|        |    9 |   162 |   547   (0)| 00:00:01 |     1 |     3 |
|* 2 |   TABLE ACCESS FULL | ORDERS |    9 |   162 |   547   (0)| 00:00:01 |
```

```
1 |   3 |


Predicate Information (IDENTIFIED BY operation id):
---------------------------------------------------

  2 - filter("TRADE_RECEIVABLE"<=11000 AND "TRADE_RECEIVABLE">=10000)
```

上述 dbms_stats.gather_schema_stats 的命令中"method_opt=>'for all columns SIZE 250'"参数的含义是对所有字段进行直方图统计，SIZE=250 表示 250 个分组桶(Bucket)。Bucket 可以理解为存储数据的容器，这个容器会按照数据的分布将数据尽量平均到各个桶里。表 14-8 进行了直方图统计前后的执行计划对比。

表 14-8　直方图统计前后的执行计划对比

直方图统计	动态采样	Rows	Bytes	Cost(%CPU)
统计前	级别 2	33	1155	547
统计后	未使用	9	162	547

从直方图统计后的结果可以看出，实际 ORDERS 表共有 5 行记录满足条件，执行计划的 Rows 为 9 行，字节数为 162，Cost 为 547，没有使用动态采样。因此可以得出结论：直观图统计显著改善了对非索引字段的查询计划。

在本例最后，通过查询 user_tables 查看每个用户表的统计情况：

```
SQL> COL TABLE_NAME FORMAT A20
SQL> COL TABLESPACE_NAME FORMAT A20
SQL> SELECT table_name, tablespace_name, num_rows FROM user_tables;
TABLE_NAME           TABLESPACE_NAME      NUM_ROWS
-------------------- -------------------- ----------
DEPARTMENTS          USERS                3
EMPLOYEES            USERS                7
PRODUCTS             USERS                9
ORDERS                                    10000
ORDER_DETAILS                             30000
ORDER_ID_TEMP
6 rows selected.
SQL>
```

从查询结果来看，各个表的总行数 NUM_ROWS 准确查询出来了，这有利于准确制订 SQL 语句的执行计划。另外，查询结果中 DEPARTMENTS、EMPLOYEES 和 PRODUCTS 的表空间都是 USERS，而其他表的表空间显示为空，这并不是说这些表没有占用表空间，而是其他表空间，比如 ORDERS 和 ORDER_DETAILS 占用了两个表空间：USERS 和 USERS02，而 ORDER_ID_TEMP 占用的是临时表空间。

14.8　应用程序开发

14.8.1　IDE 选择

"小型商品销售系统"采用 Java 语言开发，根据数据表生成实体类时需要在开发工具中配置 Datasource，有两款主流的开发工具可供选择，第一种是 Eclipse，但是由于数据库是 Oracle 12c，而目前 Eclipse 支持的 Oracle 数据库版本是 10i，不支持指定 URL 访问 Oracle 数据库的方式。第二种是 IDEA，它支持指定 URL 访问的方式，所以开发工具选择 IDEA，连接 Datasource 的 URL 是：

```
jdbc:oracle:thin:@//OracleServer:1521/pdborcl。
```

其中 OracleServer 是 Oracle 12c 的主机地址。

14.8.2　程序目录结构和通用模块

1）目录结构

程序的目录结构如图 14-7 所示。

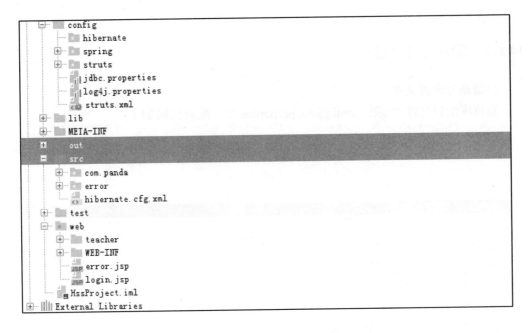

图 14-7　程序目录结构

2)配置文件

表 14-9　配置文件详述

配置项	说明
Config	用于保存系统的配置文件，主要配置文件有连接数据库的参数、struts 的 action 管理、spring 的 bean 的获取以及日志管理
Lib	库文件目录
Src	用于保存系统实现的所有 Java 代码文件
Test	用于保存测试系统功能实现情况的 Java 代码文件
Web	用于保存系统实现的所有 Jsp 文件

3)通用模块

除 login.jsp 外，每个页面都有一段判定用户是否登录的代码，如果没有登录，则直接跳转到登录页面，代码如下：

```
<s: if test="#session.user==null">
    <script type="text/javascript">
        window.location = "/login.jsp";
    </script>
</s: if>
```

14.8.3　配置文件详述

1)数据库配置文件

数据库配置文件主要在 config/jdbc.properties 中，配置代码如下：

```
jdbc.driver=oracle.jdbc.OracleDriver
jdbc.url=jdbc:oracle:thin:@//OracleServer: 1521/pdborcl
jdbc.user=study
jdbc.password=123
jdbc.checkoutTimeout=2000
jdbc.initialPoolSize=6
jdbc.idleConnectionTestPeriod=30
jdbc.maxPoolSize=30
jdbc.minPoolSize=5
jdbc.maxStatements=100
```

上述代码中的 jdbc.driver 代表的含义是 Oracle 的数据库驱动名，Mysql 的驱动名是 com.mysql.jdbc.Driver，url 代表的是连接的数据库的 url，需要指出的是，此处的 pdborcl 代表的是服务名，user 代表用户名，password 指的是密码，checkoutTimeout 指的是当连接池用完时客户端调用 getConnection()后等待获取新连接的时间，此处指定 2000 毫秒，initialPoolSize 指的是初始化数据库连接池时连接的数量，此处指定为 6，

idleConnectionTestPeriod 指的是每隔多少秒检查所有连接池中的空闲连接，minPoolSize 指的是数据库连接池中的最小的数据库连接数，maxPoolSize 指的是数据库连接池中的最大的数据库连接数，maxStatements 指的是最多可以创建多少个 Statement 对象。

2) Log4j 日志文件

Log4j 日志管理主要用于输出系统运行时的一些日志信息，便于开发人员调试程序。该配置文件在 config/log4j.properties，主要参数及含义如下：

```
#debug->info->warn->error->fetal
#定义输出级别为 debug 级别，为最高级，同时定义输出的位置为控制台和文件中
log4j.rootLogger=info, console, file
#配置输出源为控制台
log4j.appender.console=org.apache.log4j.ConsoleAppender
log4j.appender.console.Target=System.out
#配置输出源的布局
log4j.appender.console.layout=org.apache.log4j.PatternLayout
#配置输出格式
log4j.appender.console.layout.ConversionPattern=
  [%c.%p]%m%d{yyyy-MM-dd HH: mm: ss, SSS}%n
#配置输出源，指定为到达一定大小就另外生辰一个文件
log4j.appender.file=org.apache.log4j.RollingFileAppender
#指定文件名
log4j.appender.file.File=E: /Log4jLog/Oracle.log
#指定文件大小
log4j.appender.file.MaxFileSize=10MB

log4j.appender.file.Threshold=ALL
log4j.appender.file.layout=org.apache.log4j.PatternLayout
log4j.appender.file.layout.ConversionPattern=
  [%p][%d{yyyy-MM-dd HH\: mm\: ss, SSS}][%c]%m%n
```

3) Spring 配置文件

本系统所有的类都是交给 Spring 管理，Spring 配置文件在 config/spring 下，有多个配置文件，但含义基本相同，本处主要讲 applicationContext-db.xml，在该 xml 中，主要配置了 Spring 和 Hibernate 的整合以及 Spring 的一些基本配置。

```
<context: property-placeholder location="classpath: jdbc.properties"/>
```

该行代码加载刚才的数据库配置文件，用于 c3p0 数据库连接池连接。

```
<property name="hibernateProperties">
<props>
    <prop key="hibernate.dialect">
      org.hibernate.dialect.OracleDialect</prop>
```

```xml
    <prop key="hibernate.show_sql">true</prop>
    <prop key="hibernate.format_sql">true</prop>
    <prop key="hibernate.hbm2ddl.auto">UPDATE</prop>
    <prop key="hibernate.connection.autocommit">false</prop>
    <!--getCurrentSession 的配置-->
    <prop key="hibernate.current_session_context_class">
      thread</prop>
</props>
```

上述代码主要配置 Hibernate 的一些常用属性,hibernate.dialect 指数据库方言是什么,即底层用什么数据库语句,show_sql 指是否需要展示 SQL 语句,只有 true 和 false 两个值,format_sql 指是否需要有格式化的输出 SQL 语句,同样只有 true 和 false 两个值,hbm2ddl.auto 代表数据表的生成策略,此处选用如果没有数据表就生成新的数据表,有的话不再生成策略,connection.autocommit 指的是是否自动提交,此处选择 false,即手动提交事务。

```xml
<aop: config>
    <aop: pointcut expression=
      "execution(* com.panda.serviceImpl.*.*(..))"
      id="exceptionAspj"/>
    <aop: advisor advice-ref="tx" pointcut-ref="exceptionAspj"/>
    <aop: aspect ref="exceptionTransaction">
        <aop: after-throwing method="throwException"
          throwing="e" pointcut-ref="exceptionAspj"/>
    </aop: aspect>
</aop: config>
```

上述代码是 Spring 中一个很关键的知识点,即 Spring 的 AOP 操作,此处主要是利用<aop:aspect ref="exceptionTransaction">来进行的统一的异常处理。

4)Struts 配置文件

Struts 的配置文件主要在 config/struts 下,需要指明的是,struts.xml 只能放在 config 的根目录下,不能放在 config/struts,在 struts.xml 中,主要代码如下:

```xml
<struts>
    <constant name="struts.objectFactory" value="spring" />
    <constant name="struts.devMode" value="true"/>
    <constant name="struts.multipart.saveDir" value="/tmp"/>
    <include file="struts/struts-login.xml"/>
    <include file="struts/struts-employee.xml"/>
    <include file="struts/struts-product.xml"/>
    <include file="struts/struts-order.xml"/>
    <package name="struts-global" extends="struts-default">
```

```xml
        <global-results>
          <result name="errHandler" type="chain">
            <param name="actionName">errorProcessor</param>
          </result>
        </global-results>
        <global-exception-mappings>
          <exception-mapping result="errHandler"
           exception="java.lang.Exception">
          </exception-mapping>
        </global-exception-mappings>
        <action name="errorProcessor"
         class="com.panda.error.ErrorProcessor">
          <result name="error">error.jsp</result>
        </action>
     </package>
</struts>
```

在上述代码中，首先用 include 指令引入了其他 xml 文件，其次是定义了全局的异常结果处理，即<global-result></global-result>标签内的部分，指定出错之后统一跳转到 error.jsp。

14.8.4　管理主界面与登录程序设计

1）设计管理主界面

本系统的管理主界面是 index.jsp，它的功能是显示商品销售系统的管理链接。index.jsp 的主界面如图 14-8 所示。

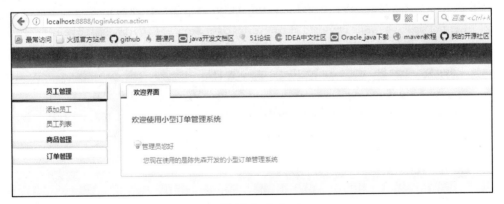

图 14-8　index.jsp

在 index.jsp 中使用框架将网页分为左右两侧，在左侧框架打开 left.jsp 网页，用于显示管理链接，在右侧框架中打开 right.jsp，用于显示链接网页的内容。left.jsp 用于显示界面的左侧部分，主要有以下链接：

表 14-10　left.jsp 的链接

管理项目	链接
添加员工	employyAdd.jsp
员工列表	employeeAction2.action
商品列表	productListAction.action
添加订单	orderAdd.jsp
订单管理	orderShow.action

2）用户登录程序设计

系统管理界面只能是管理员进入，管理员输入账号和密码时（默认为 admin，admin），后台将进行判断，判断密码和账号是否正确，如果正确，将进入管理页面，失败则返回登录页面。判断代码如下：

```java
if (username == null || password == null){
    addActionError("用户名或密码为空");
    return INPUT ;
}
if ("".equals(username)|| "".equals(password)){
    addActionError("用户名或密码为空");
    return INPUT ;
}
if ("admin".equals(username)&&"admin".equals(password)){
    request = ServletActionContext.getRequest();
    request.getSession().setAttribute("user", username);
    List<EmployeesEntity> lists = employeeService.queryEmployee();
    List<DepartmentsEntity> list2 =
      departmentService.queryDepartments();
    request.getSession().setAttribute("employees", lists);
    request.getSession().setAttribute("departments", list2);
    System.out.print(request.getSession().getAttribute("user"));
    return SUCCESS ;
} else
    return ERROR;
```

14.8.5　程序主要模块

1）员工管理模块设计

员工管理模块主要有如下功能：添加员工、删除员工、编辑员工以及查看员工详细信息。

第 14 章 小型商品销售系统

❖ 添加员工

添加员工页面为 employeeAdd.jsp,页面如图 14-9 所示。

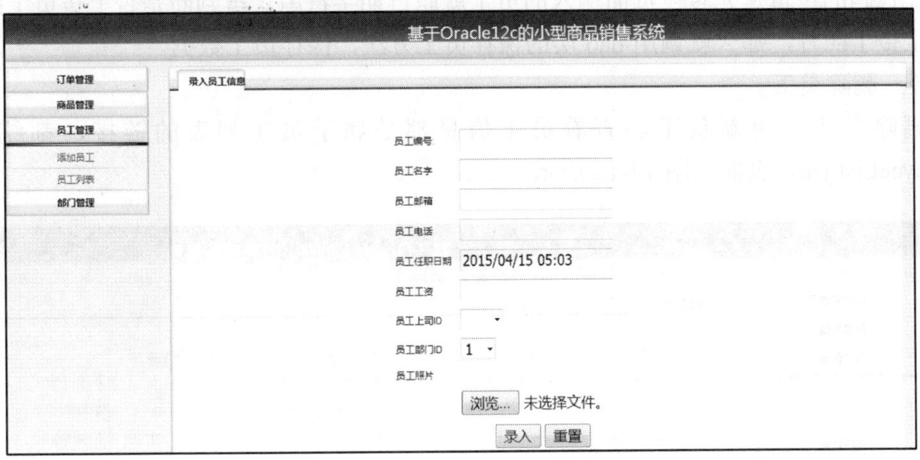

图 14-9 employeeAdd.jsp

在进行员工的添加之前,会对员工数据进行判断,查看是否合法,如果不合法,将会返回添加界面,如果合法,就将数据保存进数据库,关键代码如下:

```
Date date = new Date(hireDate);
SimpleDateFormat format = new SimpleDateFormat("yyyy-MM-dd");
format.format(date);
employeesEntity.setHireDate(date);
System.out.print(hireDate+"  " + date);
if (photo != null){
    FileInputStream in = new FileInputStream(photo);
    bytes = new byte[in.available()];
    i = in.read(bytes);
}

//如果有文件,并且成功写入 byte[]数组,那么 employeesEntity 对象的属性都赋值完成,
即可插入数据库
if (i > 0){
    employeesEntity.setPhoto(bytes);
    employeeService.addEmployee(employeesEntity);
    return  SUCCESS;
}
else {
    this.addActionError("请传入空缺参数");
```

```
        return INPUT;
}
```

此代码的逻辑是先转换页面输入的员工就职日期字符串，再判断是否上传员工照片，如果上传了照片，那么就调用 dao 层的保存员工方法，保存员工数据。

◇ 删除员工

删除员工、更新员工、查看员工信息都是属于员工列表的链接，都统一在 employeeList.jsp，页面如图 14-10 所示。

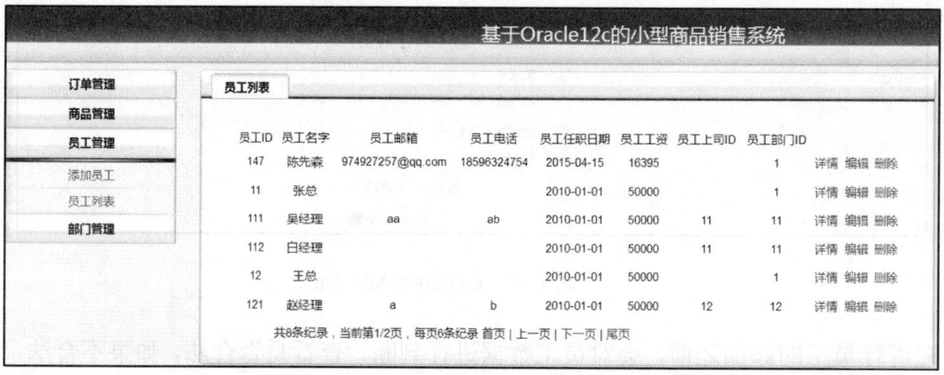

图 14-10 employeeList.jsp 的运行界面

页面首先需要分页显示企业员工，呈现的关键代码如下：

```
<table width="95%">
    <tr align="center">
    <td >员工 ID</td>
    <td >员工名字</td>
    <td >员工邮箱</td>
    <td >员工电话</td>
    <td >员工任职日期</td>
    <td >员工工资</td>
    <td >员工上司 ID</td>
    <td >员工部门 ID</td>
    </tr>
    <s: iterator value="#session.employees1" var="list">
     <tr align="center">
        <td><s: property value="#list.employeeId"/></td>
        <td><s: property value="#list.name"/></td>
        <td><s: property value="#list.email"/></td>
        <td><s: property value="#list.phoneNumber"/></td>
        <td><s: date name="#list.hireDate"
```

```
                    format="yyyy-MM-dd"/></td>
            <td><s: property value="#list.salary"/></td>
            <td><s: property value="#list.managerId"/></td>
            <td><s: property value="#list.departmentId"/></td>
            <td><a href="employeeAction3.action?
                employeeId=<s: property value="#list.employeeId"/>">
                详情</a></td>
            <td><a href="employeeAction5.action?
                employeeId=<s: property value="#list.employeeId"/>">
                编辑</a></td>
            <s: if test="#list.hasEmployee == 1">
              <td><a href="javascript: void(0)" onclick="fun1(); ">
                  删除</a></td>
            </s: if>
            <s: else>
              <td><a href="employeeAction7?employeeId=<s: property
                  value="#list.employeeId"/>">删除</a></td>
            </s: else>
          </tr>
    </s: iterator>
    <tr>
     <td colspan="6" align="center">
        共${pager.totalSize}条纪录,当前第
          ${pager.pageNow}/${pager.totalPage}页,每页
          ${pager.pageSize}条纪录
     <s: if test="#session.pager.hasPre">
            <a href="employeeAction2.action?pageNow=1">首页</a> |
            <a href="employeeAction2.action?
              pageNow=${pager.pageNow - 1}">上一页</a> |
            </s: if>
            <s: else>
             首页 | 上一页 |
            </s: else>
            <s: if test="#session.pager.hasNext">
            <a href="employeeAction2.action?
              pageNow=${pager.pageNow + 1}">下一页</a> |
            <a href="employeeAction2.action?
              pageNow=${pager.totalPage}">尾页</a>
```

```
                </s:if>
                <s:else>
                下一页 | 尾页
                </s:else>
        </td>
    </tr>
</table>
```

其中比较重要的是利用<s:iterator value="#session.employees1" var="list">来循环输出员工列表，而保存在 session 中的 employees1 主要通过后台调用存储过程得到，调用存储过程的代码如下：

```java
public void execute(Connection connection)throws SQLException {
    statement = connection.prepareCall(
       "Call MyPack.Get_EmployeeByPage_P(?, ?, ?)");
    statement.setInt(1, pageNow);
    statement.setInt(2, pageSize);
    statement.registerOutParameter(3, OracleTypes.CURSOR);
    statement.execute();
    ResultSet resultSet = (ResultSet)statement.getObject(3);
    while (resultSet.next()){
        employeesEntity = new EmployeesEntity();
        employeesEntity.setEmployeeId(
           resultSet.getLong("EMPLOYEE_ID"));
        employeesEntity.setPhoneNumber(
           resultSet.getString("PHONE_NUMBER"));
        employeesEntity.setName(resultSet.getString("NAME"));
        employeesEntity.setHireDate(
           resultSet.getDate("HIRE_DATE"));
        employeesEntity.setDepartmentId(
           resultSet.getLong("DEPARTMENT_ID"));
        employeesEntity.setEmail(resultSet.getString("EMAIL"));
        employeesEntity.setSalary(
           resultSet.getLong("SALARY"));
        employeesEntity.setManagerId(
           resultSet.getLong("MANAGER_ID"));
        employeesEntity.setPhoto(resultSet.getBytes("PHOTO"));
        list.add(employeesEntity);
    }
}
```

```
});
```
此后点击图中的删除链接之后跳转到 employeeAction3.action，代码主要如下：
```
System.out.print("Action 得到的员工 ID 是："+employeeId);
if (getEmployeeId()== 0)
    return ERROR;
try {
    employeeService.deleteEmployee(employeeId);
    return SUCCESS;
} catch (Exception e){
    return ERROR;
}
```
首先判断是否得到了参数 employeeId，如果没有得到，跳转到异常界面，如果得到了，调用 service 层的 deleteEmployee（employeeId）方法，根据员工 Id 删除员工。

◇ 更新员工

点击图 14-10 中的编辑链接跳转到编辑界面，编辑界面如图 14-11 所示。调用 employeeService.updateEmployee（employeesEntitys）更新数据。更新代码如下：
```
Query query = session.createSQLQuery(
  "UPDATE EMPLOYEES SET NAME=? , EMAIL=?,SALARY=?, " +
  " MANAGER_ID =? , DEPARTMENT_ID =? WHERE EMPLOYEE_ID = ? ");
query.setParameter(0, employeesEntity.getName());
query.setParameter(1, employeesEntity.getEmail());
query.setParameter(2, employeesEntity.getSalary());
query.setParameter(3, managerId);
query.setParameter(4, departmrntId);
query.setParameter(5, employeesEntity.getEmployeeId());
query.executeUpdate();
transaction.commit();
```

图 14-11　employeeEdit.jsp 的运行界面

❖ 查看员工详情

点击图 14-10 中的显示出的员工列表后的详情链接，跳转到 employeeShow.jsp，页面效果如图 14-12 所示：

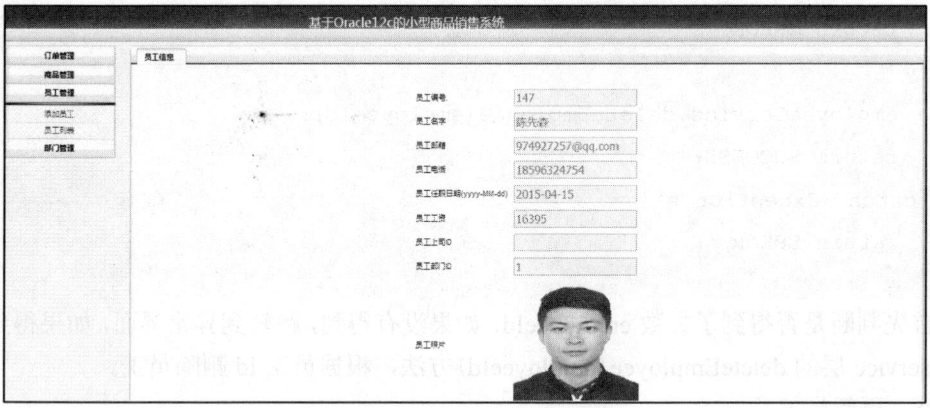

图 14-12 employeeShow.jsp

在该页面，会将保存在 session 中的 EMPLOYEE 对象的属性显示出来，重点是点击链接之后如何获得该对象？逻辑很简单，就是通过链接传入的 employeeId 去数据库中查询对应的 EMPLOYEE 对象。

```
String hql = "from EmployeesEntity WHERE  employeeId = ?";
SessionFactory sessionFactory = this.getSessionFactory();
Session session = sessionFactory.getCurrentSession();
Transaction transaction = null ;
try {
   transaction = session.beginTransaction();
   EmployeesEntity employeesEntity = (EmployeesEntity)
     session.createQuery(hql).setParameter(
     0, employeeId).uniqueResult();
   transaction.commit();
   return employeesEntity;
}
```

有一点需要补充的是，本模块与数据库交互的操作均在 src 下的 com/panda/daoImpl 包下的 EmployeeDaoImpl 类中。

2) 商品列表模块

❖ 商品列表显示

本模块对商品的操作均是在 productList.jsp 上，页面如图 14-13 所示：

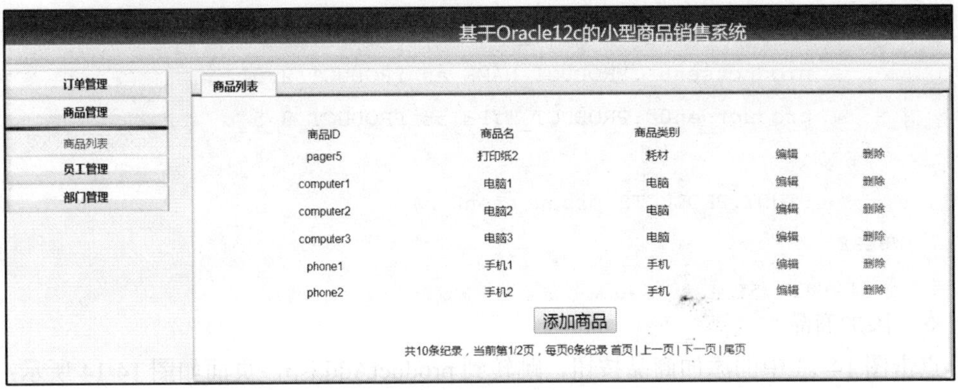

图 14-13　productList.jsp

可以看到，页面上有添加商品按钮，编辑和删除链接，由于商品的属性不多，可以直接看到所有，故不再有详情链接。该页面的关键操作是将 session 中的 productList 的 product 对象的属性显示在页面上，而 productList 又是如何得到呢？答案是分页查询数据库中的数据，此后把查询出的数据添加进 List 中，实现代码如下：

```
Session session = this.getSessionFactory().getCurrentSession();
List<ProductsEntity> list = null ;
String sql = "from ProductsEntity ";
Transaction transaction = null ;
try {
    transaction = session.beginTransaction();
    Query querys = session.createQuery(sql);
    querys.setFirstResult((pageNow-1)*pageSize); // 设置查询的起始序号
    querys.setMaxResults(pageSize); // 设置一次查询的数量
    list = querys.list();
    transaction.commit();
} catch (Exception e){
    e.printStackTrace();
    transaction.rollback();
    return null;
}
return list;
```

在这段代码中，会使用到两个关键的参数 pageNow 和 pageSize，这两个参数决定查询的数据的起始序号以及一次查询的数量。底层的 sql 语句中 rownum 后面的参数值就取决于这两个关键参数。

```
SELECT
    FROM
        ( SELECT
```

```
            productsen0_.PRODUCT_ID as PRODUCT_1_5_,
            productsen0_.PRODUCT_NAME as PRODUCT_2_5_,
            productsen0_.PRODUCT_TYPE as PRODUCT_3_5_
        FROM
            STUDY.PRODUCTS productsen0_ )
    WHERE
        rownum <= ?
```

❖ 添加商品

点击图 14-13 中的添加商品按钮，跳转到 productAdd.jsp，页面如图 14-14 所示：

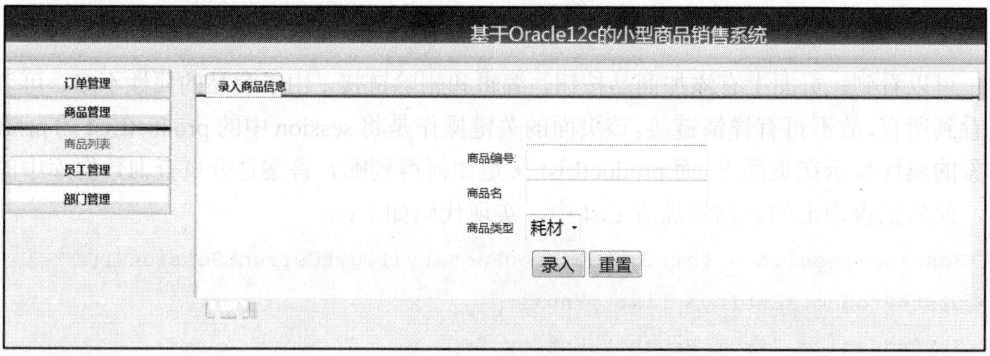

图 14-14　productAdd.jsp

点击录入按钮之后，将用户提交的数据提交到后台进行保存，由于在添加员工模块已经阐述过字段验证，故不再赘述，保存 PRODUCT 的代码如下：

```
try{
    session = this.getSessionFactory().getCurrentSession();
    transaction = session.beginTransaction();
    session.save(productsEntity);
    transaction.commit();
}catch (Exception e){
    if (transaction != null){
        transaction.rollback();
    }
```

❖ 商品编辑

点击商品编辑链接，跳转到 productEdit.jsp，页面如图 14-15 所示：

点击录入按钮，将数据提交到后台进行更新，更新的逻辑主要是通过 productId 去获取数据库中的对象，此后给该对象的属性赋值，赋值完成后，调用 update 方法进行更新。

```
transaction = session.beginTransaction();
ProductsEntity productsEntity1 = this.getHibernateTemplate().
    load(ProductsEntity.class, productsEntity.getProductId());
```

```
productsEntity1.setProductType(productsEntity.getProductType());
productsEntity1.setProductName(productsEntity.getProductName());
System.out.print(productsEntity1.getProductName()+
 productsEntity1.getProductType());
session.UPDATE(productsEntity1);
transaction.commit();
```

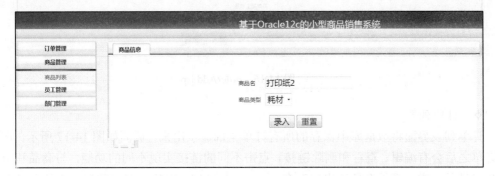

图 14-15 productEdit.jsp

✧ 商品删除

点击图 14-13 中的删除链接，传入商品 Id 到后台，根据商品 Id 删除商品。

```
String hql2 = "from ProductsEntity WHERE  productId = ?";
SessionFactory sessionFactory = this.getSessionFactory();
Session session = sessionFactory.getCurrentSession();
Transaction transaction = session.beginTransaction();
ProductsEntity productsEntity = (ProductsEntity)
 session.createQuery(hql2).setParameter(0, productId).uniqueResult();
session.delete(productsEntity);
```

3) 订单列表模块

订单的管理同其他两个模块一样，都是增删查改操作。

✧ 添加订单

点击左侧的添加订单链接，到达 orderAdd.jsp，如图所示：

在页面输入数据之后，点击录入按钮，将数据提交到后台进行保存。

```
session = getMySession();
transaction = session.beginTransaction(); //开启事务
ordersEntity.setEmployeesByEmployeeId((EmployeesEntity)
  session.get(EmployeesEntity.class, ordersEntity.getEmployeeId()));
session.save(ordersEntity);
transaction.commit();
```

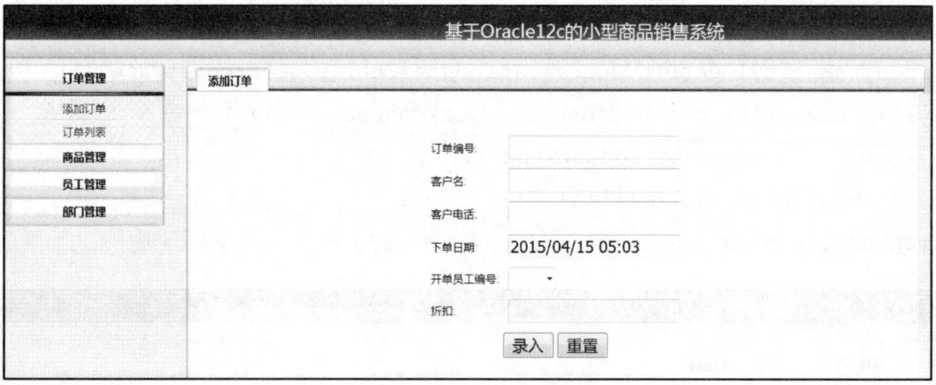

图 14-16　orderAdd.jsp

◆ 订单列表

订单列表界面将数据库中保存的所有订单全部显示出来，页面如图 14-17 所示。在每条订单之后会有编辑、查看和删除链接，点击不同的链接实现不同的功能，与商品和员工的列表显示一样，都是循环输出保存在 session 中的 List 对象。取出数据库中所有订单数据的代码如下：

```java
List<OrdersEntity> list = null ;
Session session = null ;
String sql = "from OrdersEntity ";
Transaction transaction = null ;
try {
    session = getMySession();
    transaction = session.beginTransaction();
    Query querys = session.createQuery(sql);
    querys.setFirstResult((pageNow-1)*pageSize); //设置一次查询的数量
    querys.setMaxResults(pageSize); //设置查询的起始序号
    list = querys.list();
    transaction.commit();
    return list;
} catch (Exception e){
    e.printStackTrace();
    transaction.rollback();
    return null;
}
```

第 14 章 小型商品销售系统

图 14-17 orderList.jsp

由于数据库记录很多，还需要分页查询数据库记录。在订单列表中随意选取一个订单，点击详情链接，会在数据库中查出该订单相应的订单详单表，页面如图 14-18 所示。

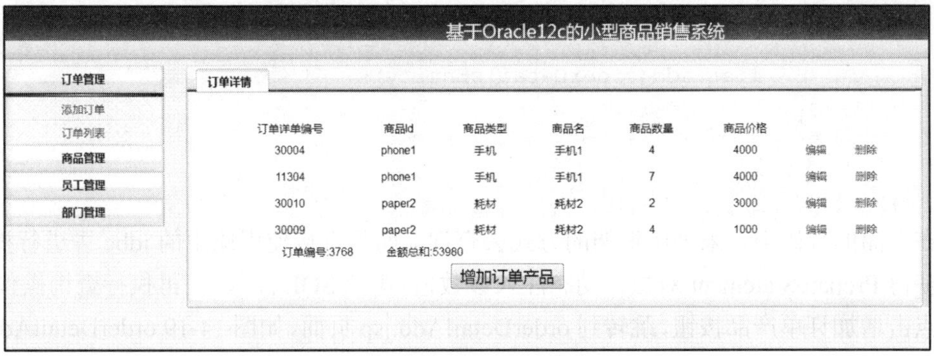

图 14-18 orderDetailsList.jsp

点击详情链接之后，会将相应的 orderId 传入后台，后台根据传入的 orderId 在 OrderDetails 表中去查询，查询订单详单产品的代码如下：

```java
String sql = "SELECT * FROM ORDER_DETAILS WHERE ORDER_ID = ?";
try {
    transaction = session.beginTransaction();
    session.doWork(new Work(){
        OrderDetailsEntity orderDetailsEntity = null;
        public void execute(Connection connection)throws
            SQLException {
            PreparedStatement statement =
              connection.prepareStatement(sql);
            statement.setInt(1, orderId);
            ResultSet resultSet = statement.executeQuery();
```

```java
            while (resultSet.next()){
                orderDetailsEntity = new OrderDetailsEntity();
                orderDetailsEntity.setId(resultSet.getInt("ID"));
                orderDetailsEntity.setOrderId(
                    resultSet.getInt("ORDER_ID"));
                orderDetailsEntity.setProductId(
                    resultSet.getString("PRODUCT_ID"));
                orderDetailsEntity.setProductName(
                    resultSet.getString("PRODUCT_NAME"));
                orderDetailsEntity.setProductType(
                    resultSet.getString("PRODUCT_TYPE"));
                orderDetailsEntity.setProductNum(
                    resultSet.getLong("PRODUCT_NUM"));
                orderDetailsEntity.setProductPrice(
                    resultSet.getLong("PRODUCT_PRICE"));
                list.add(orderDetailsEntity);
            }
        }
    });
```

在上面的代码中，未采用框架的方式去查询，而是直接使用原生的 jdbc 去进行操作，通过获得 PrepareStatement 对象，动态注入参数值（避免 SQL 注入），再执行查询操作。

点击增加订单产品按钮，跳转到 orderDetailAdd.jsp 页面，如图 14-19 orderDetailAdd.jsp 所示：

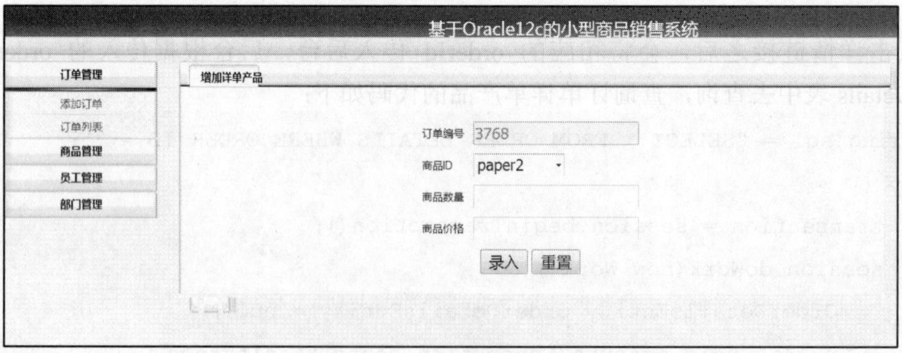

图 14-19 orderDetailAdd.jsp

由于是针对某个订单单独添加商品，所以首先会把该订单的订单编号固定的显示到页面上。

4) 部门管理模块

部门列表的界面如图 14-20 所示。点击编辑链接，可以对部门的信息进行修改，由于部门 ID 不能变动，所以只能更改部门名称。点击删除按钮，可以实现对部门的删除，但是如果部门有员工，就不允许删除这个部门。

点击新增部门，跳转至部门添加界面，如图 14-21 所示：

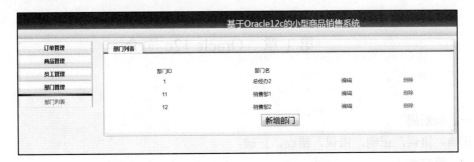

图 14-20　departmentList.jsp

图 14-21　departmentAdd.jsp

附录 练习答案与训练任务的实现

第 1 章 Oracle 12c 简介

练习

一、判断题

正确，错误，正确，错误，错误，正确

二、填空题

1. cloud/云
2. 企业版
3. CDB$ROOT
4. PDB$SEED
5. 高可用
6. 停机

第 2 章 Oracle 12c 的安装

练习

一、判断题

错误，正确，正确，错误，正确

二、单项选择题

A D A B C

三、填空题

1. 查询用户能访问的所有表
2. 用户组
3. CDB$ROOT
4. PDB$SEED
5. user_
6. ora_

训练任务

1. 在 Linux 环境下安装 Oracle 12c

略，见本教材在出版社的网上资源。

2. 连接/检测 Oracle 12c

```
--本地连接Oracle 12c
$ sqlplus hr/123@pdborcl
SQL> exit
--异地连接并访问Oracle 12c
$ sqlplus hr/123@192.168.0.50/pdborcl
SQL> SELECT * FROM hr.jobs;
JOB_ID     JOB_TITLE                        MIN_SALARY MAX_SALARY
---------- -------------------------------- ---------- ----------
AD_PRES    President                             20080      40000
AD_VP      Administration Vice President         15000      30000
...
19 rows selected.
SQL> exit
```

```
--查看环境变量、安装目录及文件
$ echo $ORACLE_HOME
/home/oracle/app/oracle/product/12.1.0/dbhome_1
$ echo $ORACLE_SID
Orcl
$ cd $ORACLE_HOME
$ ls
...
```

```
--查看Oracle进程
$ ps -ef | grep ora_
oracle   2500    1  0 06:19 ?     00:00:00 ora_pmon_orcl
oracle   2502    1  0 06:19 ?     00:00:03 ora_psp0_orcl
oracle   2504    1  2 06:19 ?     00:04:27 ora_vktm_orcl
...
```

```
--查看监听器状态
$ lsnrctl status
...
The command completed successfully

$ lsnrctl service
...
```

第 3 章 网络配置及管理工具

练习

一、判断题

错误，正确，正确，正确，错误，正确，正确，正确，正确，正确

二、单项选择题

B D D A B

三、填空题

1. 专用，共享

2. 1

3. define

4. 规则，成本

5. 定义列的别名，输出格式

6. 查看一些参数的值

7. 在命令提示符上会显示当前时间

8. 显示命令执行时间长度

9. 最快/最优

10. 没有统计过的

11. 允许哪些 IP 访问 Oracle

12. 拒绝哪些 IP 访问 Oracle

训练任务

1. 熟悉 SQL*Plus 和 SQL Developer 两个工具的使用

首先使用 sqlplus 以简单方式登录 pdborcl，连接字符串是"localhost：1521/pdborcl"。登录用户是 hr，登录后查询某个部门的部门总人数和平均工资。SQL 语句中定义了绑定变量 dn，表示需要查询的一个部门。

```
$ sqlplus hr/123@localhost: 1521/pdborcl
SQL> col department_name format a30
SQL> var dn varchar2(100)
SQL> EXEC: dn: = 'Sales';
PL/SQL procedure successfully completed.
SQL> SELECT department_name, count(e.job_id) as "部门总人数", avg(e.salary) as "平均工资" FROM departments d, employees e WHERE d.department_id=e.department_id and department_name= : dn GROUP BY department_name;
DEPARTMENT_NAME                部门总人数       平均工资
```

```
----------------------------  ----------  ----------
Sales                                 34   8955.88235
SQL> EXEC: dn: = 'IT';
PL/SQL procedure successfully completed.
SQL> SELECT department_name, count(e.job_id) as "部门总人数", avg(e.salary) as
"平均工资" FROM departments d, employees e WHERE
d.department_id=e.department_id and department_name= : dn GROUP BY
department_name;
DEPARTMENT_NAME              部门总人数    平均工资
----------------------------  ----------  ----------
IT                                     5        5760
SQL>
```

下面训练以 TNS 名称登录，在文件：

```
$ORACLE_HOME/network/admin/tnsnames.ora 中加入一个 PDBORCL 名称节：
PDBORCL =
  (DESCRIPTION =
    (ADDRESS_LIST =
      (ADDRESS = (PROTOCOL = TCP)(HOST = localhost)(PORT = 1521))
    )
    (CONNECT_DATA =
      (SERVER = DEDICATED)
      (SERVICE_NAME = pdborcl)
    )
  )
```

然后以 PDBORCL 名称直接登录成功，无需主机地址、端口号和服务名称：

```
$ sqlplus hr/123@pdborcl
SQL>
```

sqldeveloper 工具的使用省略，见教材第 3.5 节：Oracle SQL Developer。

2. 配置服务器同时工作在专用服务器模式和共享服务器模式

略，见教材第 3.2 节：服务器模式和数据库连接方式。

3. 分析 SQL 执行计划，执行 SQL 语句的优化指导

本训练任务查询两个部门('IT'和'Sales')的部门总人数和平均工资，以下两个查询的结果是一样的。但效率不相同。

查询 1：

```
SELECT d.department_name, count(e.job_id) as "部门总人数",
avg(e.salary) as "平均工资"
from hr.departments d, hr.employees e
where d.department_id = e.department_id
```

```
and d.department_name in ('IT', 'Sales')
GROUP BY department_name;
```

查询 2:

```
SELECT d.department_name, count(e.job_id)as "部门总人数",
avg(e.salary)as "平均工资"
FROM hr.departments d, hr.employees e
WHERE d.department_id = e.department_id
GROUP BY department_name
HAVING d.department_name in ('IT', 'Sales');
```

查询 1 的执行计划如下图，其中：Cost=5，Rows=20，Predicate Information（谓词信息）中有一次索引搜索 access，一次全表搜索 filter。

```
--------------------------------------------------------------------------------------
| Id  | Operation                     | Name              | Rows | Bytes | Cost (%CPU)| Time     |
--------------------------------------------------------------------------------------
|   0 | SELECT STATEMENT              |                   |    1 |    23 |     5  (20)| 00:00:01 |
|   1 |  HASH GROUP BY                |                   |    1 |    23 |     5  (20)| 00:00:01 |
|   2 |   NESTED LOOPS                |                   |   19 |   437 |     4   (0)| 00:00:01 |
|   3 |    NESTED LOOPS               |                   |   20 |   437 |     4   (0)| 00:00:01 |
|*  4 |     TABLE ACCESS FULL         | DEPARTMENTS       |    2 |    32 |     3   (0)| 00:00:01 |
|*  5 |     INDEX RANGE SCAN          | EMP_DEPARTMENT_IX |   10 |       |     0   (0)| 00:00:01 |
|   6 |    TABLE ACCESS BY INDEX ROWID| EMPLOYEES         |   10 |    70 |     1   (0)| 00:00:01 |
--------------------------------------------------------------------------------------

Predicate Information (identified by operation id):
---------------------------------------------------

   4 - filter("D"."DEPARTMENT_NAME"='IT' OR "D"."DEPARTMENT_NAME"='Sales')
   5 - access("D"."DEPARTMENT_ID"="E"."DEPARTMENT_ID")

Note
-----
   - this is an adaptive plan

Statistics
----------------------------------------------------------
          0  recursive calls
          0  db block gets
         10  consistent gets
          0  physical reads
          0  redo size
        797  bytes sent via SQL*Net to client
        552  bytes received via SQL*Net from client
          2  SQL*Net roundtrips to/from client
          0  sorts (memory)
          0  sorts (disk)
          2  rows processed
```

查询 2 的执行计划如下图，其中：Cost=7，Rows=106，Predicate Information（谓词信息）中有一次索引搜索 access，两次全表搜索 filter。

那么，查询 1 和查询 2 哪个更优呢？总的来看，查询 1 比查询 2 更优，这是因为查询 1 除了"consistent gets=10"比查询 2 的"consistent gets=9"稍差，其他参数都优于查询 2。从分析两个 SQL 语句可以看出，查询 1 是先过滤后汇总（where 子句），参与汇总与计算的数据量少。而查询 2 是先汇总后过滤（having 子句），参与汇总与计算的数据量多。

```
-------------------------------------------------------------------------------------
| Id | Operation                     | Name        | Rows | Bytes | Cost (%CPU)| Time     |
-------------------------------------------------------------------------------------
|  0 | SELECT STATEMENT              |             |    1 |    23 |    7  (29)| 00:00:01 |
|* 1 |  FILTER                       |             |      |       |           |          |
|  2 |   HASH GROUP BY               |             |    1 |    23 |    7  (29)| 00:00:01 |
|  3 |    MERGE JOIN                 |             |  106 |  2438 |    6  (17)| 00:00:01 |
|  4 |     TABLE ACCESS BY INDEX ROWID| DEPARTMENTS|   27 |   432 |    2   (0)| 00:00:01 |
|  5 |      INDEX FULL SCAN          | DEPT_ID_PK  |   27 |       |    1   (0)| 00:00:01 |
|* 6 |     SORT JOIN                 |             |  107 |   749 |    4  (25)| 00:00:01 |
|  7 |      TABLE ACCESS FULL        | EMPLOYEES   |  107 |   749 |    3   (0)| 00:00:01 |
-------------------------------------------------------------------------------------

Predicate Information (identified by operation id):
---------------------------------------------------

   1 - filter("DEPARTMENT_NAME"='IT' OR "DEPARTMENT_NAME"='Sales')
   6 - access("D"."DEPARTMENT_ID"="E"."DEPARTMENT_ID")
       filter("D"."DEPARTMENT_ID"="E"."DEPARTMENT_ID")


Statistics
----------------------------------------------------------
          0  recursive calls
          0  db block gets
          9  consistent gets
          0  physical reads
          0  redo size
        797  bytes sent via SQL*Net to client
        552  bytes received via SQL*Net from client
          2  SQL*Net roundtrips to/from client
          1  sorts (memory)
          0  sorts (disk)
          2  rows processed
```

既然查询 1 比查询 2 优，那么查询 1 就是最优的吗？通过 sqldeveloper 中的优化指导可以看出，仍然有优化的必要，下图是 sqldeveloper 给出的优化建议：

点击上图界面中的"详细资料"，sqldeveloper 给出了优化建议：

```
Recommendation (estimated benefit: 59.99%)
  - 考虑运行可以改进物理方案设计的访问指导或者创建推荐的索引。
CREATE INDEX HR.IDX$$_01220001 on
HR.DEPARTMENTS("DEPARTMENT_NAME", "DEPARTMENT_ID");
```

这个建议是让我们在 departments 表上创建一个基于 DEPARTMENT_NAME 和 DEPARTMENT_ID 字段的索引,这样就可以加快查询 DEPARTMENT_NAME 的速度。在创建这个索引之后,可以看到查询 1 的执行计划如下图,对比可以看出,谓词中已经没有全表搜索 filter,只有索引搜索 access,其他性能指标也随之全面提升了。

```
Execution Plan
----------------------------------------------------------
Plan hash value: 872036552

--------------------------------------------------------------------------------------------
| Id  | Operation                    | Name             | Rows | Bytes | Cost (%CPU)| Time     |
--------------------------------------------------------------------------------------------
|   0 | SELECT STATEMENT             |                  |    2 |    46 |     2   (0)| 00:00:01 |
|   1 |  SORT GROUP BY NOSORT        |                  |    2 |    46 |     2   (0)| 00:00:01 |
|   2 |   NESTED LOOPS               |                  |   19 |   437 |     2   (0)| 00:00:01 |
|   3 |    NESTED LOOPS              |                  |   20 |   437 |     2   (0)| 00:00:01 |
|   4 |     INLIST ITERATOR          |                  |      |       |            |          |
|*  5 |      INDEX RANGE SCAN        | IDX$$_01220001   |    2 |    32 |     1   (0)| 00:00:01 |
|*  6 |     INDEX RANGE SCAN         | EMP_DEPARTMENT_IX|   10 |       |     0   (0)| 00:00:01 |
|   7 |    TABLE ACCESS BY INDEX ROWID| EMPLOYEES       |   10 |    70 |     1   (0)| 00:00:01 |
--------------------------------------------------------------------------------------------

Predicate Information (identified by operation id):
---------------------------------------------------

   5 - access("D"."DEPARTMENT_NAME"='IT' OR "D"."DEPARTMENT_NAME"='Sales')
   6 - access("D"."DEPARTMENT_ID"="E"."DEPARTMENT_ID")


Statistics
----------------------------------------------------------
          0  recursive calls
          0  db block gets
          8  consistent gets
          0  physical reads
          0  redo size
        797  bytes sent via SQL*Net to client
        552  bytes received via SQL*Net from client
          2  SQL*Net roundtrips to/from client
          0  sorts (memory)
          0  sorts (disk)
          2  rows processed
```

第 4 章　数据库管理与配置

练习

一、判断题

错误,正确,错误,错误,正确

二、单项选择题

A B C D C

三、填空题

1. parameter

2. pdbs

3. ORACLE_SID

4. 静默

5. startup,shutdown

训练任务

1. 创建插接式数据库

由于 dbca 命令的选项较多，所以先用-help 选项查看一下帮助，再执行创建 pdborder 数据库的命令，创建 100%完成之后，用管理用户 sysorder 登录 pdborder，如果连接成功，表示创建 pdborder 成功了。注意，通过查询当前权限 session_privs 可以看出，管理用户 sysorder 权限很少，还需要通过 system 用户给它授予更多权限才行。

```
$ dbca -createPluggableDatabase -help
...
$ dbca -silent -createPluggableDatabase -sourceDB orcl -pdbName pdborder
-createPDBFrom default -pdbAdminUserName sysorder -pdbAdminPassword 123
-pdbDatafileDestination  "/home/oracle/app/oracle/oradata/orcl/{PDB_NAME}"
-createUserTableSpace true
正在创建插接式数据库
4% 已完成
12% 已完成
21% 已完成
38% 已完成
85% 已完成
正在完成创建插接式数据库
100% 已完成
$ sqlplus sysorder/123@localhost/pdborder
SQL> SELECT * FROM session_privs;
PRIVILEGE
----------------------------------------
SET CONTAINER
CREATE PLUGGABLE DATABASE
CREATE SESSION
```

注意，dbca 命令是在操作系统下运行的。Pdborder 数据库创建完成之后，如果要删除它，可以在操作系统下运行：

```
$ dbca -silent -deletePluggableDatabase -sourceDB orcl -pdbName
pdborder
```

2. 启动和关闭数据库

略。参见"4.5 数据库的启动与关闭"。

第 5 章　Oracle 12c 数据库结构

练习

一、判断题

错误，正确，正确，错误，错误，正确

二、单项选择题

A　C　D　D　A

三、填空题

1. 内存结构
2. 逻辑结构
3. 文件组
4. 块（block）
5. 数据段
6. 多线程
7. 强制性
8. 归档

训练任务

1. 增加 SGA 空间大小

本任务首先通过"show parameter sga_"查看 sga 空间的最大值 sga_max_size 和当前值 sga_target。可以看到都是 1.5G，通过修改这两个参数将 SGA 值改为 2G，然后重启数据库，重启之后，可以看出 SGA 的当前值和最大值已经变成了 2G。

```
$ sqlplus / as sysdba
SQL> show parameter sga_
NAME                                   TYPE         VALUE
------------------------------------   -----------  --------------------
sga_max_size                           big integer  1504M
sga_target                             big integer  1504M
unified_audit_sga_queue_size           integer      1048576
SQL> ALTER SYSTEM SET sga_max_size=2G scope=spfile;
System altered.
SQL> ALTER SYSTEM SET sga_target=2G scope=spfile;
System altered.
SQL> shutdown immediate
Database closed.
Database dismounted.
```

```
ORACLE instance shut down.
SQL> startup
ORACLE instance started.
Total System Global Area 2147483648 bytes
Fixed Size                  2926472 bytes
Variable Size             671090808 bytes
Database Buffers         1291845632 bytes
Redo Buffers               13848576 bytes
In-Memory Area            167772160 bytes
Database mounted.
Database opened.
SQL> show parameter sga_
NAME                                 TYPE        VALUE
------------------------------------ ----------- --------------------
sga_max_size                         big integer 2G
sga_target                           big integer 2G
unified_audit_sga_queue_size         integer     1048576
SQL>
```

要注意，sga_target 的值不能超过 sga_max_size。sga_max_size 的值不要超过物理内存。

2. 应用 In-Memory 技术创建内存表

第 1 步：本任务以 hr 用户登录，首先创建表 testd，随机生成 100 万条记录。然后查询一小部分记录，这里是 "SELECT * FROM testd WHERE name like'99999%'"；。

第 2 步：运行 "set autotrace on" 开始自动跟踪。运行同样的查询语句，可见 "2918 consistent gets"，即一致性读的数据块数量为 2918。

第 3 步：通过 "ALTER table testd inmemory；" 修改该表为 inmemory。运行同样的查询语句两次，这是由于运行第 1 次查询之后才会将表调入 in-memory 块。第二次的查询跟踪输出结果为 "inmemory("NAME" LIKE '99999%')" 以及 "8 consistent gets"。

第 4 步：对比分析。testd 修改为内存表之前的一致性读的数据块数量是 2918，之后是 8，所以查询效率大大提高了。

第 5 步：任务结束，删除这张大表 testd。

```
$ sqlplus hr/123@pdborcl
SQL> CREATE TABLE testd(id int, name varchar(50));
Table created.
SQL>
begin
for i in 1..1000000 loop
    INSERT INTO testd values(
        i, to_char(dbms_random.random()));
```

```
end loop;
commit;
end;
/
PL/SQL procedure successfully completed.
SQL> SELECT * FROM testd WHERE name like '99999%';
       ID NAME
---------- --------------------------------------------------
    940710 999999600
SQL> set autotrace on
SQL> SELECT * FROM testd WHERE name like '99999%';
...
Predicate Information (IDENTIFIED BY operation id):
---------------------------------------------------
   1 - filter("NAME" LIKE '99999%')


Note
-----
   - dynamic statistics used: dynamic sampling (level=2)


Statistics
----------------------------------------------------------
          0  recursive calls
          0  db block gets
       2918  consistent gets
          0  physical reads
          0  redo size
        622  bytes sent via SQL*Net to client
        552  bytes received via SQL*Net FROM client
          2  SQL*Net roundtrips to/from client
          0  sorts (memory)
          0  sorts (disk)
          1  rows processed
SQL> ALTER table testd inmemory;
Table altered.
SQL> SELECT * FROM testd WHERE name like '99999%';
...
SQL> SELECT * FROM testd WHERE name like '99999%';
```

附录　练习答案与训练任务的实现

```
...
Predicate Information (IDENTIFIED BY operation id):
---------------------------------------------------
   1 - inmemory("NAME" LIKE '99999%')
       filter("NAME" LIKE '99999%')


Note
-----
   - dynamic statistics used: dynamic sampling (level=2)
Statistics
---------------------------------------------------
          0  recursive calls
          0  db block gets
          8  consistent gets
          0  physical reads
          0  redo size
        622  bytes sent via SQL*Net to client
        552  bytes received via SQL*Net FROM client
          2  SQL*Net roundtrips to/from client
          0  sorts (memory)
          0  sorts (disk)
          1  rows processed
SQL> drop table testd;
Table dropped.
```

3. 维护定时作业任务

略。参见"5.3.2　定时执行作业任务"的步骤。

第 6 章　数据库存储管理

练习

一、判断题

错误，正确，错误，正确，正确，错误

二、单项选择题

C D A A A

三、填空题

1. 读写

2. 之和

3. SYSTEM，SYSAUX
4. control_files
5. 服务器

训练任务

1. 新建一个用户类型表空间

略。参见"6.2 创建表空间"中的新建用户类型表空间的步骤。

2. 在 CDB/PDB 中分别设置系统参数

本任务是在 CDB 和 pdborcl 中分别修改 open_cursors 参数。首先以 sys 登录，将 open_cursors 修改为 400，然后切换到 pdborcl，将 open_cursors 设置为 500。最后通过视图 v$pdbs 和 pdb_spfile$ 联合查看 pdborcl 的全部自定义参数。

```
$ sqlplus / as sysdba
SQL> ALTER SYSTEM SET open_cursors=400;
System altered.
SQL> show parameter open_cursors
NAME                                 TYPE         VALUE
------------------------------------ ------------ ------------------
open_cursors                         integer      400
SQL> ALTER SESSION SET CONTAINER=pdborcl;
Session altered.
SQL> ALTER SYSTEM SET open_cursors=500;
System altered.
SQL> show parameter open_cursors;
NAME                                 TYPE         VALUE
------------------------------------ ------------ ------------------
open_cursors                         integer      500
SQL> col pname format a50
SQL> col value$ format a50
SQL> SELECT b.name as pname, b.value$ as pvalue FROM v$pdbs a,
  2 pdb_spfile$ b WHERE  a.dbid=b.pdb_uid and a.name='PDBORCL';
PNAME                                             PVALUE
------------------------------------------------- ------------------
job_queue_processes                               2
optimizer_dynamic_sampling                        2
open_cursors                                      500
db_securefile                                     'PREFERRED'
```

第7章 用户及权限管理

练习

一、判断题

错误，正确，正确，正确，错误，正确，错误，正确

二、单项选择题

C B A D D

三、填空题

1. 系统权限，对象权限

2. C##

3. 设置表空间配额

4. GRANT

5. REVOKE

6. system

训练任务

1. 掌握管理角色、权根、用户的能力，并在用户之间共享对象。

第1步：以 system 登录到 pdborcl，创建角色 con_res_view 和用户 new_user，并授权和分配空间：

```
$ sqlplus system/***@pdborcl
SQL> CREATE ROLE con_res_view;
Role created.
SQL> GRANT connect, resource, CREATE VIEW TO con_res_view;
Grant succeeded.
SQL> CREATE USER new_user IDENTIFIED BY 123 DEFAULT TABLESPACE users
  2  TEMPORARY TABLESPACE temp;
User created.
SQL> ALTER USER new_user QUOTA 50M ON users;
User altered.
SQL> GRANT con_res_view TO new_user;
Grant succeeded.
SQL> exit
```

第2步：新用户 new_user 连接到 pdborcl，创建表 mytable 和视图 myview，插入数据，最后将 myview 的 SELECT 对象权限授予 hr 用户。

```
$ sqlplus new_user/123@pdborcl
SQL> show user;
```

```
USER is "NEW_USER"
SQL> CREATE TABLE mytable (id number, name varchar(50));
Table created.
SQL> INSERT INTO mytable(id, name)VALUES(1, 'zhang');
1 row created.
SQL> INSERT INTO mytable(id, name)VALUES (2, 'wang');
1 row created.
SQL> CREATE VIEW myview AS SELECT name FROM mytable;
View created.
SQL> SELECT * FROM myview;
NAME
--------------------------------------------------
zhang
wang
SQL> GRANT SELECT ON myview TO hr;
Grant succeeded.
SQL>exit
```

第3步：用户 hr 连接到 pdborcl，查询 new_user 授予它的视图 myview

```
$ sqlplus hr/123@pdborcl
SQL> SELECT * FROM new_user.myview;
NAME
--------------------------------------------------
zhang
wang
SQL> exit
```

第 8 章　数据库的对象管理

练习

一、判断题

错误，正确，错误，正确，正确

二、单项选择题

A　C　A　B　D

三、填空题

1. NOT NULL
2. PRIMARY KEY
3. ALTER TABLE

4. RENAME
5. 简单视图、连接视图、复杂视图

训练任务

1. 创建表和管理表
1) 创建 EMPLOYEES 表和 DEPARTMENTS 表

```
SQL> CREATE TABLE employees
  2  (
  3    employee_id NUMBER NOT NULL,
  4    name VARCHAR2(40) NOT NULL,
  5    email VARCHAR2(40),
  6    phone_number VARCHAR2(40),
  7    hire_date DATE NOT NULL,
  8    salary NUMBER(8,2),
  9    manager_id NUMBER,
 10    department_id NUMBER,
 11    photo BLOB,
 12    CONSTRAINT "EMPLOYEES_PK" PRIMARY KEY (employee_id)
 13  )
 14  PCTFREE 10
 15  PCTUSED 40
 16  INITRANS 1
 17  MAXTRANS 255
 18  LOGGING
 19  STORAGE
 20  (
 21    INITIAL 65536
 22    NEXT 1048576
 23    MINEXTENTS 1
 24    MAXEXTENTS 2147483645
 25  )
 26  TABLESPACE USERS;
SQL> CREATE TABLE departments
  2  (
  3    department_id NUMBER NOT NULL,
  4    department_name VARCHAR2(40) NOT NULL,
  5    CONSTRAINT DEPARTMENTS_PK PRIMARY KEY(department_id)
  6  NOLOGGING
```

```
 7  PCTFREE 10
 8  INITRANS 1
 9  STORAGE
10  (
11    INITIAL 65536
12    NEXT 1048576
13    MINEXTENTS 1
14    MAXEXTENTS UNLIMITED
15  )
16  TABLESPACE USERS;
```

2）管理 EMPLOYEES

```
--在 EMPLOYEES 表中，删除列 PHOTO 列
SQL> ALTER TABLE employees DROP COLUMN photo;
--为 EMAIL 列增加唯一约束
SQL> ALTER TABLE employees
  2  ADD CONSTRAINT EMPLOYEES_UNI1 UNIQUE(email);
--为 DEPARTMENT_ID 列增加外键约束(对应 DEPARTMENTS 表中的主键)
SQL> ALTER TABLE employees
  2  ADD CONSTRAINT EMPLOYEES_FK1 FOREIGN KEY (department_id)
  3  REFERENCES departments(department_id);
--为 MANAGER_ID 列增加外键约束(对应 EMPLOYEES 表中的主键)
SQL> ALTER TABLE employees
  2  ADD CONSTRAINT EMPLOYEES_FK2 FOREIGN KEY (manager_id)
  3  REFERENCES employees(employee_id)ON DELETE SET NULL;
--为 SALARY 列增加检查约束(值大于 0)
SQL> ALTER TABLE employees
  2  ADD CONSTRAINT EMPLOYEES_ CHK1 CHECK (salary>0)
--为 MANAGER_ID 列增加检查约束(不等于当前的 EMPLOYEE_ID 值)
SQL> ALTER TABLE employees
  2  ADD CONSTRAINT EMPLOYEES_CHK2 CHECK (employee_id<>manager_id)
```

2. 创建分区表

本任务参考教材【示例 8-7】完成操作。

3. 创建索引

本任务参考教材 8.3.1 完成操作。

第 9 章 表数据维护

练习

一、判断题

正确，正确，正确，正确，错误

二、单项选择题

C A D B A A

三、填空题

1. 原子性、一致性、隔离性、持久性
2. RETURNING
3. COMMIT
4. ROLLBACK TO SAVEPOINT
5. 错读

训练任务

1. 表数据的增加、修改、删除

1）创建 EMPLOYEES 表和 DEPARTMENTS 表（参见第 8 章的训练任务）

2）对 EMPLOYEES 表和 DEPARTMENTS 表中数据进行增删改

```
--在 DEPARTMENTS 表中插入 1 条记录
SQL> INSERT INTO departments(department_id, department_name)
  2  VALUES(60, 'IT 部门');
--在 EMPLOYEES 表中插入 2 条记录
SQL> INSERT INTO employees(
  2  employee_id, name, email, phone_number, hire_date,
  3  salary, manager_id, department_id, photo)
  4  VALUES( 1, '张三', 'zs@qq.com', '123456789',
  5  to_date('2016-5-25', 'YYYY-MM-DD'), 8000, null, 60, null);
SQL> INSERT INTO employees(
  2  employee_id, name, email, phone_number, hire_date,
  3  salary, manager_id, department_id, photo)
  4  VALUES( 2, '李四', 'ls@qq.com', '123456789',
  5  to_date('2017-3-20', 'YYYY-MM-DD'), 4000, 1, 60, null);
SQL>COMMIT;
--使用 INSERT INTO 语句插入数据，SALARY 填充默认值，PHOTO 为空值
SQL> INSERT INTO employees(
  2  employee_id, name, email, phone_number, hire_date,
```

```
  3  manager_id, department_id, photo)
  4  VALUES( 3, '高七', 'gq@qq.com', '123456789',
  5  to_date('2017-5-25', 'YYYY-MM-DD'), 1, 60, null);
-- 创建 EMPLOYEES1 表, 并复制 EMPLOYEES 数据到 EMPLOYEES1 中
SQL> CREATE TABLE employees1
  2  (
  3     employee_id NUMBER NOT NULL,
  4  name VARCHAR2(40)NOT NULL,
  5  email VARCHAR2(40),
  6  phone_number VARCHAR2(40),
  7  hire_date DATE NOT NULL,
  8  salary NUMBER(8, 2),
  9  manager_id NUMBER,
 10  department_id NUMBER,
 11     photo BLOB,
 12  CONSTRAINT "EMPLOYEES1_PK" PRIMARY KEY (employee_id)
 13  )
 14  TABLESPACE USERS;
SQL>INSERT INTO employees1 SELECT * FROM employees;
SQL>COMMIT;
--更新员工编号为 3 的记录, SALARY 增加 1000
SQL> UPDATE employees SET salary=salary+1000 WHERE employee_id=3;
--删除员工编号为 3 的记录
SQL>DELETE FROM employees WHERE employee_id=3;
```

2. 熟悉事务的 ACID 特性

本任务参考教材 9.5.4 完成操作。

3. 熟悉事务锁

本任务参考教材 9.5.5 完成操作。

第 10 章　SQL 语言基础

练习

一、判断题

正确，正确，正确，错误，正确

二、单项选择题

A C D B A

三、填空题

1. 右连接、全连接
2. 自连接
3. 子查询
4. 单行子查询
5. 树查询

训练任务

1. 熟悉分页查询

```
SELECT * FROM employees
  OFFSET 10*(1-1)+1 ROWS FETCH NEXT 10 ROWS ONLY;

SELECT * FROM employees
  OFFSET 10*(2-1)+1 ROWS FETCH NEXT 10 ROWS ONLY;

SELECT * FROM employees
  OFFSET 10*(3-1)+1 ROWS FETCH NEXT 10 ROWS ONLY;
```

2. 熟悉递归查询

本任务参考教材 10.9 完成操作。

第 11 章 使用函数

练习

一、判断题
正确，正确，错误，正确，正确

二、单项选择题
B B C B C

三、填空题

1. acef
2. World
3. TRUNC()
4. CONCAT()
5. SYSDATE

训练任务

1. 使用单行函数

◇ 在 employee 表中查询出在(任何年份)2 月受聘的所有员工

```
SELECT * FROM employees WHERE TO_CHAR(hire_date,'mm')='02';
```

✧ 在 employee 表中查询满 10 年服务年限的员工的姓名和受雇日期

```
SELECT first_name, last_name, hire_date
FROM employees WHERE hire_date<=ADD_MONTHS(sysdate, -12*10);
```

✧ 在 employee 表中查询 50 号部门的雇员工作的月数

```
SELECT employee_id, first_name, last_name,
MONTHS_BETWEEN(sysdate, hire_date)
FROM employees WHERE department_id=50;
```

2. 略

3. 略

参照教材"11.3 SQL 语句优化"。对比参数 cost，filter，access，phiscal access，consistant get 等参数值，分析执行效率。

第 12 章 PL/SQL 语言

练习

一、判断题

正确，错误，正确，错误，错误

二、单项选择题

C B A C D

三、填空题

1. CALL

2. trigger

3. all_source

4. 读取游标

5. OTHERS

训练任务

1. 编写程序块

```
DECLARE
   num_sal EMPLOYEES.SALARY%TYPE;
BEGIN
   SELECT salary INTO num_sal FROM employees
      WHERE employee_id=200; --查询 employees 表中职工号 200 的工资
   IF(num_sal<3000) --判断工资是否小于 3000
   THEN
      UPDATE employees SET salary=3000  --把小于 3000 的工资改为 3000
         WHERE employee_id=200;
```

```
    END IF;
    DBMS_OUTPUT.PUT_LINE('更新'||SQL%ROWCOUNT||'条记录');
    DBMS_OUTPUT.PUT_LINE('工资是: '||num_sal);
END;
```

2. 创建游标

```
set serveroutput on;
DECLARE
    cur_id fruit.f_id%TYPE;
    cur_name fruit.f_name%TYPE;
    cur_price fruit.f_price%TYPE;
    CURSOR frt_cur
        IS SELECT f_id , f_name, f_price FROM fruit WHERE f_price>10;
BEGIN
    OPEN frt_cur;
    LOOP
        FETCH frt_cur INTO cur_id, cur_name, cur_price;
        IF frt_cur%FOUND THEN
            INSERT INTO fruitage VALUES(cur_id, cur_name, cur_price);
        ELSE
            dbms_output.put_line('已经取出所有数据，共有'
            ||frt_cur%ROWCOUNT ||'条记录');
            EXIT;
        END IF;
    END LOOP;
    CLOSE frt_cur;
END;
```

3. 创建函数和过程

1) 创建函数

```
CREATE OR REPLACE FUNCTION get_emp_name (v_no in NUMBER)
RETURN VARCHAR2
IS
    v employees.first_name%type;
BEGIN
    select first_name into v from EMPLOYEES where EMPLOYEE_ID=v_no;
    return v;
END get_emp_name;
/
--调用:
```

```sql
select get_emp_name(200)from dual;
```

2）创建过程

```sql
CREATE OR REPLACE PROCEDURE COUNT_SCH
AS
   cur_count number(6);
   cur_sum number(6);
BEGIN
   SELECT COUNT(*)INTO cur_count FROM sch;
   SELECT SUM(id)INTO cur_sum FROM sch;
   IF SQL%FOUND THEN
      DBMS_OUTPUT.PUT_LINE('记录总数: '||cur_count);
      DBMS_OUTPUT.PUT_LINE('记录总数: '||cur_sum);
   END IF;
END;
```

在 Oracle SQL Developer 中执行存储过程 COUNT_SCH，语句如下：

```sql
EXEC COUNT_SCH;
```

执行结果如下：

记录总数：2

id 总和：3

4．创建包和触发器

1）创建包

```sql
--创建程序包规范
CREATE OR REPLACE PACKAGE pack_op
IS
   PROCEDURE pro_print_ename(id number);
   PROCEDURE pro_print_sal(id number);
   FUNCTION fun_re_date(id number)return date;
END pack_op;
--创建程序包主体
CREATE OR REPLACE PACKAGE body pack_op
IS
   PROCEDURE pro_print_ename(id number)IS
   name  employees.first_name%type;
BEGIN
   SELECT first_name INTO name FROM employees WHERE employee_id=id;
   dbms_output.put_line('职员姓名: '||name);
END pro_print_ename;
```

```
PROCEDURE pro_print_sal(id number)IS
   em_salary employees.salary%type;
     BEGIN
        SELECT salary INTO em_salary
        FROM employees WHERE employee_id=id;
        dbms_output.put_line('职员工资：'||em_salary);
END pro_print_sal;

FUNCTION fun_re_date(id number)return date IS
   bedate employees.hire_date%type;
     BEGIN
        SELECT hire_date into bedate
        from employees where employee_id=id;
      RETURN bedate;
END fun_re_date;
END pack_op;

--调用程序包中创建的过程和函数
EXEC pack_op.pro_print_ename(200);
Exec pack_op.pro_print_sal(200);
SELECT pack_op.fun_re_date(200)FROM dual;
```

2) 触发器的创建和使用

```
CREATE TRIGGER NUM_SUM
AFTER INSERT
ON persons
BEGIN
   IF INSERT THEN
      INSERT INTO sales VALUES (NEW.name, 7*NEW.num);
   END IF;
END;
```

--测试语句

向 persons 表中插入记录，代码如下：

```
SQL>INSERT INTO persons VALUES ('xiaoxiao', 20);
SQL>INSERT INTO persons VALUES ( 'xiaohua', 69);
```

查询 persons 表中的记录，代码如下：

```
SQL> SELECT * FROM persons;
NAME           NUM
----------   ------
```

```
    xiaoxiao      20
    xiaohua       69
```

查询 sales 表中的记录，代码如下：

```
SQL> SELECT *FROM sales;
NAME              NUM
----------      --------
 Xiaoxiao         140
 Xiaohua          483
```

从执行的结果来看，在 persons 表插入记录之后，NUM_SUM 触发器计算插入到 persons 表中的数据，并将结果插入到 sales 表中相应的位置。

第 13 章　备份与恢复

练习

一、判断题

正确，错误，正确，错误，错误

二、单项选择题

A　B　C　D　A

三、填空题

1. run
2. 归档
3. 物理备份　逻辑备份
4. 数据文件　控制文件　归档日志文件
5. 关闭

训练任务

1. 脱机备份和恢复

首先查询所有数据文件，日志文件以及控制文件

```
$ sqlplus / as sysdba
SQL> SELECT NAME FROM v$datafile
  2  UNION ALL
  3  SELECT MEMBER AS NAME FROM v$logfile
  4  UNION ALL
  5  SELECT NAME FROM v$controlfile;
```

参数文件 spfileorcl.ora 和初始化文件 init.ora，initorcl.ora 存储在目录 $ORACLE_HOME/dbs 中。编写的脚本如下：

```
$ cat cold_cpfiles.sh
```

```bash
#!/bin/bash
if [ $# -lt 1 ]; then
    echo "cold_cpfiles.sh -b -r"
    echo "-b backup"
    echo "-r restore"
    return
fi
export ORACLE_HOME=/home/oracle/app/oracle/product/12.1.0/dbhome_1
if [ $1 = "-b" ]; then
echo "backup..."
    cp                      /home/oracle/app/oracle/oradata/orcl/system01.dbf  /home/oracle/cold_bak_files/system01.dbf
    cp                      /home/oracle/app/oracle/oradata/orcl/sysaux01.dbf  /home/oracle/cold_bak_files/sysaux01.dbf
    cp                      /home/oracle/app/oracle/oradata/orcl/undotbs01.dbf  /home/oracle/cold_bak_files/undotbs01.dbf
    cp           /home/oracle/app/oracle/oradata/orcl/pdbseed/system01.dbf  /home/oracle/cold_bak_files/pdbseed/system01.dbf
    cp                      /home/oracle/app/oracle/oradata/orcl/users01.dbf  /home/oracle/cold_bak_files/users01.dbf
    cp           /home/oracle/app/oracle/oradata/orcl/pdbseed/sysaux01.dbf  /home/oracle/cold_bak_files/pdbseed/sysaux01.dbf
    cp           /home/oracle/app/oracle/oradata/orcl/pdborcl/system01.dbf  /home/oracle/cold_bak_files/pdborcl/system01.dbf
    cp           /home/oracle/app/oracle/oradata/orcl/pdborcl/sysaux01.dbf  /home/oracle/cold_bak_files/pdborcl/sysaux01.dbf
    cp /home/oracle/app/oracle/oradata/orcl/pdborcl/SAMPLE_SCHEMA_users01.dbf  /home/oracle/cold_bak_files/pdborcl/SAMPLE_SCHEMA_users01.dbf
    cp           /home/oracle/app/oracle/oradata/orcl/pdborcl/example01.dbf  /home/oracle/cold_bak_files/pdborcl/example01.dbf
    cp   /home/oracle/app/oracle/oradata/orcl/pdborcl/pdbtest_users02_1.dbf  /home/oracle/cold_bak_files/pdborcl/pdbtest_users02_1.dbf
    cp   /home/oracle/app/oracle/oradata/orcl/pdborcl/pdbtest_users02_2.dbf  /home/oracle/cold_bak_files/pdborcl/pdbtest_users02_2.dbf
    cp                      /home/oracle/app/oracle/oradata/orcl/redo03.log  /home/oracle/cold_bak_files/redo03.log
    cp                      /home/oracle/app/oracle/oradata/orcl/redo02.log
```

```
/home/oracle/cold_bak_files/redo02.log
    cp                       /home/oracle/app/oracle/oradata/orcl/redo01.log
/home/oracle/cold_bak_files/redo01.log
    cp                    /home/oracle/app/oracle/oradata/orcl/control01.ctl
/home/oracle/cold_bak_files/control01.ctl
    cp          /home/oracle/app/oracle/fast_recovery_area/orcl/control02.ctl
/home/oracle/cold_bak_files/control02.ctl
    #备份初始化文件和参数文件
    cp $ORACLE_HOME/dbs/init.ora /home/oracle/cold_bak_files/init.ora
    cp                                   $ORACLE_HOME/dbs/initorcl.ora
/home/oracle/cold_bak_files/initorcl.ora
    cp                                   $ORACLE_HOME/dbs/spfileorcl.ora
/home/oracle/cold_bak_files/spfileorcl.ora
    elif [ $1 = "-r" ]; then
        echo "restore..."
    cp                          /home/oracle/cold_bak_files/system01.dbf
/home/oracle/app/oracle/oradata/orcl/system01.dbf
    cp                          /home/oracle/cold_bak_files/sysaux01.dbf
/home/oracle/app/oracle/oradata/orcl/sysaux01.dbf
    cp                          /home/oracle/cold_bak_files/undotbs01.dbf
/home/oracle/app/oracle/oradata/orcl/undotbs01.dbf
    cp                       /home/oracle/cold_bak_files/pdbseed/system01.dbf
/home/oracle/app/oracle/oradata/orcl/pdbseed/system01.dbf
    cp                          /home/oracle/cold_bak_files/users01.dbf
/home/oracle/app/oracle/oradata/orcl/users01.dbf
    cp                       /home/oracle/cold_bak_files/pdbseed/sysaux01.dbf
/home/oracle/app/oracle/oradata/orcl/pdbseed/sysaux01.dbf
    cp                       /home/oracle/cold_bak_files/pdborcl/system01.dbf
/home/oracle/app/oracle/oradata/orcl/pdborcl/system01.dbf
    cp                       /home/oracle/cold_bak_files/pdborcl/sysaux01.dbf
/home/oracle/app/oracle/oradata/orcl/pdborcl/sysaux01.dbf
    cp         /home/oracle/cold_bak_files/pdborcl/SAMPLE_SCHEMA_users01.dbf
/home/oracle/app/oracle/oradata/orcl/pdborcl/SAMPLE_SCHEMA_users01.dbf
    cp                      /home/oracle/cold_bak_files/pdborcl/example01.dbf
/home/oracle/app/oracle/oradata/orcl/pdborcl/example01.dbf
    cp             /home/oracle/cold_bak_files/pdborcl/pdbtest_users02_1.dbf
/home/oracle/app/oracle/oradata/orcl/pdborcl/pdbtest_users02_1.dbf
    cp             /home/oracle/cold_bak_files/pdborcl/pdbtest_users02_2.dbf
```

```
/home/oracle/app/oracle/oradata/orcl/pdborcl/pdbtest_users02_2.dbf
    cp                          /home/oracle/cold_bak_files/redo03.log
/home/oracle/app/oracle/oradata/orcl/redo03.log
    cp         .                /home/oracle/cold_bak_files/redo02.log
/home/oracle/app/oracle/oradata/orcl/redo02.log
    cp                          /home/oracle/cold_bak_files/redo01.log
/home/oracle/app/oracle/oradata/orcl/redo01.log
    cp                          /home/oracle/cold_bak_files/control01.ctl
/home/oracle/app/oracle/oradata/orcl/control01.ctl
    cp                          /home/oracle/cold_bak_files/control02.ctl
/home/oracle/app/oracle/fast_recovery_area/orcl/control02.ctl
    #恢复初始化文件和参数文件
    cp /home/oracle/cold_bak_files/init.ora $ORACLE_HOME/dbs/init.ora
    cp                          /home/oracle/cold_bak_files/initorcl.ora
$ORACLE_HOME/dbs/initorcl.ora
    cp                          /home/oracle/cold_bak_files/spfileorcl.ora
$ORACLE_HOME/dbs/spfileorcl.ora
    else
        echo "wrong parameter"
    fi
```

该脚本有两个参数，一个是"-b"表示备份，另一个是"-r"表示恢复。最后根据脚本创建存储备份文件的一些目录：cold_bak_files、cold_bak_files/pdbseed 和 cold_bak_files/pdborcl。然后就可以运行脚本了。在运行脚本之前应该关闭数据库，运行脚本之后，应该打开数据库。

2. 自动联机备份

首先创建一个用存储备份文件和日志的目录：/home/oracle/rman_backup，然后新建两个 shell 文件：/home/oracle/rman_level0.sh 和/home/oracle/rman_level1.sh。rman_level0.sh 是 0 级备份的脚本文件，rman_level1.sh 是一级备份的脚本文件。要保证这两个文件可执行。

rman_level0.sh 内容如下：

```
#rman_level0.sh
#!/bin/sh
export NLS_LANG='SIMPLIFIED CHINESE_CHINA.AL32UTF8'
export ORACLE_HOME=/home/oracle/app/oracle/product/12.1.0/dbhome_1
export ORACLE_SID=orcl
export PATH=$ORACLE_HOME/bin: $PATH

rman    target   /    nocatalog    msglog=/home/oracle/rman_backup/lv0_`date
+%Y%m%d-%H%M%S`_L0.log << EOF
    run{
```

```
    configure retention policy to redundancy 1;
    configure controlfile autobackup on;
    configure controlfile autobackup format for device type disk to
'/home/oracle/rman_backup/%F';
    configure default device type to disk;
    crosscheck backup;
    crosscheck archivelog all;
    allocate channel c1 device type disk;
    backup as compressed backupset incremental level 0 database format
'/home/oracle/rman_backup/dblv0_%d_%T_%U.bak'
        plus archivelog format '/home/oracle/rman_backup/arclv0_%d_%T_%U.bak';
    report obsolete;
    delete noprompt obsolete;
    delete noprompt expired backup;
    delete noprompt expired archivelog all;
    release channel c1;
    }
    EOF

exit
```

rman_level1.sh 内容如下：

```
#rman_level1.sh
#!/bin/sh

export NLS_LANG='SIMPLIFIED CHINESE_CHINA.AL32UTF8'
export ORACLE_HOME=/home/oracle/app/oracle/product/12.1.0/dbhome_1
export ORACLE_SID=orcl
export PATH=$ORACLE_HOME/bin: $PATH

rman target / nocatalog msglog=/home/oracle/rman_backup/lv1_`date
+%Y%m%d-%H%M%S`_L0.log << EOF
    run{
    configure retention policy to redundancy 1;
    configure controlfile autobackup on;
    configure controlfile autobackup format for device type disk to
'/home/oracle/rman_backup/%F';
    configure default device type to disk;
    crosscheck backup;
```

```
    crosscheck archivelog all;
    allocate channel c1 device type disk;
    backup as compressed backupset incremental level 1 database format
'/home/oracle/rman_backup/dblv1_%d_%T_%U.bak'
        plus archivelog format '/home/oracle/rman_backup/arclv1_%d_%T_%U.bak';
    report obsolete;
    delete noprompt obsolete;
    delete noprompt expired backup;
    delete noprompt expired archivelog all;
    release channel c1;
    }
    EOF

    Exit
```

上述两个 shell 文件中包含有 rman 的一些格式化参数%F，%d，%T，%U，目的是自定义备份文件的文件名称。

下面通过 linux 的 crontab 命令创建自动执行脚本文件的调度：

```
    $crontab -e
    #每个星期天 23：00 执行 rman_level0.sh
    00 23 * * 0 /home/oracle/rman_level0.sh >>
/home/oracle/rman_backup/rman_level0_`date +\%Y\%m\%d-\%H\%M\%S`.log 2>&1
    #星期一，二，三，四，五，六 23：00 执行 rman_level1.sh
    00 23 * * 1,2,3,4,5,6 /home/oracle/rman_level1.sh >>
/home/oracle/rman_backup/rman_level0_`date +\%Y\%m\%d-\%H\%M\%S`.log 2>&1
```

所有工作做完之后，等待 Linux 系统自动执行脚本，然后每天定期检查目录中的自动生成的各类备份文件和日志文件，查看备份状态。比如：

```
    $ pwd
    /home/oracle/rman_backup
    $ ls
    arclv0_ORCL_20170502_4ps3896e_1_1.bak
dblv0_ORCL_20170502_4ns38948_1_1.bak    lv0_20170502-160802_L0.log
    c-1392946895-20170502-15
dblv0_ORCL_20170502_4os38951_1_1.bak    lv1_20170502-160342_L0.log
    dblv0_ORCL_20170502_4ms38934_1_1.bak        lv0_20170502-155630_L0.log
rman_level0_20170502-160802.log
```

3. 数据迁移

数据迁移的过程是先在源服务器上导出(使用命令 expdp)一个 pdb 的数据，然后在目的服务上导入(使用 impdp)数据。

难点在于数据的导入。这是由于数据的存储方式多种多样，比如数据可能存储在多个表空间中，有些表还可能是分区存储的。因此在执行导入命令之前，必须了解源数据库的空间存储结构，要在目标数据库创建和源数据库相同数目的表空间，并且要在目标数据库上在创建一个可以访问这些表空间的用户，并设置足够的表空间配额(QUOTA)。

典型在导出与导入命令是：

```
$ expdp study/***@pdborcl directory=expdir dumpfile=study.dmp
$ impdp study/***@pdborcl directory=expdir dumpfile=study.dmp
```